Protein Targeting

Protein Targeting

Anthony P. Pugsley
Unité de Génétique Moléculaire
Institut Pasteur
Paris, France

Academic Press, Inc.
Harcourt Brace Jovanovich, Publishers
San Diego New York Berkeley
Boston London Sydney Tokyo Toronto

ACADEMIC PRESS, INC.
San Diego, California 92101

United Kingdom Edition published by
ACADEMIC PRESS LIMITED
24-28 Oval Road, London NW1 7DX

Library of Congress Cataloging-in-Publication Data

Pugsley, Anthony P.
Protein targeting.

Bibliography: p.
Includes index.
1. Proteins—Physiological transport. 2. Proteins
—Secretion. I. Title
QP551.P83 1989 574.19′245 88-7522
ISBN 0-12-566770-1 (alk. paper)

PRINTED IN THE UNITED STATES OF AMERICA
89 90 91 92 9 8 7 6 5 4 3 2 1

For C, G, and B,
the not-so-silent majority

Contents

Preface xi

CHAPTER I

An overview of protein targeting 1

A. Introduction 1
B. Protein traffic in eukaryotic cells 1
C. Protein traffic in bacterial cells 5
D. Terminology and basic principles 8

CHAPTER II

Basic principles and techniques 13

A. Protein identification and function 13
B. Primary structure of targeted proteins 17
C. Posttranslational modification 17
D. Secondary structure of targeted proteins 19
E. Expression systems and model proteins 28
F. The targeting pathway 29
Further reading 43

CHAPTER III

Early stages in the secretory pathway 45

A. Secretory signal peptides 46
B. Secretory signal sequences 62
C. How do secretory routing signals work? 65
D. Signal peptidases 90
E. Protein translocation 94
F. Membrane protein topogenesis 99
Further reading 111

CHAPTER IV

Later stages in the prokaryotic secretory pathway 112

A. Secretory proteins without sorting signals 112
B. Outer membrane proteins 115
C. Secreted proteins of Gram-negative bacteria 120
Further reading 125

CHAPTER V

Later stages in the eukaryotic secretory pathway 126

A. General concepts 126
B. Protein modification, proofreading, and retention in the endoplasmic reticulum 128
C. Transport from the endoplasmic reticulum to the Golgi 136
D. Protein modification in the Golgi 137
E. Organization of the Golgi cisternae 141
F. Intra-Golgi movement of secretory proteins 144
G. Post-Golgi sorting of secretory proteins 147
H. Secretory pathway-independent export and secretion 166
I. Concluding remarks 167
Further reading 168

CHAPTER VI

The targeting of mitochondrial, chloroplast, and peroxisomal proteins 169

A. Mitochondrial and chloroplast organization 169
B. Protein import into mitochondria and chloroplasts 170
C. Sorting of imported and endogenous mitochondrial and chloroplast proteins 190
D. Import of proteins into peroxisomes 195
Further reading 198

CHAPTER VII

The targeting of nuclear proteins 199

A. The structure of the nucleus 199
B. Targeting and retention of nuclear proteins 201
Further reading 210

CHAPTER VIII

Endocytosis

Endocytosis 211

A. Receptor-mediated endocytosis 212
B. Other modes of protein uptake 222
Further reading 228

CHAPTER IX

Applications of protein targeting

Applications of protein targeting 229

A. Applications of protein export and secretion 229
B. Applications of organelle targeting 239
C. Applications of protein import and endocytosis 240
Further reading 240

References 241

Index 269

Preface

Protein targeting is one of the most actively researched and rapidly developing aspects of cell biology. Knowledge of the subject has increased greatly through the use of molecular genetics and through the development and refinement of *in vitro* techniques for studying the complex processes involved in intracellular protein transport.

Different aspects of protein targeting have already been the subject of many specialized reviews. One aim of this book is to update these reviews and to discuss the various views and opinions expressed by their authors. More importantly, almost all the various aspects of the subject are reviewed here in a single volume, allowing similarities and differences among protein targeting pathways to be presented and discussed. For example, the early stages of protein export and secretion in prokaryotes and eukaryotes have been reviewed separately in the past, and yet the underlying principles involved are very similar. Here, the two pathways are discussed in a single chapter, which not only highlights principles common to both groups of organisms but also points to significant differences between the groups. Likewise, the targeting of mitochondrial and chloroplast proteins is reviewed in a single chapter that reveals striking similarities as well as some important differences in the biogenesis of the two organelles. Extensive cross-referencing among chapters and a chapter describing general principles and techniques serve to increase the continuity of the book and to emphasize the underlying themes of the subject as a whole.

This book will broaden the interest and knowledge of scientists already familiar with specific aspects of protein targeting, for whom the extensive and up-to-date reference list will prove particularly useful. It will also be useful to students of cell biology, biochemistry, microbiology, and molecular biology, for whom the specialized reviews listed at the end of each chapter will provide more extensive coverage and discussion of certain aspects of the subject. Finally, this book will direct the attention of scientists interested in fundamental aspects of the subject to the multiple applications of protein targeting; conversely, it will direct the attention of biotechnologists to the basic principles of the techniques that they are employing.

Anthony P. Pugsley

CHAPTER I

An overview of protein targeting

A. Introduction

Cells contain a large number of mainly cytoplasmic proteins with widely differing structures and functions. The ability to insert proteins into lipid bilayer membranes was undoubtedly fundamental in the evolution of the complex cell structure. The earliest membrane proteins were almost certainly derived from cytoplasmic proteins which had acquired properties enabling them to insert spontaneously into cell membranes. These primordial membrane proteins need not have been catalytically active; they could instead have evolved diverse functions thereafter. Whatever characteristics distinguished the cytoplasmic protein from its membrane counterpart might be regarded as primitive targeting signals. The idea that relatively simple changes in protein structure could have led to the evolution of the first membrane proteins is supported by the existence of several known targeting signals which turn out to be relatively short sequences of amino acids. As we shall see, even minor alterations to these targeting sequences can abolish or reduce targeting efficiency, whereas their inclusion in the sequence of a normally cytoplasmic protein can convert it into a targeted protein. Proteins are targeted with high specificity and efficiency to a variety of locations within and outside the cell. The central theme of this book is that protein targeting is principally a property of signals or codes within the structure of the protein and of cytoplasmic and membrane proteins which "decode" these signals and facilitate specific protein targeting and transmembrane movement.

B. Protein traffic in eukaryotic cells

The complexity of protein targeting in eukaryotic cells is illustrated in Fig. I.1. The cell is bordered by a lipid bilayer plasma membrane which pre-

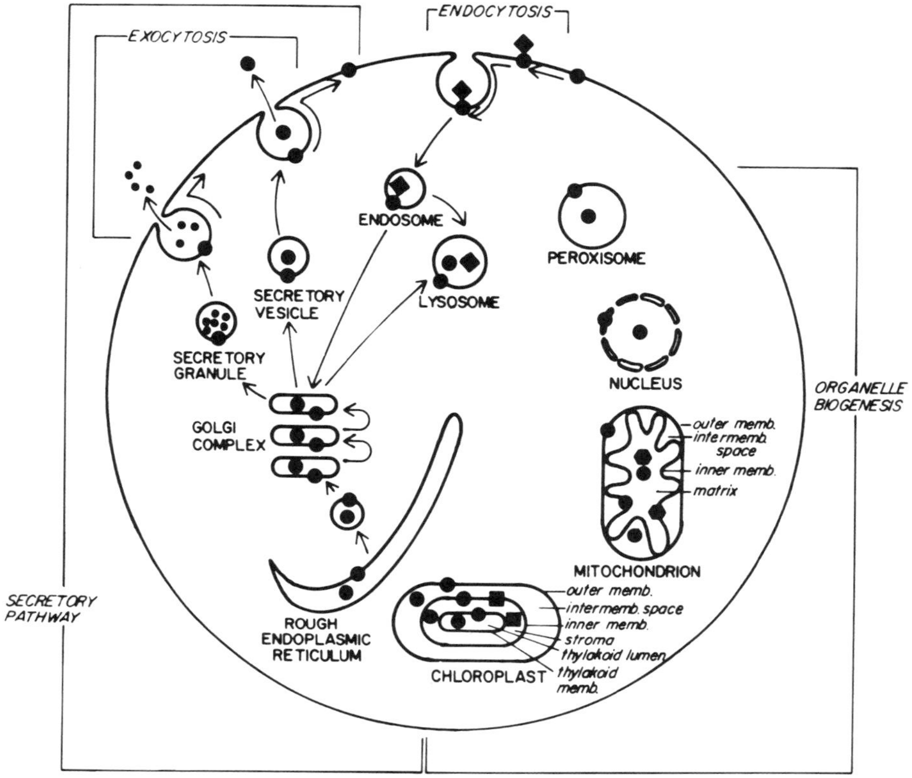

Fig. I.1. Simplified representation of a eukaryotic cell showing the sites to which targeted proteins are directed. Nuclear-encoded proteins (●) are directed to a variety of sites. Rough endoplasmic reticulum-dependent protein traffic (the secretory pathway) represents the major protein route by which nuclear-encoded proteins are directed into and through the endoplasmic reticulum to lysosomes and via secretory vesicles (constitutive secretion) or secretory granules (regulated secretion) to the plasma cell membrane or to the outside of the cell. Four types of organelle are shown. Chloroplasts and mitochondria have small genomes coding for a limited number of proteins localized to the chloroplast stroma and inner membrane (■) or to the mitochondrial matrix and inner membrane (⬢). Other chloroplast and mitochondrial proteins, and all nuclear and peroxisomal proteins, are encoded by the nucleus and are imported from the cytoplasm. Proteins synthesized by other cells (◆) are imported via endocytosis, as are plasma membrane proteins. Some of these proteins may be redirected to the cell surface, or they may be routed into lysosomes or other intracellular targets.

vents cytosolic material, including proteins, from leaking out into the medium. Specific transport systems are required for the movement of small solutes into and out of the cell, and for the exit and entry of specific proteins. Other functions of cell surface proteins vary according to cell type and growth conditions. Some cells (e.g., epithelial cells) may grow as polarized sheets in which plasma membrane composition is different on opposing cell surfaces (the apical and basolateral membranes). Furthermore, certain proteins are selectively secreted into the extracellular milieu, and secretion may occur specifically from either the apical or basolateral membranes in polarized cells (not shown in Fig. I.1; see Chapter V). A further distinction is made between two groups of secreted proteins on the basis of whether or not they are released constitutively. Some secreted proteins may be stored in highly concentrated form in a special class of vesicle called *secretory granules*. Their release from the cell is triggered by external stimuli: this is the regulated secretory pathway (Chapter V).

Very few cytoplasmically synthesized cell surface proteins or secreted proteins reach the cell surface directly. As illustrated in Fig. I.1, proteins are released to the outside of the cell, and concomitant insertion of proteins into the plasma membrane occurs as vesicles carrying the proteins fuse with the cell surface. These vesicles are the end products of the secretory pathway, which starts at the rough endoplasmic reticulum (RER) and continues through the Golgi complex. Secretory proteins are extensively posttranslationally modified (e.g., by proteolysis, glycosylation, or fatty acylation) in both the RER and the Golgi complex. The RER is a specialized domain of an extensive vesicular–endothelial network, the endoplasmic reticulum (ER), which in most cells is present almost throughout the cytoplasm and is continuous with the nuclear envelope. The RER is studded with ribosomes exclusively engaged in the synthesis of secretory proteins, giving it its characteristic rough morphology as revealed by electron microscopy (Section III.C.2). The ER is also the major site of lipid synthesis, which may have profound and as yet largely ignored implications for protein targeting.

The Golgi apparatus is usually represented as a series of three or more stacked discs as in Fig. I.1. It is usually located close to the nucleus. Enzyme activities are frequently located in specific parts of the stack, which is orientated from the *cis side* (closest to and receiving material from the ER) through the *medial zone* to the *trans side,* from which released material is directed to the cell surface and probably also to the lysosome (Chapter V). The lysosome (or the functionally equivalent vacuole in plant and fungal cells) is the terminus of a branch of the main secretory pathway (Chapter V). Like the Golgi and ER, it is bordered by a single proteolipid bilayer membrane. The soluble proteins contained within the lumen of the lysosome are mainly enzymes which degrade material imported from the outside of the cell by endocytosis (see below).

Secreted and plasma membrane proteins, as well as proteins retained in secretory organelles, are synthesized in the cytoplasm and are referred to as *secretory proteins*. All of them cross only one membrane, the RER membrane (RERM), during transit through the cell. They are targeted to and cross the RERM by the same mechanisms, irrespective of their final destination (Chapters III and V). Subsequently, proteins in the secretory pathway are sorted to their final destinations by complicated signal–receptor interactions, which are far from being fully understood. Bulk flow through the secretory pathway carries soluble proteins to the cell surface via the RER and Golgi and the terminal, constitutive branch of the pathway, but some proteins take longer than others to reach the cell surface. Proteins remain embedded in membranes or are encapsulated within them throughout transit through the secretory pathway (Fig. I.1); they are never released in free form into the cytosol. Vectorial interorganelle or organelle-to-surface transport occurs via vesicles which bud from and fuse with donor and receptor membranes, respectively (Chapter V).

Most of the proteins located in nonsecretory organelles (the nucleus, mitochondria, chloroplasts, peroxisomes, and glyoxisomes) are also nuclear-encoded and made in the cytoplasm. Only the mitochondria and chloroplasts have endogenous genomes, which allow them to synthesize some of their own proteins. These proteins either stay at their site of synthesis (the mitochondrial matrix or the chloroplast stroma) or are targeted to the mitochondrial inner membrane or intermembrane space, or to the chloroplast inner membrane, thylacoid membranes, or thylacoid lumen. By analogy with secretory protein targeting systems, targeting of these endogenous proteins almost certainly involves the decoding of targeting signals within the sequence of the targeted protein, but these signals have not yet been extensively characterized. All other mitochondrial and chloroplast proteins are nuclear-encoded, are imported from the cytosol, and are known to contain specific routing signals (Chapter VI). The proteins have a wide variety of functions; they contribute to almost all of the known enzymatic activities of these organelles and often associate with endogenous proteins to form enzyme complexes. One of the most interesting features of protein targeting into mitochondria and chloroplasts is that the proteins may have to cross up to three membranes to reach their final destinations (Chapter VI). These organelles also synthesize some specific classes of lipids, such as those in the chloroplast thylacoid membrane, but the bulk of the lipids in these organelles is probably synthesized in the ER.

As discussed in Chapter VI, *peroxisomes* and *glyoxisomes* (microbodies) contain specific enzymes such as catalase, alcohol oxidase, and luciferase. Unlike mitochondria and chloroplasts, these organelles usually have only single lipid bilayer membranes. Very little is known about peroxisomal or glyoxisomal protein targeting, but it is clear that it is not a branch of the secretory pathway.

Finally, many elegant studies show that proteins are imported into the nucleus by distinctly different mechanisms from those involved in the biogenesis of other organelles (Chapter VII). The DNA is retained by a double lipid bilayer membrane containing pores through which only small proteins can diffuse. Specific nuclear targeting of DNA-binding proteins including histones and regulatory proteins also occurs through the pores, and once again specific targeting signals are involved.

The development of true protein secretion (i.e., release of proteins into the medium) probably represented another significant leap in evolution, since it allowed cells to release enzymes which could furnish nutrients, to communicate with each other via polypeptide messages (hormones), and to develop defense systems (immunoglobulins) and even methods for attacking other cells (toxins). Several of these processes require a system, *endocytosis,* for recognizing and internalizing proteins from the outside (Chapter VIII). Endocytosed proteins carry signals which are recognized by receptors on cell surfaces and are directed to appropriate targets within the cell, often in association with the cell surface receptor. In some cases, a protein entering a cell at one site will exit at another (*transcytosis*) (see Chapter VIII).

Figure I.1 is not an accurate representation of the internal organization of a eukaryotic cell, but it illustrates the routes used by targeted proteins. Organelles such as the endoplasmic reticulum and the Golgi might be represented as continuous reticula rather than as discrete bodies. The nuclear membrane is also in close juxtaposition to the endoplasmic reticulum and share components with it. Organelles may also be tethered to each other by microtubules or other filamentous structures, which may also be used for the directed and energy-driven displacement of organelles and possibly also of small vesicles involved in protein targeting throughout the cell.

C. Protein traffic in bacterial cells

Bacteria generally belong to one or two morphologically distinct groups distinguished by their reaction in a staining test developed by Christian Gram. In the Gram stain, crystal violet is precipitated on the cell surface and in the cytoplasm by a mordant, iodine. The surface precipitate is washed off by ethanol. Gram-negative bacteria also release their cytoplasmic crystal violet, probably through small holes formed in the plasma membrane, when extracted with ethanol, whereas Gram-positive bacteria retain the dye during ethanol extraction (81). Dye retention is a property of cell envelope structure. The *cytoplasmic membrane* (BCM) of Gram-positive bacteria is surrounded by a layer of interwoven peptidoglycan and teichoic–teichuronic acid strands which form the cell wall. Gram-negative bacteria have a second lipid bilayer membrane (the *outer membrane*) outside the peptidoglycan, which is enclosed within the *periplasm*. The outer membrane and peptidoglycan layer of Gram-negative bacteria seem to be more sensitive to ethanol than the Gram-positive cell wall, allowing ethanol action on the plasma membrane to release more of the precipitated cytoplasmic crystal violet (Fig. I.2).

All bacterial membranes contain proteins, as does the periplasm of Gram-negative bacteria. Bacteria also secrete proteins into the medium, from which they may be either released or locally assembled into proteinaceous coats (surface layers) or appendages. Most bacteria do not have membrane-limited organelles, but some have specialized regions of the plasma membrane involved in photosynthesis (not shown in Fig. I.2).

Bacteria export proteins to their cell envelope and secrete them into the medium by a process which is quite similar to the early stages of the secretory pathway in eukaryotic cells. The cytoplasmic membrane is the first and in some cases the only membrane which the proteins will cross. As in the eukaryotic secretory pathway, polysomes synthesizing exported proteins are attached to the cytoplasmic membrane (Chapter III). A protein which is integrated into the cytoplasmic membrane will remain there unless the cell forms and releases cytoplasmic membrane vesicles analogous to those budding from the ER (Fig. I.1). There is only very tentative evidence that proteins can be released from the BCM in this way (see below). Proteins extruded across the BCM are analogous to eukaryotic proteins which are extruded into the lumen of the RER and will be released into the medium (Gram-positive bacteria) or into the periplasm (Gram-negative bacteria). Three different models explain how proteins reach the Gram-negative outer membrane: (1) export through fixed regions of contact between the inner and outer membranes; (2) release from the cytoplasmic membrane in vesicles which fuse with the outer membrane (see above); and (3) spontaneous or assisted assembly into the outer membrane via a soluble periplasmic intermediate. These models are discussed in further detail in Chapter IV. Gram-negative bacteria use a variety of different mechanisms to secrete proteins into the growth medium, some of which are extensions of the secretory pathway whereas others bypass the secretory pathway altogether (Chapter IV).

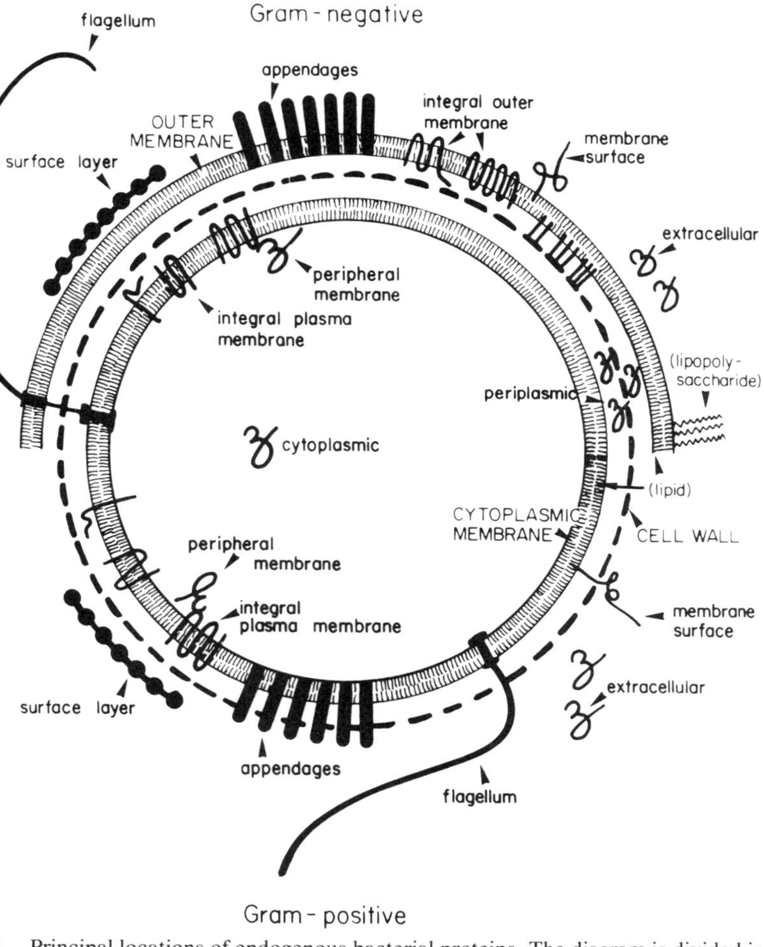

Fig. I.2. Principal locations of endogenous bacterial proteins. The diagram is divided into two parts representing Gram-negative and Gram-positive cell structures. Proteins are indicated as cytoplasmic, integral plasma membranous (many different configurations are possible), peripheral membranous (associated with integral membranous proteins), periplasmic, outer membranous (several configurations), or extracellular. Additional proteinaceous structures (i.e., appendages, surface layers, and flagella) are also indicated, as are membrane lipids, the cell wall (peptidoglycan and other polymers), and lipopolysaccharide, an outer membrane-specific glycolipid.

There are fewer known cases of posttranslational modification of secreted or exported proteins in bacteria than in eukaryotes. In particular, there are few authenticated cases of glycosylated bacterial proteins. Certain bacterial proteins may be fatty-acylated by enzymes in the cytoplasmic membrane (Section III.E.3).

D. Terminology and basic principles

With such a massive amount of literature being devoted to the subject of protein targeting, the nomenclature used to describe even fundamental aspects of the subject has inevitably become confused. The following "standard" terminology is used throughout this book, irrespective of the terminology used in the article being cited.

1. Targets

As already explained, the secretory route directs proteins to a variety of different locations both within and outside the cell. The names applied to these proteins and their targets are relatively straightforward, but there is some confusion over the distinction between secreted and exported proteins, particularly in bacteria. A *secreted* protein can be defined as one which exists in the medium surrounding the cell, having been released from the cell without any alteration of cell structure greater than the maximum compatible with the cell's normal processes of growth and reproduction. This definition excludes proteins which are inserted into the plasma membrane, which will be referred to as *exported*. Many workers consider that a protein should be regarded as secreted if it is released into the periplasm between the two membranes of Gram-negative bacteria, but strictly speaking, such proteins are not outside the cell envelope boundary. Periplasmic and outer membrane proteins are therefore also referred to as exported. The term *excreted* has also been applied to extracellular (secreted) bacterial proteins but is more widely used for the release of low-molecular-weight compounds (toxic metabolites, antibiotics, amino acids, etc.) for which it should be specifically reserved.

Membrane proteins are referred to as *integral* if the polypeptide spans the lipid bilayer at least once. Integral membrane proteins are subdivided into three types depending on the number of times the polypeptide crosses the membrane (*type I–II* or *bitopic* and *type III* or *polytopic*), and, in the case of bitopic polypeptides, which only cross the membrane once, subdivided on the basis of their orientation (types I and II) (Fig. I.3). *Peripheral* membrane proteins are only loosely anchored in the mem-

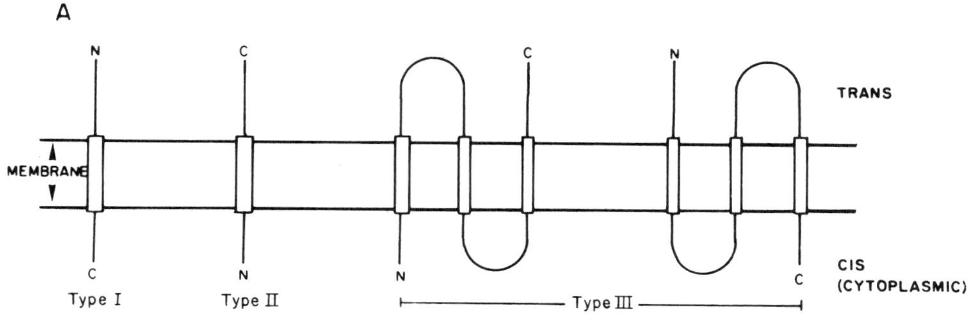

Fig. I.3. Nomenclature applied to different types of transmembrane proteins according to the orientation of their N and C termini and the number of transmembrane segments. Types I and II, bitopic; type III, polytopic with two or more membrane segments (see Section I.D.1).

brane via interactions with integral membrane proteins or with other membrane components or by partial insertion of the protein into the membrane bilayer (*monotopic*); they can usually be released by chaotropic agents or by other mild extraction procedures (Section II.A.1).

Proteins which are internally routed to different organelles are said to be *compartmentalized*. If the organelle actively participates in the uptake of soluble proteins from the cytoplasm, the process is called *protein import*. Proteins entering the cell from the outside are also *imported*. The term *endocytosis* is reserved for protein import by the invagination of surface-derived vesicles.

2. Targeting signals

The basic principle behind most studies on protein targeting is that proteins (other than those remaining at their site of synthesis, i.e., the cytoplasm, the mitochondrial matrix, and the chloroplast stroma) have signals which tell them where they are to go and how they are to fold and interact with other cell components once they reach their destination. These signals may be divided into the following four basic groups:

(i) *Routing signals* are concerned with the earliest stages of protein targeting; they direct the specific interaction between targeted polypeptides and the surface of the first membrane with which they come into contact. The various types of routing signals, together with alternative names used in the literature, are listed in Table I.1. Note that routing signals need not be located at the extreme N-terminus of the polypeptide and need not be proteolytically cleaved off during the early stages of polypeptide translocation. In those

Table I.1. Characteristics of protein targeting signals[a]

Group	Route or target	Name	Alternative names	Processing
Routing signals	Secretory route	Signal peptide	Leader peptide[b]	+
		Secretory signal sequence		–
	Mitochondria	Prepeptide[c]	Presequence Leader peptide Signal peptide Signal sequence	+
		OM routing signal[c]		–
		Matrix export signal[d]		+(?)
	Chloroplasts	Transit peptide[c]	Signal peptide	+
		Stromal export signal[d]		+(?)
		Thylacoid routing signal[c,d]		?
	Peroxisome	Peroxisome routing signal		–
	Nucleus	Karyophilic signal	Nuclear targeting signal	–
Sorting signals	Vacuole	Vacuolar propeptide		+
	Lysosome	Lysosomal sorting signal[e]		–
	Basolateral membrane	Basolateral polarity signal		–
	Apical membrane	Apical polarity signal		–
	Secretory granules	Granule sorting signal		?
	Bacterial OM	OM sorting signal		–
	Culture medium	Secretion signal		–
	Mitochondrial IM	IM sorting signal[f]		+(?)
	Chloroplast IM	IM sorting signal[g]		?
	Thylacoid	Thylacoid sorting signal[g]		+(?)

[a] IM, inner membrane; OM, outer membrane.

[b] Leader peptides are also the short peptides translated from leader transcripts in bacterial operons regulated by attenuation.

[c] In nuclear-encoded polypeptides.

[d] In polypeptides encoded by mitochondrial or chloroplast DNA.

[e] Mannose-6-phosphate in lysosomal enzymes.

[f] In nuclear-encoded proteins; may be the same as or equivalent to matrix export signal.

[g] In nuclear-encoded proteins; may be the same as or equivalent to stromal export or thylacoid routing signals.

cases in which the routing signal is proteolytically processed, the initial form of the polypeptide is called the *precursor* (or *pre*-form), and the processed polypeptide is the *mature* form. Some precursors may be processed more than once, in which case the first precursor is the *prepro*polypeptide, and the product of the first processing step (removal of the routing signal) the *pro*polypeptide. In many cases, the second processing step removes another N-terminal segment of the polypeptide, which is then referred to as the *propeptide*. Some propeptides function as sorting signals (Table I.1).

(ii) *Sorting signals* direct proteins into different branches of a targeting pathway or redirect them once they have reached their initial targets (Table I.1).

(iii) *Retention or salvage signals* prevent proteins from passing along the entire length of a targeting pathway or from being released from organelles to which they have been targeted. This is particularly important for soluble and membrane proteins in the ER, the Golgi apparatus, and lysosomes or vacuoles because these organelles are extremely dynamic transit points in different protein targeting pathways and, as such, are likely to lose their own complement of specific proteins as well as the proteins which are normally passing through them (Chapter V). Retention signals may also be important in nuclear proteins, for although routing signals may direct them into the nucleus, nuclear pores are large enough to allow many of them to diffuse out again (Chapter VII).

(iv) *Membrane topology–stop transfer signals*. Proteins are probably "threaded" through membranes as unfolded polypeptide chains. For membrane proteins, this process is halted at a particular point along the length of the polypeptide and, in the case of polytopic membrane proteins, is reinitiated at a site further down the polypeptide, and so on. Sequences which initiate the translocation of the polypeptide across the membrane are called routing signals (see above). If these routing signals are not cleaved off, they usually remain embedded in the membrane and may be called *membrane anchor sequences*. Sequences which halt the translocation of the polypeptide through the membrane are referred to as *stop transfer signals*.

The activity of some signals may change or be completely lost when moved to other parts of the polypeptide or into other, foreign sequences. As emphasized in Section II.D.5, targeting signal function may be lost when sequences are placed out of context simply because they are no longer exposed on the surface of the polypeptide. It is also important to note that targeting and other signals are not necessarily linear stretches of amino acids. For example, secretory proteins fold in the lumen of the ER and yet are faithfully targeted along appropriate branches of the secretory pathway. These folded polypeptides may have *targeting patches* rather than targeting sequences. These patch signals may be recognized by appropriate receptors in much the same way as epitopes are recognized by antibodies (47).

CHAPTER II

Basic principles and techniques

Before reviewing our present understanding of protein targeting, it seems pertinent to describe some of the techniques employed. Space does not allow full methodological details to be given, but it is important to understand their basic principles, their limitations, and the interpretation of results.

A. Protein identification

The location of a protein should be known before studies on its targeting are initiated. In some cases, the protein is selected as a marker for a particular structure, and its function need not be known. Proteins may be identified by their characteristic migration in polyacrylamide gel electrophoresis (PAGE) in the presence of the denaturing detergent sodium dodecylsulfate (SDS) and by their reactivity with specific antibodies. In many cases, proteins are chosen for study because they are relatively abundant and therefore easier to detect by SDS-PAGE, to purify, and to characterize.

Antibodies are probably the most widely used tool in studies on protein targeting. Antibody specificity is obviously of paramount importance. Other important features might be the antibodies' ability to differentiate between native and denatured or precursor forms of the protein and their ability to recognize specific epitopes. Increasing use is therefore being made of antibodies which recognize a specific site or sequence in a polypeptide (*epitope-specific antibodies;* monoclonal antibodies or antibodies raised against synthetic peptides corresponding to defined parts of the proteins being studied).

Some proteins are only poorly immunogenic, making it difficult to obtain high titers of antibodies for use in immunodetection studies. A novel

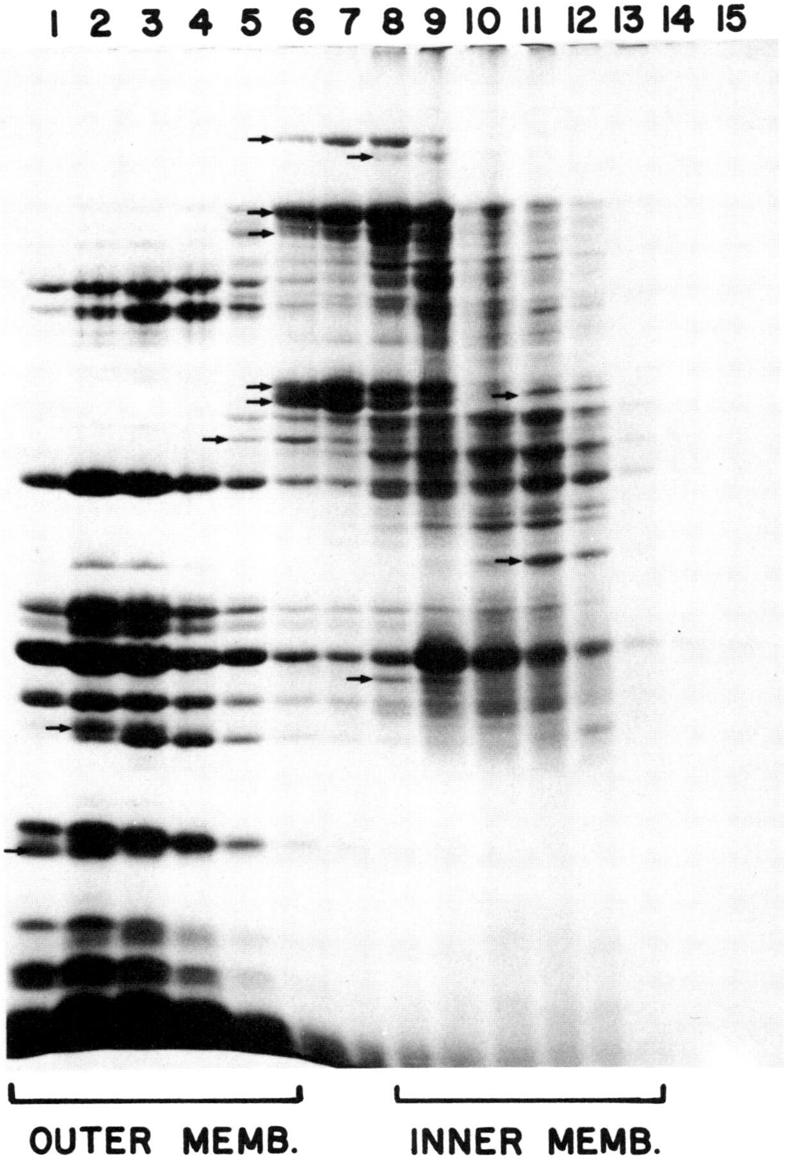

Fig. II.1. Sodium dodecyl sulfate (SDS) polyacrylamide gel electrophoresis (PAGE) of proteins in membrane vesicles of *Escherichia coli* separated by isopycnic sucrose density centrifugation (1019). Proteins in fractions collected from the gradient were dissolved in SDS and heated to 100°C before being loaded onto the gel. Proteins in the heavier outer membrane vesicles were identified by their characteristic profiles [major 30–36-kDa proteins way to overcome this problem is to fuse a highly antigenic peptide to the targeted protein in such a way as not to affect its localization. This can be done by *in vitro* recombinant genetics such that DNA coding for the antigenic peptide is inserted in frame into the coding sequence of the gene for the targeted peptide (358,757). Another application of recombinant genetics which allows proteins to be detected more easily is to fuse the proteins being studied to proteins which have a readily detectable enzymatic activity (Section II.D.4).

1. Cell fractionation

The development of techniques for separating cell components into relatively pure subfractions has greatly assisted studies on protein targeting. Cells are usually lysed by sonication, freeze-thawing, osmotic shock, or homogenization. The resulting mixture is centrifuged at high speed to sediment the particulate fraction, which is then resuspended and centrifuged to equilibrium through a density gradient of suitable material [e.g., sucrose and silica particles (Percoll™)] in which individual cell components will be distributed through the gradient according to their density. Fractions collected from the gradients are examined by biochemical methods for known characteristics of different cell components such as enzymes or other proteins, or by electron microscopy. In many cases, gradient centrifugation gives good separation of cell components, as, for example, in the case of the two membranes of Gram-negative bacteria (Fig. II.1). In other cases, however, organelles, and particularly subparticles within organelles, such as different parts of the Golgi apparatus, are not well separated from each other because their densities are very similar. Cell fractionation is also used to prepare organelle-derived membrane vesicles for use in *in vitro* studies on protein targeting (797) (Section II.F.3).

Additional procedures are used to extract proteins in particular regions of the cell. For example, proteins located on organelle or cell surfaces may sometimes be released by chaotropic agents, high pH, or limited proteolysis. Proteases and epitope-specific antibodies have also been used to determine the extent to which proteins have been translocated across membranes (Section II.D.2). The differential detergent solubility of certain membrane proteins has also been used to determine their location. One example of this is the preferential solubility of inner membrane

(porins and OMP A) in center left of gel]. Arrows indicate proteins present in minor vesicle fractions migrating mainly between inner and outer membrane vesicles and indicative of possible minor membrane domains. (Samples prepared by Sylvie Chauvaux and Claire Salmeron.)

proteins of many Gram-negative bacteria in the detergent Triton X-100 (1019). In this case, the protein profiles obtained by polyacrylamide gel electrophoresis of material in the detergent-soluble and -insoluble extracts of an envelope preparation are very similar to those shown in peak fractions in Fig. II.1. Another detergent with properties useful in cell fractionation is Triton X-114. Membrane proteins are generally taken up into Triton X-114 micelles and are precipitated when the detergent is heated past its cloud point (the temperature at which microscopic phase separation occurs). They can thus be readily separated from hydrophilic proteins by centrifugation. This provides a simple method for detecting and separating hydrophobic proteins in cell homogenates (101).

Although physical fractionation has undoubtedly proven very useful, the location of some proteins as determined by cell fractionation does not correspond to the actual location as determined by other methods. For example, overproduction or structural alterations may cause incorrect sorting of targeted proteins or may cause them to form cytoplasmic aggregates with buoyant densities or other properties similar to those of cell membranes or organelles.

2. *In situ* detection

The earliest *in situ* protein detection studies were based on the identification of cell- or organelle-surface proteins by nonpenetrating reagents (e.g., proteases or antibodies), by iodination (with lactoperoxidase) or by other protein-modifying agents. Immunofluorescence is one of the simplest methods available. Cells are first mixed with a specific antibody, then washed and mixed with a fluoresceine- or rhodamine-tagged antibody specific for immunoglobulins of the animal used to raise the first serum, or with fluoresceine-tagged staphylococcal protein A, which specifically binds to most G class immunoglobulins. The fluorescence exhibited by cells with a surface antigen recognized by the first antibody can be detected by ultraviolet light microscopy. Light microscopy does not permit the sites of protein accumulation in some organelles or other structures to be clearly identified, however, and protein concentrated in specific regions of the cell can only be distinguished from protein dispersed in the cytoplasm (general background fluorescence).

The electron microscope provides much greater definition of cell structures. *In situ* protein detection by immunocytochemistry uses antibodies or protein A coupled to colloidal gold or some other electron-dense material of uniform size and shape. The cells are usually embedded in an appropriate resin at low temperature under conditions which cause minimal structural perturbation. They are then sectioned and treated with an appropriate antibody followed by protein A–gold (956). Immunogold complexes appear as small dark dots on electron micrographs, and structural features can be highlighted by staining with uranyl acetate or lead citrate. This provides a very powerful tool for studying protein targeting and is being increasingly used in preference to physical fractionation methods.

B. Primary structure of targeted proteins

Knowledge of the primary structure of a protein can be invaluable in the study of protein targeting. Comparisons of the complete or partial amino acid sequence of the purified protein with the same sequence as predicted from the nucleotide sequence of its structural gene allows the sites of posttranslational proteolytic processing steps to be determined. One of the commonest proteolytic processing steps is the removal of N-terminal routing signals (Table I.1; Section I.D.2). However, secondary proteolytic processing of some exported or secreted proteins has been reported, making it difficult to identify the initial processing site. Cleavage steps such as signal peptide processing may sometimes be detected by comparing the molecular weights of the mature protein and that produced under conditions such that the signal peptidase is inactive. However, errors inherent in molecular weight determination by SDS-PAGE, as well as other posttranslational modifications, may make such comparisons unreliable.

Amino acid sequences can be used to predict protein secondary structure (Section II.D.3). In addition, they can be scanned for sequences implicated in protein targeting, especially if specific targeting sequences have been identified in other proteins from the same location. Furthermore, sites of posttranslational proteolytic cleavage events can also be predicted from the primary sequence of a targeted protein because processing peptidases have relatively well-defined specificities. Sites for other posttranslational processing events (glycosylation or fatty acylation) may also be predicted because they too occur at relatively well-defined sites within the polypeptide.

C. Posttranslational modification

Side chains of 14 of the 20 amino acids can be covalently modified, and over 120 posttranslationally modified amino acids have been identified in proteins (1196). Many targeted proteins are posttranslationally modified in a variety of different ways in addition to proteolysis and oligomerization (e.g., N- and O-linked glycosylation and trimming, fatty acylation, lipid attachment, or disulfide bridge formation). These changes may be important in protein targeting in two ways. First, they may confer targeting competence on the protein or may themselves be targeting signals. Second, posttranslational modifications demonstrate that proteins have passed through organelles in which the modifying enzymes are located.

Other types of modification are often involved in the assembly of enzyme complexes in the cytoplasmic membrane or in organelles. These modifications include the noncovalent addition of heme groups, the formation of iron–sulfur clusters, and the binding of various ions and coenzymes. These changes are often relatively simple to detect as changes in protein mass or by the incorporation of radioactive groups, and they may be useful indicators of correct protein targeting and assembly.

In some cases, covalently linked groups can be detected by the metabolic incorporation of appropriate radioactive substrates such as fatty acids. Enzymes may also be used to detect covalent modifications. For example, phospholipase C treatment of phosphatidylinositol-modified cell surface proteins releases them from the cell surface (Section III.F.4.b). Endoglycosidases will cleave off different N-linked carbohydrates according to their specificity for different sugar linkages. Endoglycosidases do not cleave O-linked carbohydrates, but these oligosaccharides are the substrates of O-glycanase. Methanolic alkali or hydroxylamine have been used to distinguish between ester- and amide-linked fatty acids in lipoproteins (sensitive and resistant, respectively, to both agents; Section III.F.4.c). Glycoproteins can be stained with periodate–Schiff reagent when separated by SDS-PAGE (296), providing a simple test for the unambiguous identification of covalent glycosylation. Posttranslational modification can also be prevented by specific inhibitors or competitors. For example, N-glycosylation acceptor peptides (Asn-X-Thr or Ser or Gly) compete with natural glycosylation sites to prevent glycosylation (Section V.B.1), and N-linked glycosylation is prevented by D,L-threo-β-fluoroasparagine (1112) or by the antibiotic tunicamycin (Section II.F.2).

Oligomers may be detected by a variety of techniques including molecular weight and sedimentation velocity measurements and chemical cross-linking. Cross-linking agents can be used *in situ* or with purified proteins. Cleavable cross-linkers are preferred because they allow cross-linked complexes to be resolved by two-dimensional SDS-PAGE, in which the cross-links are cleaved between the first and second gels. The disadvantage of cross-linking reagents is that their efficiency and specificity is determined by their ability to cross-link only certain amino acids, by the proximity of these amino acids in neighboring polypeptides, and by the length of the spacer arm. Antibodies may also be able to distinguish between proteins in different states of oligomerization because different epitopes will be exposed. Furthermore, different proteins in heterooligomers will usually be immunoprecipitated by antibodies against one of the component proteins, provided that antibody binding does not cause other proteins to be released from the complex. Disulfide bridges are disrupted by reducing agents.

D. Secondary structure of targeted proteins

In trying to understand how proteins are targeted, it is essential to learn as much as possible about the proteins themselves, including their structure and organization when they reach their destinations and the differences between precursor and mature forms. Features which are important for targeting may thus be distinguished from other structural or functional features.

Localized regions of a protein may adopt a number of different conformations including β-strands, α-helices, turns (induced principally by proline or glycine residues, but also by some longer sequences of amino acids), or random structure. Different regions of a polypeptide interact with each other through hydrophobic and electrostatic interactions and also by formation of salt bridges, disulfide bridges, etc. to form the secondary and tertiary structures. Defined conformations can be formed within the complex organization, including β sheets (an array of parallel β strands), β barrels, and clusters of α helices.

It is sometimes possible to determine the shape of protein molecules or particles by crystallography or by other physical or chemical techniques. Much of what we know about protein structure is based on a number of physical parameters, especially on X-ray diffraction studies of purified protein crystals and other crystallographic analyses. X-ray diffraction patterns can be correlated with the known sequence of residues in a protein to enable three-dimensional maps to be drawn. In principle, X-ray diffraction can define the position of every atom in a protein or protein complex. Data obtained by these methods can be correlated with the presence of particular sequences of amino acids in the polypeptide to allow predictions to be made regarding the structure of uncharacterized proteins which contain the same or similar sequences. It is often important to know how membrane proteins are organized, and, in particular, which parts of the polypeptide are exposed on the inner or outer surfaces of the membrane, in order to determine how the protein is inserted into the membrane and how it adopts its final configuration.

1. Physical analysis of membrane proteins

The proteins for which we have precise structural information are almost all soluble proteins. Membrane proteins are more difficult to study because they are generally insoluble in water or other polar solvents. Some measurements can be obtained with membrane proteins dissolved in nonionic detergent micelles. Circular dichroism and other spectroscopic measurements provide information on the extent of α-helices and β structures in such proteins but cannot predict the arrangement of the various structures.

Membrane proteins are notoriously difficult to crystallize in a form suitable for X-ray diffraction. Only one membrane–protein complex (the photosynthetic center of the bacterium *Rhodopseudomonas viridis*) (239) has been successfully analyzed in this way, but even this limited information has been useful in validating structural predictions based on the primary sequence.

Some membrane proteins form regular two-dimensional crystalline arrays (planar crystals) amenable to electron microscopic analysis. Highly sensitive scanning transmission electron microscopy coupled with image processing can provide a great deal of information on the surface structure of these arrays, allowing such features as pores to be detected and providing at least some information on the organization of the protein in the membrane. Large amino acids can also be resolved by electron diffraction of planar crystals.

2. Epitope and domain mapping

Information on membrane protein organization can be obtained by defining parts of the polypeptide chain which are exposed on the two faces of the membrane. The range of reagents used includes proteases, membrane-impermeant protein-modifying agents, antibodies, and other ligands (for proteins with receptor activities). Some protein-modifying agents react with specific amino acids. Many such reagents cannot cross the membrane and therefore react only with residues exposed on the cell surface. Modified proteins can be purified, subjected to peptide mapping to identify the affected parts of the protein, and ultimately sequenced to define which residues are affected. This process is laborious, but it may be useful if the protein has only a limited number of potentially reactive sites. Proteases are used with whole or permeabilized cells, membrane vesicles, or protein–detergent micelles. The resulting fragments are mapped and partially sequenced to determine their location and hence the cleavage site in the primary sequence (1005). Membrane proteins are often protease-resistant, necessitating the use of unusually high amounts of protease to obtain efficient cleavage.

Whole cells of wild-type or mutant strains can be screened for their reactivity with antibodies raised against *in vitro*-synthesized peptides corresponding to different sequences within the protein, or with monoclonal antibodies. An interesting example of this approach is the study by Sayre *et al.* (999) of the topology of a chloroplast thylacoid membrane protein. In this case, the epitope map differs markedly from that predicted from the assumption that all hydrophobic segments span the thylacoid membrane (see Section II.D.3.). In bacteria, mutants which no longer bind surface-reactive antibodies can be obtained by selecting for resistance to the antibody-mediated bactericidal effects of complement (231). The genes coding for these altered proteins can then be sequenced to determine which residues have been changed and hence which regions of the protein are normally accessible to antibodies. Other agents such as viruses (called *bacteriophages* in the case of bacteria) and toxins which bind to protein receptors on the outside of the cell may also be used to select resistant mutants. In all cases, results should be interpreted with caution because antigen or receptor determinants may be composed of discontinuous amino acids in the polypeptide chain (47). These approaches are well suited to defining sites which are exposed on the cell surface but are less useful for mapping sites on the inner face of the membrane. A recent development in this area has been the use of gene fusions to construct membrane proteins containing novel epitopes at sites predicted to be exposed on the cell surface (Section II.D.4).

Intron–exon boundaries in eukaryotic genes for cytoplasmic proteins often correspond to the borders between distinct domains in the polypeptide. It remains to be seen whether introns in genes for membrane polypeptides also mark the borders between different protein domains.

It is often important to establish the stage at which a protein adopts its final conformation during transport through the cell. Unfolded or partially folded polypeptides are often more sensitive to proteases than fully folded proteins and may lack some epitopes recognized by specific monoclonal antibodies. These biochemical tools therefore provide simple tests for measuring the kinetics of protein folding (1222).

3. Structure predictions

Many different programs have been devised for predicting protein secondary structure, the most commonly used being those of Chou and Fasman (173) and of Garnier *et al.* (350). All are based on the physical and chemical properties of individual amino acids (hydrophobicity, size, charge, and presence of side chains) and on the amino acids' collective tendency to form identifiable structures in proteins whose secondary structure has been determined. Properties may be weighted according to their supposed influence in known structures. For example, amino acids such as proline or glycine are often present where polypeptides have sharp turns, long stretches of hydrophobic amino acids (as determined by a hydropathy plot) (605) usually have a strong affinity for lipids, and charged residues are rarely found in transmembrane segments. These transmembrane stretches of amino acids have a strong tendency to form α helices with a low hydrophobic moment (a measure of the periodicity of hydrophobic residues along the length of a given sequence). "Membrane-seeking" stretches of amino acids may have lower overall hydrophobicity but a considerably higher hydrophobic moment (276).

It is unlikely that structure predictions will ever be 100% accurate. They are more applicable to small polypeptides or short lengths of a protein which function independently of the bulk of the polypeptide (e.g., routing signals). Predicting the structure of membrane proteins is even more difficult because of the unknown effects of the lipid environment, which are not taken into consideration in programs such as that of Chou and Fasman. In conclusion, structure predictions backed up by at least some physical characterization may be valuable for determining membrane protein structure and should be useful for studying how proteins enter and cross lipid bilayers, but, as pointed out by Wallace *et al.* (1158), currently available methods based on limited data bases from soluble proteins are inherently inaccurate.

4. Genetic techniques

Targeting or other signals may be tentatively identified by sequence analyses. Their function can be verified by changing them so that they no longer function, or by placing them in a different context to see whether they can target another polypeptide to the same location as the protein from which they were derived. Although this can sometimes be achieved by manipulating the protein itself, it is much simpler to do so by genetic methods.

a. *Gene fusions*

Gene fusions are widely used in studies of protein targeting. Basically, the object is to create gene fusions coding for part of the sequence of a targeted protein coupled to the sequence of another protein. Gene fusions are usually constructed by recombinant genetic techniques [reviewed in (898)], although *in vivo* techniques have also been successfully used in bacteria to create banks of essentially random gene fusions (147,688). One advantage of gene fusions is that "reporter" or "passenger" proteins may be easier to detect and assay than the targeted protein itself. Table II.1

Table II.1. Desirable features of reporter proteins for studying protein targeting

1.	Simple genetic construction
2.	Simple screening or selection for in-phase fusions
3.	Simple assay and histological or immunocytochemical detection
4.	Hybrid proteins must be stable
5.	Should be a low-molecular-weight, monomeric polypeptide
6.	Should retain activity when N-terminus is replaced
7.	Should have no "irreversible" secondary structure
8.	Should not aggregate when overproduced
9.	Should not have "membrane-jamming" sequences
10.	Should not have intrinsic targeting signals

lists some of the features desirable in reporter proteins. It is very unlikely that any protein could embody all of these properties, and the choice of reporter is usually a compromise. Cytoplasmic proteins are often preferred because they should not contain targeting signals, but they may be unable to cross certain membranes (see Tables II.2 and II.3). Some reporter proteins are active only when present in a certain cell compartment, permitting the direct selection of appropriate targeting signals or screening for mutations affecting protein targeting.

Some examples of reporter proteins which have been used in protein targeting studies are listed in Tables II.2 and II.3. Other systems are being continually developed and researched (see also Chapter IX). The most commonly used gene fusion is one which replaces the N-terminus of the reporter protein by various lengths of a targeted protein, retaining the initiation codon of the targeted protein gene and, wherever possible, the enzymatic activity of the reporter (Fig. II.2). Many enzymes remain active when their extreme N-terminus is drastically altered.

Different reporters have been used with varying degrees of success, and none is ideal for all purposes. One drawback to the use of gene fusions is that the inclusion of supposed targeting signals in the sequence of a normally cytoplasmic protein places it in a totally different context, in which it may not be able to function correctly (935). Negative results (failure to target the hybrid protein) are not as easy to interpret as positive results because of the often drastic conformational changes which might be introduced when sequences from two polypeptides are fused. Patch signals (Section I.D.2) cannot be detected by gene or protein fusion analyses because they are not composed of colinear residues.

A novel development has been the recent use of gene fusions for determining the configuration of membrane proteins. Sandwich gene fusions (Fig. II.2), in which short oligonucleotides are inserted within the coding

Table II.2. Proteins with potential applications as reporter proteins

Protein	Normal location	Size	Active form	Simple to screen or select	Colorimetric assay
β-Galactosidase (*E. coli*)	Cytoplasm	117 kDa	Tetramer	+	+
Alkaline phosphatase[a] (*E. coli*)	Periplasm	47 kDa	Dimer with intramolecular S–S & Zn^{2+}	+	+
Neomycin phosphotransferase (bacterial)	Cytoplasm	28 kDa	Monomer	+[b]	–
Chloramphenicol transacetylase (bacterial)	Cytoplasm	24 kDa	Tetramer	+[b]	–
Dihydrofolate reductase (mouse)	Cytoplasm	18 kDa	Monomer	–[c]	–
Pyruvate kinase (chicken)	Cytoplasm	57 kDa	Tetramer	–	–
β-Lactamase[a] (bacterial)	Periplasm	32 kDa	Monomer	+[b]	+
α Globin (monkey red blood cells)	Cytoplasm	15 kDa	Monomer	–	–
Invertase[a] (yeast)	Secreted & cytoplasm	60 kDa	Dimer or octamer	+	+
Acid phosphatase[a] (yeast)	Secreted	56 kDa	Monomer	+	+
Galactokinase (*E. coli*)	Cytoplasm	40 kDa	?	+	+

[a] Lacking endogenous signal peptide. The natural cytoplasmic invertase isomer lacks a signal peptide.
[b] In-frame gene fusions may confer antibiotic resistance.
[c] Selection for in-frame fusions may be possible in bacteria lacking Dihydrofolate reductase.

sequence of a gene, can be used to determine the topology of membrane proteins (Fig. II.3). If the region in which the additional amino acids are exposed is on the surface of the membrane, the residues encoded by the oligonucleotide should also be exposed and consequently detectable either by epitope-specific antibodies or by changes in proteolytic cleavage patterns (153). An alternative procedure is to use 3′ gene fusions. In integral membrane proteins, the orientation of the inserted epitope or other sequence depends on the position of the site in the membrane protein into which it has been inserted, as illustrated in Fig. II.3. Excellent examples of this technique come from studies on the topology of cytoplas-

Table II.3. Examples of the use of reporter proteins in protein targeting studies

Protein	Examples of use in potentially targeted proteins[a]							References[b]
	Bacterial export	RER sec, route	Mitochondria	Chloroplast	Nucleus	Vacuole or lysosome	Peroxisome	
β-Galactosidase	–[c]	–	+	NT	+	NT	NT	76,255,280,281,416 459
Alkaline phosphatase	+	NT	NT	NT	NT	NT	NT	
Neomycin phosphotransferase	NT	NT	NT	+	NT	NT	NT	129
Chloramphenicol acetyltransferase	–[c,d]	+[e]	+	+	NT	NT	NT	111,505,640
Dihydrofolate reductase	–[c,d]	NT	+	NT	NT	NT	+	361,392,494,495
Pyruvate kinase	NT	+	NT	NT	+	NT	NT	450a,933
β-Lactamase	+	+	NT	NT	NT	NT	NT	369,1049
α-Globin	NT	+	NT	NT	+	NT	NT	219,220,1049
Invertase	NT	+	+	NT	NT	+	NT	279,282,541
Acid phosphatase	NT	+	NT	NT	NT	NT	NT	1041
Galactokinase	NT	NT	NT	NT	+	NT	NT	667

[a] +, Targeted correctly in at least some cases; –, not targeted or incorrectly targeted; NT, not tested.
[b] The reference list is intended as a guide only; see text for further details and references.
[c] Often toxic to producing cells.
[d] B. Dobberstein and I. Palva, personal communications.
[e] *In vitro* using dog pancreatic microsomes.

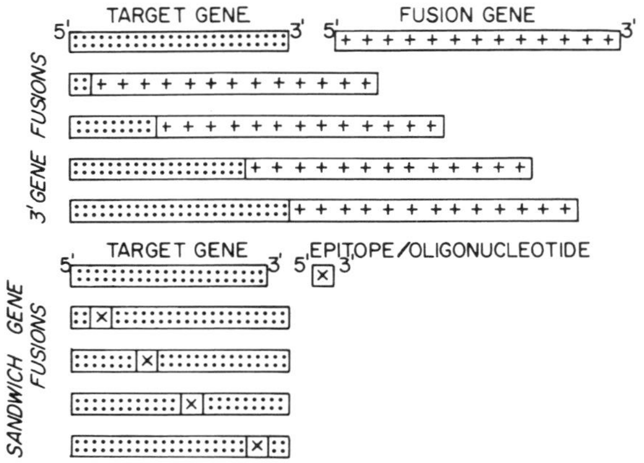

Fig. II.2. Two methods for creating in-frame fusions. In 3′ gene fusions, a foreign sequence (i.e., fusion gene) is fused at the 3′ end of varying lengths of the target gene. The encoded protein will contain varying lengths of the sequence of the targeted polypeptide together with the almost complete sequence of the reporter polypeptide (lacking its N-terminus) as a C-terminal tail. In sandwich gene fusions, foreign sequences (e.g., an oligonucleotide coding for an antigenetic peptide or for a presumed targeting signal) are inserted at different sites within the reading frame of the target gene.

mic membrane proteins in the bacterium *Escherichia coli*. The reporter proteins used are alkaline phosphatase, which is only active when exposed on the outside of the cytoplasmic membrane (689); β-lactamase, which confers high-level penicillin resistance only when it is exposed on the periplasmic face of the membrane (272); and β-galactosidase, which usually cannot cross the cytoplasmic membrane, so that only hybrids in which the β-galactosidase part of the polypeptide is free to tetramerize in the cytoplasm exhibit high enzymatic activity. However, a report by Boyd *et al.* (113) indicates that hybrid proteins need not necessarily adopt the same transmembrane configuration as the membrane protein from which they are derived.

b. Mutations affecting targeting signals

Targeting signals may also be identified and further characterized by studying mutations which alter their primary sequence. The most direct method for obtaining such mutants is site-directed mutagenesis. In general, the changes are designed such that the new amino acid has drastically different properties from those of the residue it replaces. However, such changes may affect overall protein configuration, and subtle amino acid substitutions may be more useful in defining the precise nature of a targeting sequence.

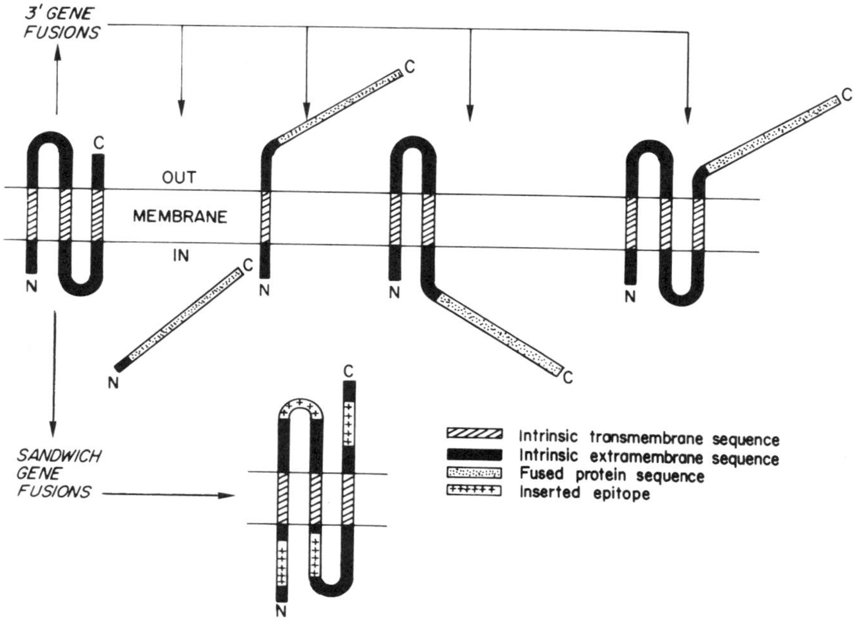

Fig. II.3. Consequences of 3′ gene fusions and sandwich gene fusions into a type III polytopic membrane protein. In this example, the polypeptide has three transmembrane sequences and four hydrophilic sequences, two each exposed on either side of the membrane. 3′ gene fusions with junctions at sites within sequences coding for the hydrophilic loops or the extreme N- and C-termini will encode hybrid polypeptides in which the reporter sequence is exposed accordingly on the membrane surface. The disposition of the epitope inserted by sandwich gene fusions will be similarly determined. Antibody or protease accessibility can be used to determine the location of the reporter protein or the inserted epitope.

Mutations affecting protein targeting may also arise spontaneously or following more generalized mutagenesis. They may occasionally be identified in mutants producing a protein which does not function correctly due to incorrect localization, and in some cases may be selected directly. For example, aborted export of hybrid polypeptides containing nonexportable β-galactosidase fused to the N-terminus of an exported protein (including the signal peptide) may eventually kill the cells, whereas mutants producing hybrids with a defective signal peptide survive (Section III.A.4). Overproduction of some exported proteins, or the production of artificially truncated exported proteins, may also be toxic to bacteria, again facilitating the selection of mutations affecting signal peptide structure. Other mutations may be obtained by the same process, and signal peptide mutations themselves are useful for the selection of mutations in genes involved in the export pathway (see Fig. II.4, Section II.F.1). A similar strategy, this time based on the lack of essential function of a normally cytoplasmic protein when it is localized to the lumen of the RER, has been developed for use in the yeast *Saccharomyces cerevisiae* by Deshaies and Schekman (233).

5. Chemically cross-linked polypeptides

Chemical methods can also be used to construct hybrid polypeptides. Segments of two polypeptides can be cross-linked using chemical reagents and their distribution followed after microinjection into suitable cells (Section II.F.2). Although the applications of this technique are clearly rather limited, it has been successfully used in the study of nuclear protein targeting (Chapter VII). One advantage is that proteins can be radiolabeled *in vitro* before they are injected.

6. Amino acid analogs

Amino acid analogs may be incorporated into proteins during translation, when they replace the natural amino acid in the translation mixture. *In vitro*, it is possible to ensure that the analog is incorporated at every site which would normally be occupied by the natural residue. Analogs often have very different properties (hydrophobicity, charge, etc.) from their natural counterparts and so, as with substitutions introduced by genetic techniques, it is possible to study the effects of these changes on protein targeting in an *in vitro* system (472,473,939). The obvious disadvantage is that several sites will be modified throughout the length of the polypeptide, and it is therefore difficult to determine the importance of individual residues.

E. Expression systems and model proteins

A further application of molecular genetics has been the development of expression systems enabling proteins to be produced in large quantities either *in vitro* or *in vivo*. *In vivo* systems are based on autonomously replicating fragments of DNA (plasmids, viruses) or DNA fragments which can integrate into the chromosome. Additional elements incorporated into these vectors include genes which permit their maintenance in host cells, and a regulated promoter and other signals for efficient expression of the inserted gene.

The advantages of expression systems include the ability to produce a targeted protein at higher than normal levels, thereby simplifying detection and analysis, although overproduced targeted proteins often aggregate or are misrouted. A further advantage is that proteins can be produced in cells from which they are normally absent. This may be particularly important in the study of protein targeting in eukaryotic cells, in which the choice of cells is limited by the ease with which they can be grown in culture. The advantage of studying the targeting of viral proteins is that host cell protein synthesis is often drastically reduced as a result of viral infection, so virus-encoded proteins predominate.

In vitro reconstitution systems are being increasingly used to study protein targeting (Section II.F.3). These systems require precursor polypeptide, which can be produced in radiolabeled form by *in vitro* synthesis primed with the appropriate structural gene together with factors necessary for its transcription and translation (RNA polymerase and transcription activating factors, ribosomes, and translation initiation and elongation factors). Radioactive amino acids, usually [^{35}S]methionine or [^{35}S]cysteine, are incorporated into the reaction mixture to label the precursor protein.

Mitochondria do not use the universal genetic code. Therefore, mitochondrial genes cannot be expressed in "standard" *in vitro* translation systems. Site-directed mutagenesis can be used to change the "nonuniversal" codons to make them compatible with these translation mixtures.

F. The targeting pathway

Targeted proteins interact directly or indirectly with a variety of other proteins en route to their final destination. These include peptidases involved in processing and maturation, other enzymes which introduce posttranslational modifications, receptors, and proteins which mediate vesicle formation and other membrane fusion events. These proteins can be analyzed in the same way as the targeted protein itself, i.e., identification and characterization of enzymatic activity and cellular location or alteration or inactivation by mutagenesis. Such studies can indicate the route taken by a targeted protein as it moves through the cell. Mutations affecting steps in the secretory pathway may cause the accumulation of normally targeted proteins at the stage immediately preceding that affected by the mutation. There is an unfortunate tendency to overlook other cell components which might contribute to protein targeting, including lipids in the various donor and acceptor membranes. Some of the methods described below might also be used to study these components of the targeting pathway.

1. Mutations affecting protein targeting

Increasing use is being made of mutations affecting protein targeting in bacteria and in some eukaryotes. The first group of mutations generally affect very specific classes of proteins, those whose targeting does not depend on general targeting pathways or those which reach their final destination via a minor branch of a general targeting pathway. Mutations which block the targeting of these proteins are therefore less likely to have a deleterious effect on cell metabolism than the second group of mutations, which directly or indirectly affect general targeting pathways. Methods for selecting the latter mutations are discussed in the following sections.

a. *Bacteria*

The methods used to select mutations affecting the function of the secretory pathway in bacteria (mainly *E. coli*) (Fig. II.4) are essentially extensions of the techniques described in Section II.D.4 for selecting mutations affecting targeting signals. The basic idea is that mutations affecting one component of the secretory pathway can be suppressed by a second mutation affecting another component of the same pathway, provided that the two act together. The essential features of the selection procedures are mutations which prevent or reduce the efficiency of general protein export and which are consequently deleterious to the cell (probably because of defective cell surface assembly), and mutations or conditions which specifically prevent or reduce the export of proteins whose activity is required only under certain conditions. Four classes of mutations can be envisioned:

(i) Mutations which specifically suppress the effect of a defective signal peptide (Sds) can be obtained by selecting for restored export and processing of the affected precursor protein. This is the simplest type of interaction revealed by suppressor selection and implies that the product of the gene affected in the Sds mutant interacts directly with the signal peptide (42,277,278).

(ii) Mutations which prevent protein export (Pex^-) can be obtained by screening banks of cold- or temperature-sensitive mutants for strains which fail to process secretory protein precursor signal peptides at restrictive temperatures. Pex^- mutations will also overcome the lethal effects of or restore full activity to β-galactosidase hybrids by preventing them from entering the secretory pathway. Pex^- mutations generally have pleiotropic effects on all proteins targeted along the same pathway, and most are lethal. The mutated genes code for components of the secretory pathway which function more

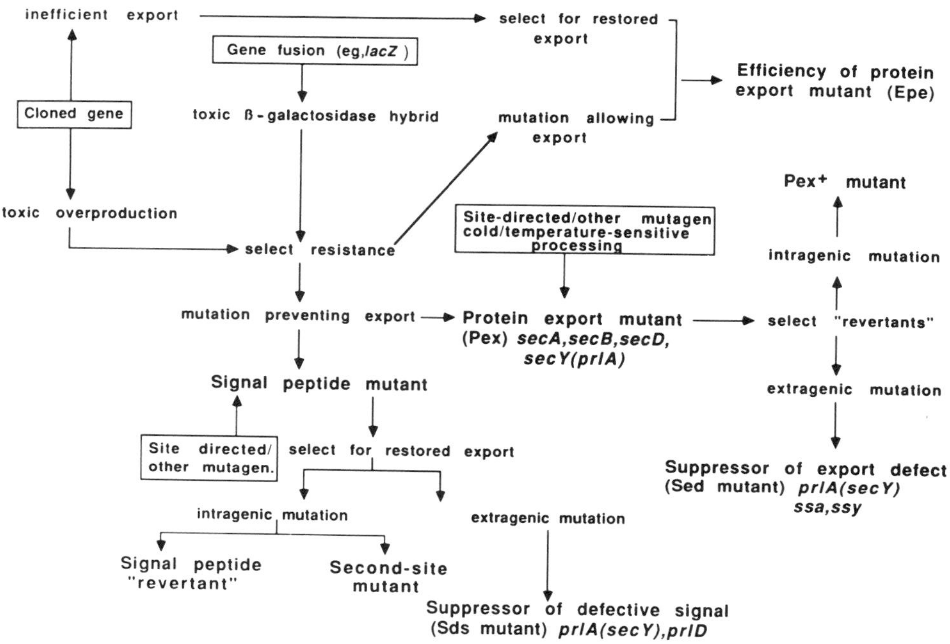

Fig. II.4. Flow diagram illustrating methods for selecting mutations affecting protein export functions in bacteria. The production of some hybrid proteins containing a signal peptide and a normally cytoplasmic sequence, or the overproduction of some exported proteins, may saturate or block the export machinery, causing cell death. Survivors (resistant) have mutations which either specifically block the entry of the particular protein into the export pathway (usually signal peptide mutations), or exert a general block on the export of some or all other exported proteins (Pex mutants). These mutations may be lethal unless the export of only a specific subclass of nonessential proteins is affected; therefore, mutations provoking the loss of export functions at a restrictive temperature (e.g., 42°C), but not at a permissive temperature (e.g., 30°C), are selected. Similar mutations may be detected following mutagenesis and screening (e.g., temperature-sensitive mutants which accumulate secretory protein precursors at the nonpermissive temperature). Likewise, site-directed or localized mutagenesis can be used to change the sequence of DNA coding for the signal peptides of secretory protein precursors.

Revertants of signal peptide mutations may be selected by virtue of the restoration of wild-type phenotype associated with renewed export of the protein. For this reason, it is usual to choose proteins whose absence confers a growth defect on the cells under certain circumstances. These mutations may be either true revertants of the original mutation or second-site mutations affecting another region of the signal peptide or, occasionally, amino acids in the mature part of the polypeptide. Alternatively, extragenic suppressors of signal peptide mutations can be selected (Sds mutants), Pex$^-$ mutants also revert to growth at the restrictive temperature. Some of these reversions may affect the gene originally mutated to give the Pex$^-$ phenotype, but pseudorevertants affecting a second gene may also be selected (Sed mutants). Finally, it may be possible to select for mutations with increased efficiency of protein export (Epe) in cases in which overproduction of normally exported proteins saturates the export pathway and, for example, prevents the export of another, essential protein. Further details of the types of mutants selected using these approaches are given in Chapter III.

or less normally at a permissive temperature but which are nonfunctional (or not produced) at the restrictive temperature. Since the mutations may have some effect even at permissive temperatures, this procedure may be biased towards the selection of mutations which have greater effect on the export and processing of nonessential proteins or proteins which are not toxic when they accumulate in the unprocessed form (349,524,809).

(iii) Mutants exhibiting increased efficiency of protein export (Epe) could theoretically be obtained by counterselection against the toxicity or other selective disadvantages of the overproduction of secretory protein precursors and consequent general or specific defects in export and processing. These mutations might affect the production or action of proteins interacting directly with the precursor(s) during transit through the cytoplasmic membrane. No mutations of this type have been reported so far.

(iv) Revertants of Pex^- mutants could be obtained either by selecting for reversal of the cold or heat sensitivity or for renewed export of an essential protein at partially or totally restrictive temperatures. The mutations in these revertants may be either in the gene affected in the original Pex^- mutant (Pex^+ revertants), or in another gene coding for a component of the secretory pathway (Sed mutant) (808).

An additional selection procedure which is not illustrated in Fig. II.4 is based on the observation that synthesis of SEC A protein, the product of a gene which mutates to give Pex^- clones, is increased when protein export is blocked (810). Thus, direct selection for overexpression of *secA* [in a *secA*$^+$, ϕ(*secA–lacZ*) gene fusion diploid strain] by virtue of increased β-galactosidase production produces Pex^- mutants including *secA, secD, secE,* and *secY* (928).

Several types of Pex, Sed, and Sds mutants have been isolated and characterized (Sections III.C and II.E.2). Interestingly, mutations in the *secY* gene have been isolated by several different selection procedures, which is strong evidence that SEC Y protein is directly involved in the secretory pathway. Many Sed mutants, however, carry mutations which reduce the overall rate of protein synthesis (616,1034). They are thought to suppress the effects of Pex^- mutations by increasing the time during which precursor proteins can interact with components of the secretory pathway while remaining attached to the ribosome.

b. Saccharomyces cerevisiae

The budding yeast *S. cerevisiae* is particularly amenable to genetic analysis of protein targeting because of the ease with which it can be grown and manipulated by molecular genetics. In theory, the techniques described below are equally applicable to the yeast *Schizosaccharomyces pombei* (which divides by a fission process more akin to cell division in other eukaryotes) and to some other single-celled eukaryotes.

Schekman and his collaborators have developed a simple procedure for identifying mutants affected in the targeting of proteins by the secretory pathway (*sec* mutants). The selection procedure is based on the fact that protein export and secretion are coupled to cell surface growth. A bank of mutants which failed to grow at the restrictive temperature (37°C) was screened for clones which exported active invertase and acid phosphatase (both periplasmic proteins) only at the permissive temperature (25°C) (794,795). This selection procedure yielded two mutants. More mutants were obtained following a refinement of the screening technique which was based on the observation that *sec*-mutant cells were denser than wild-type cells when incubated at 37°C.

Different *sec* mutants affect different stages in the secretory pathway, as judged by the sites of invertase or acid phosphatase accumulation, by the extents to which these enzymes were posttranslationally modified, and by morphological changes in secretory organelles. Rather surprisingly, very few mutations affecting the early stages of protein secretion were obtained, possibly because strains carrying these mutations died during the 3-hr incubation at 37°C prior to density centrifugation. Two new selection procedures were therefore developed to overcome this bias. The first, developed by Deshaies and Schekman (233), is similar to one used to obtain mutations affecting protein export in *E. coli*. Signal peptides from yeast secretory proteins were genetically fused to the normally cytoplasmic protein histidinol dehydrogenase. The gene fusions were introduced into a mutant strain lacking the histidinol dehydrogenase gene *HIS4*, which is essential for growth in the absence of histidine. The hybrids were targeted into the RER, and the strains therefore grew only in the presence of histidine. This provided a simple method for RER-targeting selective mutants which could grow in the absence of histidine. These mutations could specifically affect targeting of the hybrid proteins (e.g., mutations preventing signal peptide function) or could have a general effect on RER targeting. The latter mutants were identified among those exhibiting temperature-sensitive growth (discussed previously).

The second selection procedure is based on the idea that mutations blocking early stages in the secretory pathway should drastically reduce the normally extensive covalent mannose modification, which occurs mainly in the Golgi apparatus (292). A modification of the [^{3}H]mannose suicide selection procedure originally developed for the obtainment of glycosylation-defective mutants (490) allowed Newman and Ferro-Novick (775) to obtain further mutants which were blocked in a stage in protein secretion between the RER and the Golgi at the restrictive growth temperature. The results of studies on these mutations are summarized in Fig. II.5 and are discussed in more detail in Chapter III.

Further novel mutations affecting the secretory process were reported by Smith *et al.* (1065). Increased secretion of normally poorly secreted hybrid proteins comprising the invertase signal peptide and bovine chymosin was detected by the larger zones of prochymosin activity produced by the mutants growing on agar plates. The colonies were overlayered with skimmed milk agar at low pH, which activated prochymosin, resulting in casein hydrolysis and the appearance of clear zones around the colonies. This simple approach could be developed into a suppressor selection procedure similar to that used in *E. coli*. Attempts are now being made to select mutations which affect secretory protein targeting in com-

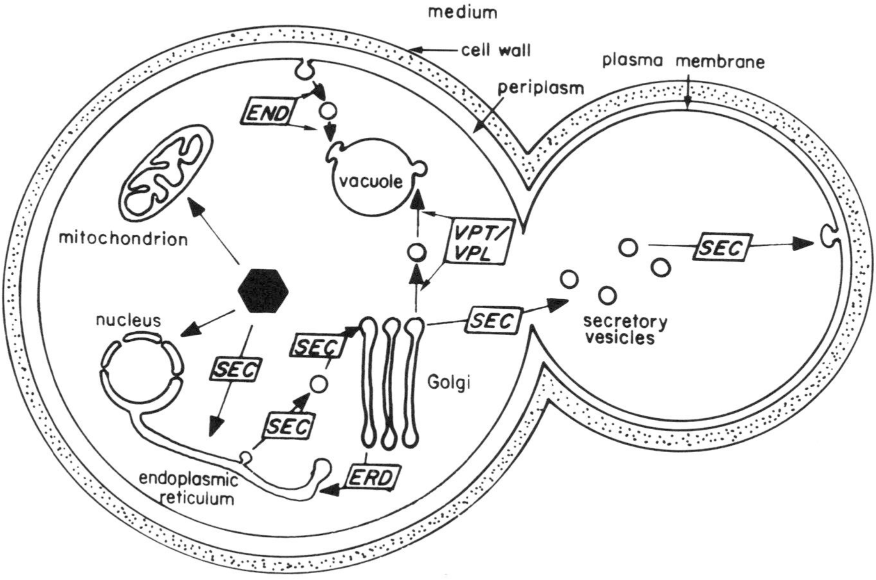

Fig. II.5. Categories of mutations which affect protein traffic in the budding yeast *Saccharomyces cerevisiae*. The hexagon represents a protein whose synthesis is initiated on free ribosomes in the cytoplasm. Classes of genes known to affect various stages in protein targeting are indicated in boxes. Additional mutations affecting protein processing and modification in the Golgi apparatus and mutations affecting intra-Golgi movement have also been isolated. The *ERD* genes, of which only one has been identified so far, appear to affect the recycling of escaped ER proteins (858) (see Section V.B.4). Additional *SEC*-type genes necessary for protein translocation into the ER have been called *PTL* by D. Meyer. Some *sec* mutations also affect endocytosis and the vacuolar branch of the secretory pathway.

plex eukaryotic cell lines (1128). These mutants may be useful for identifying key components of the secretory pathway in complex eukaryotes.

Mutations affecting the vacuolar branch of the *S. cerevisiae* secretory pathway (*vpt* and *vpl*) (Fig. II.5) have been isolated by Rothman and Stevens (971) and Bankaitis *et al.* (44). The selection procedures used were based on the observation that overproduction of either the normally vacuolar carboxypeptidase Y (CPY) (1083) or carboxypeptidase Y–invertase hybrids (CPY–INV) (44) led to their secretion into the periplasm, probably because the vacuolar sorting was saturated by the overproduced proteins. Mutations which caused the misrouting of CPY–INV to the periplasm when it was not overproduced allowed a strain which was otherwise devoid of invertase to grow on sucrose as a sole carbon source. Similarly, mutations which caused the secretion (and fortuitous activation) of CPY allowed a leucine auxotroph to utilize the N-blocked dipeptide L-phenylalanine-L-leucine (Section V.G.5.c).

The endocytotic pathway in yeast cells is also amenable to genetic analysis. Chvatchko *et al.* (175) screened a bank of temperature-sensitive mutants for their ability to accumulate lucifer yellow, which is taken up by fluid-phase endocytosis (Chapter VIII) at the restrictive growth temperature. Two mutants which were endocytosis-defective (*end*) but which continued to secrete invertase normally (SEC^+) were obtained (Fig. II.5). Both mutants were defective in their response to pheromone α, which is taken up by receptor-mediated endocytosis. However, both mutants turned out to be endocytosis-defective at the permissive growth temperature (24°C), suggesting that the defect in endocytosis did not itself cause the temperature-sensitive phenotype (Section VIII.B.3). This result underlines the need for caution in studies on temperature-sensitive mutants which are defective in protein targeting.

2. *In vivo* studies

Protein targeting pathways can be studied *in vivo* in a variety of organisms. First, specific stages in the pathway can be blocked or reduced in efficiency. For example, different stages in the secretory pathway of complex eukaryotes exhibit different degrees of cold sensitivity, allowing secreted proteins to be "trapped" at different stages along the route by changing the temperature (398,995). Second, inhibitors of protein secretion including the ionphores monesin and drefeldin A, which act at different stages of the secretory route and alter secretory organelle (RER and Golgi) morphology (457), can be used. Third, the role of pH in protein targeting can be studied in cells treated with weak bases such as NH_4Cl or chloroquinone (18). Anderson *et al.* (19) have developed a simple technique for measuring the pH of secretory organelles which is based on the preferential accumulation of a dinitrophenol derivative in acidic compartments and its immunocytochemical detection with monoclonal antibodies. Results obtained by this technique are comparable to those reported by others using techniques based on the quenching of fluoresceine at low pH (807).

The antibiotic tunicamycin inhibits UDP-GlcNAc : dol-P GlcNAc transferase, the enzyme located on the cytoplasmic side of the RER membrane which catalyzes the initial step in the assembly of the core complex which is eventually added onto Asn residues to produce N-glycosylated proteins (Section V.B.1). Consequently, tunicamycin can be used to study the importance of N-linked glycosylation on protein targeting because tunicamycin-treated cells produce exclusively nonglycosylated secretory proteins (Section II.C).

Membrane perturbants have also been extensively used in *in vivo* studies of protein targeting, but their effects are often nonspecific. Membrane perturbants alter the state of membrane lipids, making them more "fluid" or more "crystalline." Temperature changes may have the same effect if they occur around the phase transition temperature. Cells may compensate for these effects by changing the fatty acid composition of their lipids. Similarly, changes in the fluid–crystalline state of membrane lipids can be achieved by growing fatty acid auxotrophic mutants in medium supplemented with fatty acids having different ratios of saturated to unsaturated bonds. Membrane lipid composition can also be changed by using inhibitors of enzymes involved in specific stages of fatty acid biosynthesis or in the synthesis of particular phospholipids, glycolipids, or other lipids, or by mutations affecting these enzymes. These studies may be useful in defining the role of the physical state and chemical composition of membrane lipids in protein targeting.

The requirement for and source of energy used for translocating proteins across membranes can also be studied *in vivo* (Section II.F.4).

Another useful approach to the *in vivo* study of protein targeting is pulse labeling of proteins with radioactive amino acids. Amino acids with high specific activity are employed in order to obtain labeled proteins within periods as short as 10 sec. Incorporation can be stopped by adding cold trichloroacetic acid (TCA) or SDS, and the proteins are then examined by SDS-PAGE. The radioactivity can be chased by washing away the labeled amino acid, replacing it with an excess of unlabeled amino acids, and continuing incubation before adding TCA or hot SDS. Further protein synthesis can be arrested by adding specific inhibitors (e.g., chloramphenicol in bacteria, cyclohexamide in eukaryotes). Changes in the molecular weight of the labeled protein can provide information on the processing steps involved. Cells can also be fractionated at different times after pulse labeling to determine potential sites of transient protein accumulation.

A novel technique which is being increasingly used to study protein targeting is microinjection. The pioneering work in this field used frog oocytes to study the fate of microinjected proteins, particularly nuclear proteins derived from *Xenopus laevis* oocytes and radiolabeled with ^{125}I *in vitro* or with ^3H-labeled amino acids *in vivo* (411) (Section VII.B.1). Briefly, the mixture of labeled proteins is injected into the oocyte cytoplasm. The oocytes are then incubated, and the fate of the protein is followed either by autoradiography (Fig. II.6) or by lysing the cells, separating the nucleus from other cell components, determining the amount of radioactivity it contains, and examining protein content by SDS-PAGE and/or autoradiography. Fluoresceine-tagged proteins can also be injected, in which case fluorescence microscopy is used to follow routing to the nucleus (Fig. II.6). This approach has been further refined by using

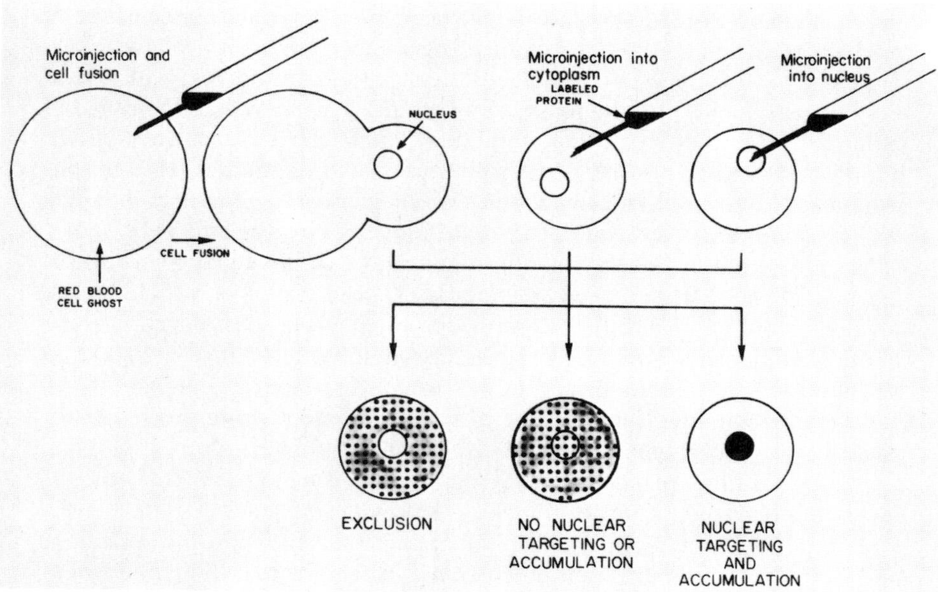

Fig. II.6. An example of the use of microinjection and cell fusion techniques for studying protein targeting. A solution of a suitably tagged protein (e.g., radioactive or fluorescent label) is injected into red blood cell ghosts or into the cytoplasm or nucleus of the cells being studied, using a glass microsyringe. If red blood cell ghosts are used, these are then fused with the cells under study. The cells are incubated and protein movement is followed either by fluorescence microscopy or autoradiography or by lysing the cells and separating the nuclear and postnuclear fractions, which are then examined by SDS-PAGE to study their protein contents.

karyophilic signals chemically cross-linked or genetically fused to other proteins (611) or to inert particles coated with nuclear proteins (306) (Section VII.B.3). Although technically more demanding, it is also possible to microinject proteins into nuclei and to study their retention or diffusion into the cytoplasm (Fig. II.6) (242). Another approach to introducing proteins into cells which are not amenable to microinjection is to inject the proteins into red blood cell ghosts, which are then fused with cells such as HeLa or other tissue culture cell lines (Fig. II.6) (257,1127).

Microinjection and cell fusion can be used to study the targeting of proteins other than those destined for the nucleus. For example, Rothman *et al.* (969,970) used cell fusion techniques to study protein movement through the Golgi apparatus, and Koren *et al.* (575) studied the effects of microinjected signal peptides on secretory protein routing in oocytes. Reports of the use of microinjection to introduce antibodies recognizing specific epitopes on virus-encoded proteins into virus-infected cells (25,590) provide an elegant example of other ways in which microinjection might be used to study protein targeting. Other potential applications of microinjection are its use to change the ionic composition of the cytoplasm or to introduce macromolecules other than antibodies, which cannot normally penetrate the plasma membrane. This increases the range of potential inhibitors of protein targeting which can be used *in vivo*. In addition, DNA or mRNA coding for targeted proteins can also be microinjected into cells, offering a possible alternative to genetic techniques for producing targeted proteins in heterologous cell systems (1187). Localized production of secretory proteins can be achieved by microinjecting mRNA into defined areas of oocytes. These proteins, which can be pulse labeled, enter into the secretory pathway via the RER near the site of mRNA injection, and their progress through the secretory pathway can be monitored by biochemical analysis of cell sections (151,263).

Another technique for studying protein movement along the secretory route involves permeabilizing cells by gentle homogenization or by using a membrane filter to peel off the apical surface membrane from monolayers of polarized tissue culture cells. The cells lose most of their cytosolic components, including nuclei, but retain their endoplasmic reticulum and Golgi apparatus. Proteins can be pulse-labeled with radioactive amino acids and then "frozen" in transit through the secretory pathway by chilling the cells before permeabilization. Protein movement can then be reinitiated by adding back cytosolic extracts and ATP and then warming the cells to 37°C (62,1051).

There is also considerable interest in the study of protein targeting in intact polarized cells, in which different proteins go to different poles of the cell. Polarized cells are relatively easy to study because they maintain their polarized organization when grown on membrane filters, on which they form electrically tight monolayers (Fig. II.7). Different agents can be added to or samples withdrawn from the medium bathing the apical or basolateral surfaces (Fig. II.7). Three cell types have been extensively used: Manine Darby canine kidney (MDCK) cells, hepatocytes, and Caco-2 human colon carcinoma cells. All three cell types mimic *in situ* polarized protein-sorting phenomena when grown in tissue culture. Increasing use is now being made of another intestinal carcinoma cell line, HT29, which can be induced to switch from nonpolarized to polarized morphology by removing glucose from the growth medium (489).

3. *In vitro* studies

One of the most exciting developments in the study of protein targeting has been the use of *in vitro* systems for reconstituting specific steps in targeting pathways. Briefly, these systems comprise a specific mRNA coding for a targeted protein or, in the case of posttranslational targeting, the purified precursor protein, together with the target membrane such as microsomes (closed vesicles derived from the RER), chloroplasts, mitochondria, nuclei, BCM vesicles, or liposomes. These mixtures are used to determine which additional factors, if any, are required for protein–membrane interaction or translocation across the membrane. To determine whether translocation has occurred, the vesicles or organelles are pelleted by centrifugation and treated with a broad-specificity protease such as

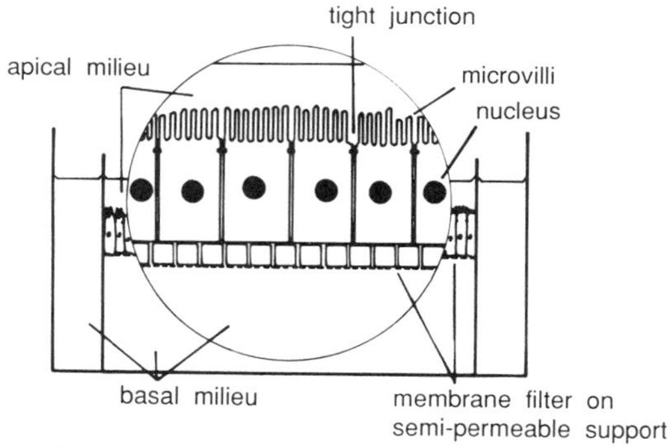

Fig. II.7. Chamber for growing polarized cells in culture on membrane filters. The filters (e.g., Nucleopore™) are coated with collagen and placed on a semipermeable support. The medium is separately supplied to the basal and apical surfaces of the cells in the monolayer. Tight junctions maintain the electrical resistance of the monolayer, which serves as an indication that the filter surface is completely covered with cells. The central region is enlarged to show details of the filter and the cells. See Section V.G.3 for further details on the organization of these cells.

proteinase K. Proteins exposed on the surface of the target membrane will usually be completely degraded, whereas fully or partially translocated proteins will be protected unless the target vesicle or organelle is first lysed by detergents or by other methods (Fig. II.8). Additional indicators of at least partial translocation include the removal of the cleavable N-terminal routing signal (see Table I.1) and glycosylation of proteins translocated across the RER membrane in microsomes (Fig. II.8).

As discussed in Section III.C.2, wheat germ extracts were used to translate mRNA coding for secretory proteins, and dog pancreatic microsomes were used as target membranes in early studies on protein targeting to the RER (90,91). Dog pancreatic microsomes were selected because they contain very low levels of RNase, which would otherwise degrade mRNA in the translation–translocation mixture. There was no reason to assume that dog and wheat systems would be compatible, which raises doubts about the validity of some of the results obtained (714). The recent development of a yeast-derived, coupled translation–translocation system (961,962,1172) and the increasing use of reticulocyte or HeLa tissue culture cell lysate–dog pancreatic microsome systems should resolve some of these difficulties. Several groups have developed similar systems for studying the early stages of bacterial protein export *in vitro* (Section III.C). One of the aims of these studies is to correlate defects observed *in vivo* in bacterial or yeast mutants affected in protein targeting with translocation activity in reconstituted systems.

In vitro tests have been devised to study protein movement from the RER to the Golgi in extracts from yeast cells carrying a *SEC* mutation blocking this step in the secretory pathway (428). Several groups have developed similar *in vitro* systems based on the detection of posttranslational modifications occurring in different Golgi compartments. Donor Golgi vesicles are prepared by fractionation of cells producing the protein being studied but lacking a particular modifying enzyme. Recipient Golgi vesicles are prepared from cells that do not produce the protein being studied, but do contain the modifying enzyme (39,40,966,969) (Chapter V). Movement of the protein from one Golgi cisterna to another is monitored by the incorporation of prosthetic groups (usually from radioactive donors) present only in the acceptor Golgi. Vesicle-mediated protein movement from the ER to Golgi-derived vesicles has also been studied using vesicles immobilized on cellulose nitrate strips (797). Woodman and Edwardson (1204) have developed an *in vitro* assay based on the removal of neuraminic acid residues from a viral envelope protein by a viral neuraminidase present in the recipient membrane as an indicator of protein insertion into the cytoplasmic membrane. Several groups have devised *in vitro* assays for the fusion of endocytic vesicles (115,237). Gruenberg and

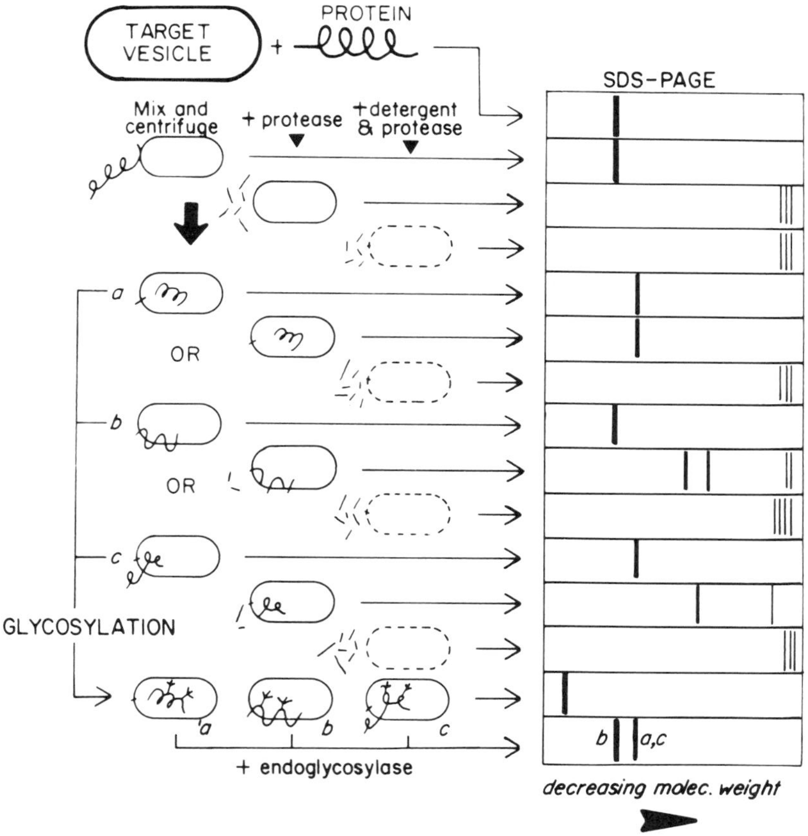

Fig. II.8. Procedures for studying protein targeting *in vitro*. In the simplest system, protein and target vesicle (or organelle) are mixed and incubated at an appropriate temperature. The vesicles are then centrifuged to separate them from free protein, lysed with SDS, and examined by SDS-PAGE. The protein is identified either by autoradiography, in which case radioactive amino acids are incorporated into the protein *in vivo* or *in vitro*, or by immunoblotting with antibodies specific for the protein. Processing of the protein [e.g., removal of a transitory targeting signal (signal peptide, presequence, transit peptide) (*a* and *c*), or glycosylation (*a, b, c*)] is detected by changes in the migration pattern of the protein in the SDS-PAGE gels. Parts of the polypeptide which are exposed on the outside of the vesicle are susceptible to prior treatment with broad spectrum proteases such as proteinase K or pronase. Samples of the same vesicles are lysed with detergents or by other means such as membrane-destructive toxins to check that the parts of the polypeptide protected by the vesicle membrane are in fact susceptible to protease action.

Howell (404,405) used antibody affinity chromatography to purify vesicles containing newly imported vesicular stomatatitis virus glycoprotein G which had been endocytosed after insertion into the plasma membrane by fusion with the virus particles at low pH.

Studying protein targeting *in vitro* has the further advantage that the mixtures are much more amenable to modification than the interior of living cells. Translocation can be "frozen" shortly after initiation (e.g., after binding to receptor under conditions such as low temperature or absence of energy source) and adjacent proteins cross-linked with cleavable cross-linking reactions. Complexes can then be purified and resolved into their component parts. A more elegant approach adopted by Wiedmann *et al.* (1188) and Krieg *et al.* (591) is to introduce modified amino acids into the polypeptide chain of the targeted protein in a suitably modified *in vitro* translation mixture. Some such modified amino acids can be photo-cross-linked to neighboring proteins, allowing specific protein–protein interactions to be detected by SDS-PAGE of radiolabeled material and their subsequent identification by immunoblotting.

4. Energy

Energy is probably required for polypeptide movement into and through a lipid bilayer membrane and the formation of vesicles from the surface of one secretory organelle and subsequent fusion with the surface of the next organelle in the secretory route. Three sources of energy have been proposed: (i) intrinsic energy liberated by the initial protein–membrane interaction (e.g., induced conformational changes in the polypeptide), which might be used to drive the polypeptide through the membrane (288,441); (ii) a hydrolyzable high-energy bond such as that in ATP or other nucleotide triphosphates; or (iii) a transmembrane energy potential comprising electrical and chemical components. The relative importance of these sources of energy in different protein targeting systems are discussed in the relevant sections of this book; here we are concerned with the methods employed in these studies.

ATP may be generated by glycolysis or from the transmembrane potential via membrane ATPases. Conversely, the electrochemical gradient may be generated by ATP hydrolysis via membrane ATPases, or by oxidative phosphorylation via electron transfer through membrane respiratory cytochromes. The two sources of energy are intimately coupled. Cells normally have a pool of ATP and maintain an electrochemical gradient across their plasma membranes. Both sources of energy are used to drive active uptake of nutrients, etc., and possibly also for excretion.

The basic approach to studying the role of energy in protein targeting is to remove all potential sources of exogenous energy and to determine the effect of these changes on protein localization. Although various agents which dissipate membrane potential may be used *in vivo* [e.g., 2,4-dinitrophenol (DNP), cyanide-*m*-chlorophenylhydrazone (CCCP), potassium cyanide, sodium azide], ATP levels may remain high for quite some time, allowing cells to maintain a transmembrane potential via ATP hydrolysis. ATPase activity can be blocked by using mutants or the specific ATPase inhibitor *N*-*N*′-dicyclohexylcarbodiimide. Alternatively, ATP levels may be depleted by growing cells in the presence of arsenate, by prolonged starvation with or without treatment with DNP, by flushing the cells with nitrogen (41), or by apyrase treatment (1172). ATP can then be added back to cells under conditions which do not allow them to regenerate the electrochemical potential. Alternatively, the culture can be supplemented with a source of energy such as lactate, which can only be utilized by respiration, under conditions which prevent ATP formation. Transmembrane potentials can also be artificially imposed by treating cells with specific ionophores and the ion which they transport (e.g., valinomycin, a K^+-specific ionophore).

Energy sources are more readily controlled *in vitro*. In the case of vesicles or liposomes, a source of energy must be supplied, either by continually adding ATP or by adding an ATP-regenerating system such as creatine phosphate plus creatine phosphokinase or phosphoenolpyruvate plus pyruvate kinase (961,1216). Vesicles can be prepared from mutants lacking ATPase activity, or ATPase can be inactivated as described above to prevent NTP hydrolysis from being used to generate a transmembrane proton gradient (360). Apyrase, glucose plus hexokinase, or alkaline phosphatase can remove ATP sources *in vitro*. The requirement for a cleavable high-energy phosphate bond in nucleoside triphosphates can be studied by using nonhydrolyzable analogs (158), and photoaffinity labeling can be used to determine whether nucleoside triphosphates bind to proteins in membrane vesicles (159). In the case of mitochondria, uncouplers can be used *in vitro* in the same way as they can be used *in vivo* in whole cells. The transmembrane chemical potential across bacterial cytoplasmic membrane vesicles can be collapsed with valinomycin plus K^+, whereas nigericin dissipates the transmembrane pH gradient.

Further reading

Fleischer, S., and Fleischer, B. (eds.) (1983). "Methods in Enzymology," Vols. 96–98. Academic Press, New York.

Fleischer, S., and Fleischer, B. (eds.) (1986). "Methods in Enzymology," Vol. 125. Academic Press, Orlando, Florida.

Garoff, H. (1985). Using recombinant DNA techniques to study protein targeting in the eukaryotic cell. *Annu. Rev. Cell Biol.* **1,** 403–445.

Pugsley, A. P. (1988). The use of hybrid proteins in the study of protein targeting signals. *In* "Membrane Biogenesis" (J. A. F. Op den Kamp, ed.), pp. 399–418. Springer-Verlag, Berlin.

CHAPTER III

Early stages in the secretory pathway

Most proteins reach the cell surface and the medium outside the cell via the secretory pathway. In eukaryotic cells, proteins enter the secretory pathway at the rough endoplasmic reticulum membrane (RERM) and travel through the Golgi apparatus to the cell surface or to secretory (storage) granules, lysosomes, or vacuoles (see Fig. I.1 and Section I.B). In this chapter we are concerned only with the earliest stages in this pathway, i.e., secretory protein interaction with the RERM and the events after the protein has crossed or inserted into the RERM. Later stages in the secretory pathway are discussed in Chapter V.

Proteins translocated into the RER lumen remain enclosed within vesicles throughout their passage through the cell. Similarly, most integral membrane proteins probably adopt their final topography as they insert into the RERM, although conformational changes may occur in lumenal and possibly in cytoplasmic loops due to posttranslational modifications occurring in the ER or Golgi (Chapter V). Only at the earliest stage in the pathway can cells distinguish between secretory and nonsecretory proteins. However, as discussed in Chapter V, incorrectly folded or "foreign" proteins may be "rejected" at later stages in the secretory pathway because they lack the correct conformation for transport through the secretory pathway.

The cytoplasmic membrane (CM) is the point of entry into the bacterial secretory pathway. The bacterial CM (BCM), or specific regions thereof, is therefore the functional equivalent of the RER membrane. In this chapter, we will also consider the extent to which the initial stages of the eukaryotic secretory pathway and the export and secretion of proteins in bacteria represent variants of the same basic process.

A. Secretory signal peptides

What is the secretory routing signal which directs proteins to the RERM or BCM? Milstein *et al.* (722) began to answer this question when they observed that secreted immunoglobin light chains are shorter than precursor proteins synthesized *in vitro*. This observation was extended and developed by Blobel and Sabatini (92,93) and by Blobel and Dobberstein (90,91), who found that precursors of secretory proteins synthesized *in vitro* in the presence of RER-derived microsomes were (i) processed to the mature polypeptide and (ii) imported into the lumen of the microsomes (note that microsomes are not naturally occurring cell components; they appear as small ER-derived vesicles when cells are broken open and can be easily purified for use in *in vitro* assays).

The transitory sequences were located at the N-terminus of the immunoglobulin light chain precursors and, significantly, similar sequences were found at the N-termini of other secretory protein precursors but not in cytoplasmic proteins. The signal hypothesis developed and subsequently refined by Blobel and others proposes that these sequences, called *signal peptides,* are primarily concerned with the initial stages in RERM–secretory protein interaction and in catalyzing the translocation of the precursor polypeptide across the RERM. There are numerous examples of secretory proteins which remain largely or entirely cytoplasmic if deprived of a functional signal peptide by mutations affecting the 5′ end of their structural gene (see Section III.A.4 for examples). Signal peptides were thus the first examples of transitory targeting signals to be characterized. The demonstration that bacterial secretory proteins also have signal peptides soon followed Milstein's studies with immunoglobulin light chains, although other techniques such as inhibition of signal peptidases with membrane perturbants and pulse–chase labeling experiments were substituted for the *in vitro* tests used by Milstein.

Many secretory proteins do not have signal peptides. As discussed in Section III.B, these typically membrane proteins have one or more sequences resembling signal peptides in all but the absence of processing by signal peptidase. These sequences target proteins into the RERM or BCM, and are therefore referred to as (noncleavable) secretory signal sequences (Table I.1). Secretory signal sequences usually remain embedded in the membrane, where they act as a membrane anchor (Section III.B).

1. Signal peptide primary structure

What do signal peptides look like? They might be expected to have strictly defined and highly conserved sequences which could be recognized by a specific receptor on the cytoplasmic face of the RERM or BCM. In fact, almost no signal peptides from among the several hundred examples which have been characterized (1174) have exactly the same sequence. However, detailed analyses of large numbers of signal peptides, principally by von Heijne (433–436) and by Perlman and Halvorson (861), have defined a number of characteristics which are shared to greater or lesser extents by all signal peptides (Fig. III.1). These features are

(i) an exclusively N-terminal location
(ii) a total length of 15–35 residues
(iii) a net positively charged region spanning the first 2–10 residues at the extreme N-terminus which usually contains one or more Lys or Arg residues
(iv) a central core region of nine or more residues containing uniquely hydrophobic or neutral residues and with a predicted strong tendency to form an α helix (Physical measurements confirm this to be the case in apolar environments; see Section III.C.1.)
(v) a turn-inducing amino acid (e.g., proline or glycine) next to the central core region and approximately six residues downstream from the cleavage site
(vi) a specific cleavage site recognized by a signal peptidase (Section III.D) and preceded in most cases by small, apolar amino acids (alanine in 50% of known examples) at positions −1 and −3 (−1 being the position immediately in front of the cleavage site)

The length and overall hydrophobicity of the core region may be the critical factor allowing signal peptides to span the 4-nm-thick lipid bilayer

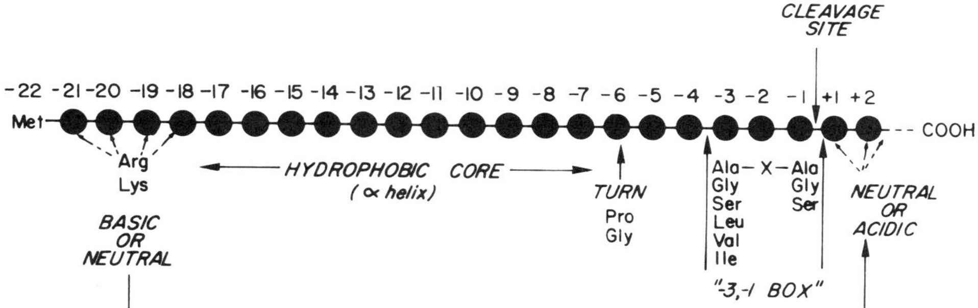

Fig. III.1. General features of a typical bacterial signal peptide. By convention, amino acid residues in the signal peptide and in the mature part of the polypeptide are numbered starting from the cleavage site.

of the RERM or BCM, probably as α helices (29). The turn in front of the cleavage site may break this helix and bring the cleavage site close to the catalytic site of signal peptidase, a transmembrane protein (or protein complex) of the RERM or BCM with its active site on the luminal or periplasmic side of the membrane (Section III.D).

Only some signal peptides exhibit all of these characteristics. Two studies have attempted to catalog features which distinguish signal peptides from precursor proteins of different origins. The detailed analyses of Gascuel and Danchin (351) clearly define 17 independent or semi-independent features called *descriptors* which distinguish between signal peptides of prokaryotic (*E. coli*) and eukaryotic (human) origin. The study by Sjöström *et al.* (1055) suggests other patterns which might be useful for distinguishing between signal peptides of precursors destined to different locations. From these and other studies, we can draw some general conclusions regarding differences in signal peptide sequences which might be important when studying their activity in heterologous systems (either linked to heterologous polypeptides or in different cell lines). For example, signal peptide lengths are quite variable, with longer signal peptides usually having a longer N-terminal basic region. This is particularly striking in the case of signal peptides from secreted (extracellular) proteins of Gram-positive bacteria, which are often longer and have a more basic N-terminus than their periplasmic counterparts in Gram-negative bacteria. Bacterial signal peptides generally have more basic changes at the N-termini and, in particular, are more likely to have a net basic charge difference between the N-terminus of the signal peptide and the N-terminus of the mature polypeptide than are eukaryotic signal peptides (a negatively charged amino acid is often found at position +2 of bacterial secretory proteins) (see Fig. III.1). Von Heijne has suggested that this feature may be related to the unique dependence of bacterial protein export on a transmembrane energy potential (435) (see Section III.E.2.c). Another characteristic feature of bacterial signal peptides is the much higher incidence of turns starting around position −6, which might indicate different requirements for processing by eukaryotic and prokaryotic signal peptidases (Section III.D).

Future analyses using sophisticated computer programs may indicate whether any of these or other emerging patterns have any real significance. However, present knowledge allows signal peptides and signal peptidase cleavage sites to be predicted with relatively high precision (351,436). Specific changes in signal peptide structure introduced by mutagenesis have been extensively used to define the limits of sequence variation which can be tolerated in signal peptides (Section III.A.4).

2. Are signal peptides interchangeable and universal?

According to the signal hypothesis, a signal peptide from one secretory protein should be able to replace the signal peptide of another protein which is targeted to the same site without affecting the efficiency or specificity of protein targeting. Furthermore, one might expect signal peptides to function normally in heterologous environments, e.g., bacterial signal peptides should interact with the RERM in the same way as eukaryotic signal peptides. These predictions have been fulfilled in some but not all cases. For example, the signal peptides of *E. coli* periplasmic and outer and inner membrane proteins (888,1119), of vacuolar and secreted proteins of *S. cerevisiae* (541), and of proteins secreted by the constitutive and regulated pathways in multicellular eukaryotes (135) can be exchanged without affecting the targeting of the mature protein. This is not surprising, however, as signal peptides are proteolytically processed at a very early stage in the secretory pathways, well before proteins are sorted into different branches of the pathway. The translocation of bacterial secretory proteins across the cytoplasmic membrane in heterologous strains of bacteria has also been reported (985,1126), but there are also reports of bacterial secretory proteins which are not exported or secreted when produced by bacteria belonging to a different species (911), and protein export or secretion is often only partially effective in many such recombinant strains (897).

There are relatively few reports of the expression of genes for eukaryotic secretory proteins in bacteria. Among the secretory proteins which were successfully translocated across the BCM are insulin (1105) and ovalbumin (324), both of which were to some extent exported to the periplasm in *E. coli*. In other studies, eukaryotic proteins were overproduced and formed inclusion bodies or aggregates in the bacterial cytoplasm, possibly because of deficiencies in signal peptide function in the new host resulting from differences in signal peptide primary structures in different organisms (Section III.A.1; see also Chapter IX). There are also few examples of full-length DNA clones for bacterial secretory proteins being expressed in eukaryotic cells. *Escherichia coli* pre-β-lactamase is not efficiently processed by signal peptidase or secreted when it is produced by cells of the yeast *S. cerevisiae* (946), whereas processing is efficient in the cases of a hybrid of β-lactamase and the precursor of the *E. coli* outer membrane major lipoprotein (LPP) (877) and of a bacterial α-amylase (976). Yeasts can also correctly process and secrete secretory proteins expressed from cloned DNA from complex eukaryotes (see Chapter IX), but at least one yeast signal peptide does not function in a mammalian cell line (84).

It is difficult to define the features which determine whether a signal peptide will function in a heterologous cell. As we shall see later in this chapter, signal peptides interact with several cell components. Deficiencies in any of these interactions or in signal peptide processing often drastically reduce the efficiency of protein export across the RERM or BCM. *In vitro* systems may be less subject to incompatibility involving the mature part of the polypeptide and the membrane through which it is being translocated; they also avoid the need for the protein to travel through the entire eukaryotic secretory pathway. Pre-β-lactamase (753), hybrid proteins carrying the *E. coli* heat-labile enterotoxin B subunit signal peptide (505), a pre-β-lactamase–α globin hybrid polypeptide (638), and bacteriophage M13 precoat protein (1176) are translocated across or into the RERM and processed by eukaryotic signal peptidase *in vitro*. This suggests that signal peptides may be more universal than is indicated by some *in vivo* experiments.

3. Signal peptide–cytoplasmic protein chimeras

When the signal hypothesis was originally proposed, the signal peptide was thought to be the only requisite for successful translocation of a protein across the RER or bacterial cytoplasmic membrane. Although there are undoubtedly many secretory proteins in which the signal peptide is indeed the only targeting signal (Chapters IV and V), the basic idea that signal peptides can promote the transmembrane movement of *any* protein is clearly not always tenable. The role played by the mature part of secretory proteins is discussed in greater detail in Section III.C.3; at this point, a brief review of studies with hybrid proteins containing signal peptides fused to cytoplasmic proteins will suffice to illustrate the extent to which signal peptides can function correctly irrespective of the polypeptide sequence to which they are attached.

The first *in vivo* studies on such hybrid proteins were with bacterial signal peptides fused to the N-terminus of *E. coli* cytoplasmic β-galactosidase [see Table II.2 and (76)]. Results from studies show that signal peptides can only rarely lead β-galactosidase out of the cytoplasm (331a,737). However, hybrids containing the signal peptide together with a small N-terminal segment of the mature part of periplasmic or outer membrane proteins do at least partially penetrate into the BCM, apparently with the β-galactosidase part of the hybrid exposed on the cytoplasmic face of the membrane (1118), and may be processed by signal peptidase (911a) or fatty-acylated (285) by enzymes located on the periplasmic face of the membrane. In the two cases which have been extensively studied, large amounts of the hybrid protein also accumulated in the cytoplasm either in aggregates (1120) or in membranous structures in the cytoplasm (1152), even though earlier reports suggested that some of them might reach the cell surface (see Chapter IV). These hybrids all have low specific β-galactosidase activity, probably because they cannot tetramerize efficiently. Synthesis of these hybrids is toxic to the cells, probably because they become jammed in the BCM at specific sites involved in protein export. The consequent failure to export enzymes involved in cell wall biosynthesis probably causes the cells to die (285,523,720,1120).

Many explanations can be offered for the failure of β-galactosidase to cross the BCM:

(i) The signal peptide could be buried within the hybrid polypeptide and unable to interact with the protein translocation machinery in the BCM.
(ii) The β-galactosidase could oligomerize before it crosses the membrane.
(iii) The monomeric form of β-galactosidase could fold into a conformation which is incompatible with translocation through the BCM, possibly because it is not recognized by cytoplasmic proteins which prevent normally exported proteins from folding prior to their translocation across the cytoplasmic membrane (Section III.C.3).
(iv) β-Galactosidase could contain so-called stop transfer signals [stretches of mainly hydrophobic amino acids with a strong tendency to remain embedded in the membrane (Section III.E)], the major one being located immediately after residue 97. This would prevent its complete translocation across the BCM.

There are very few reports of attempts to transport other cytoplasmic proteins across the bacterial cytoplasmic membrane. However, chicken triose phosphate isomerase (879), phage T4 protein 37 (673), and superoxide dismutase (1104) are all reported to cross the *E. coli* cytoplasmic membrane when fused to the signal peptide and, in some cases, to part of the mature sequence of an *E. coli* exported signal. These results indicate that there is no insurmountable barrier to the transmembrane movement of cytoplasmic polypeptide sequences in bacteria. Major factors may be (i) the absence of significant secondary structure in the protein to be translocated, (ii) the ability of components of the translocation machinery to prevent folding (competence factors), or (iii) tight translation–translocation coupling (331a) (Section III.C.3).

Signal peptide–β-galactosidase hybrids have also been studied in the yeast *S. cerevisiae* (see Table II.2). Emr *et al.* (280) found that hybrids comprising the signal peptide from secreted invertase, various lengths of the mature invertase polypeptide, and an almost complete β-galactosidase polypeptide were not secreted. Instead, the hybrids accumulated in the lumen of the ER, where they were core-glycosylated. These hybrids were not toxic to the cells, did not prevent normal secretory processes, and did not cause the accumulation of secretory protein precursors. These observations may point to a fundamental difference between the earliest stages in bacterial and eukaryotic secretory pathways; i.e., β-galactosidase can readily cross the RERM but not the BCM when attached to a signal peptide. The explanation for this difference may lie in the tighter coupling between protein synthesis and translocation in eukaryotes, which could ensure that the β-galactosidase polypeptide crosses the RERM well before its synthesis is complete (Section III.C.3). Somewhat different results were obtained with acid phosphatase–β-galactosidase hybrids, which were found to reduce the export of normal acid phosphatase (but not other secretory proteins) across the RERM, probably because they formed oligomers which bound to the RERM surface and could not be translocated (see Section III.C.3) (1198).

In vitro studies indicate that chloramphenicol acetyltransferase (CAT) and the normally cytoplasmic dihydrofolate reductase from mouse can be translocated across the RERM when fused to the signal peptide of *E. coli* heat-labile enterotoxin subunit B (361,505).

4. Genetic studies on the limits of signal peptide sequence variation

Molecular genetic studies have been used to determine the effects of sequence variations on bacterial and eukaryotic signal peptide function and hence to define key elements in the structure of these peptides.

a. *Bacterial signal peptides*

Most mutations reducing the activity of signal peptides in precursors of bacterial secretory peptides were selected by their ability to relieve signal peptide–β-galactosidase hybrid protein toxicity or to restore normal β-galactosidase activity in strains producing these hybrids (see Sections II.D.4 and III.A.3. and Fig. II.4) (63,76,898). Here we will consider one extensively characterized signal peptide, that of the periplasmic maltose-binding protein (MAL E), a vital component of the maltose transport system. Cultures of strains carrying gene fusions coding for preMAL E–β-galactosidase hybrids lost viability and died within 1–3 hr when the gene fusion was expressed. Occasional survivors often carried a mutation in the part of the gene fusion coding for the signal peptide; this caused the unprocessed hybrid to accumulate in the cytoplasm, where it was no longer toxic and could accrue normal, high-level specific β-galactosidase activity. Sequence analysis of these mutated genes indicated that the changes invariably introduced charged or turn-inducing residues into, or deleted a segment from, the hydrophobic core region (64) (Fig. III.2, lines 2–6 and 8). In every case, the resulting hydrophobic core was shorter than that of any known signal peptide. PreMAL E proteins encoded by a recombinant gene carrying the same signal peptide mutations were at best only slowly processed by signal peptidase and were inefficiently exported, as judged by the cells' ability to utilize maltose as a sole carbon source (maltose transport efficiency is directly proportional to periplasmic MAL E concentration) or by cell fractionation. The deletions and the charged residue insertions reduced signal peptide function to a greater extent than the introduction of a polar Pro residue, a phenomenon also observed for other signal peptide mutations (see below for a further example). These were among the first studies showing the critical importance of the signal peptide for the movement of proteins across the BCM.

One advantage of the use of MAL E is the ability to select for improved export of the export-defective protein by growing bacteria with maltose as the sole carbon source. This led to the identification of both extragenic suppressors of signal peptide mutations (see Fig. II.5 and Section II.F.1) and intragenic suppressors which, in most cases, regenerated a functional signal peptide (978). For example, mutations restoring the function of a defective signal peptide with a Met → Arg substitution at position −8 replaced this Arg residue by Ser, Gly, or Met, while other mutations increased the overall hydrophobicity or length of the hydrophobic core (Fig. III.2, lines 9–18). Of particular interest is the substitution of the basic residue Arg at position −19 by the hydrophobic Cys. This mutation lengthens the hydrophobic core at the expense of the basic, hydrophilic extreme N-terminus, which is longer in preMAL E than in most other *E. coli* secretory protein precursors. Two other interesting classes of intragenic suppressor mutations are those affecting position +1 of mature MAL E (Fig. III.2, lines 17 and 18), and a single, inefficient suppressor mutation causing a Gly → Cys mutation at the +19 position in mature MAL E [not shown in Fig. III.2; (978)]. The former class of mutations was proposed to extend the hydrophobic core into the mature part of the polypeptide without affecting the processing event, whereas the mutation at position +19 indicates an interaction between the signal peptide and the N-terminus of the mature part of MAL E [(978), see below].

Similar mutations were obtained as revertants of the other MAL E signal peptide mutations shown in Fig. III.2, lines 1–5 and 8 (978). Of particular interest are revertants of the deletion which removed seven residues from the hydrophobic core [(43), Fig. III.2, lines 19–23]. Two of these mutations replaced Arg at −12 (originally Arg at −19) by Cys or Leu. The Leu substitution (line 21) restored function sufficiently to permit the selection of novel types of mutations in a now considerably shorter MAL E signal peptide (symbolized as pre*MAL E) by the β-galactosidase

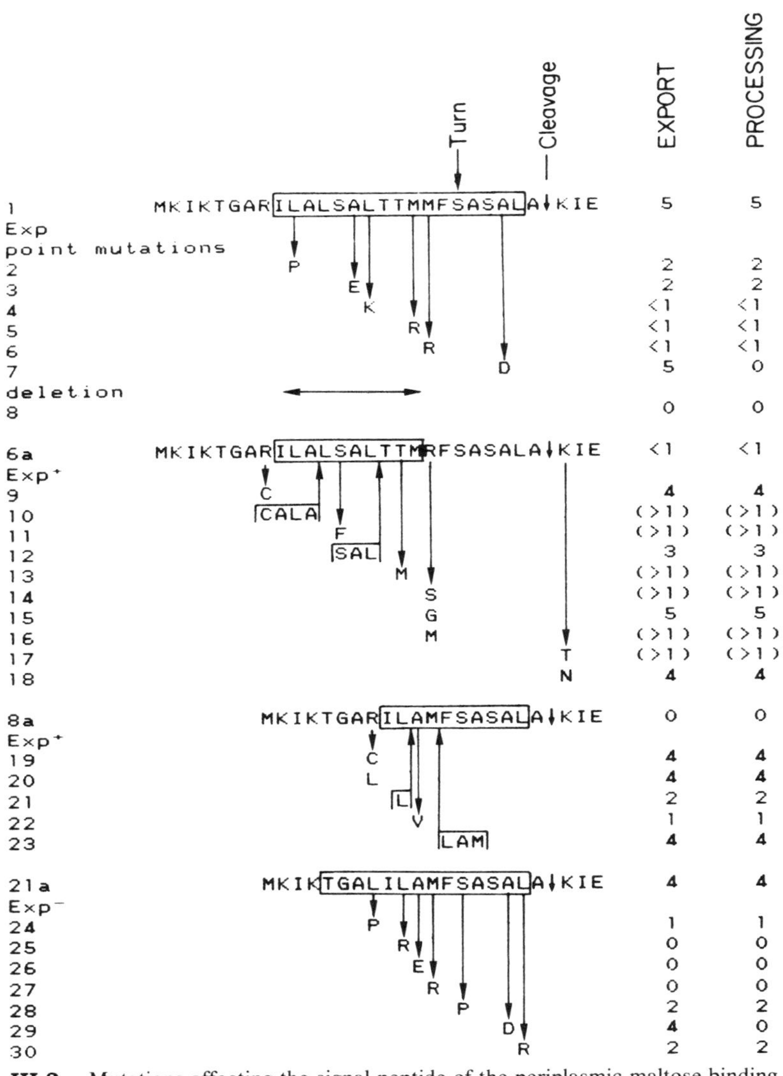

Fig. III.2. Mutations affecting the signal peptide of the periplasmic maltose binding protein of *E. coli*. Amino acids are shown in the conventional single-letter code. Line 1 shows the sequence of the wild-type signal peptide and the first three residues of the mature polypeptide. The cleavage site is indicated by the small arrow. Structure predictions indicate that there is a β-turn around position −6 of the signal peptide. The boxed region defines the probable limits of the hydrophobic core region of the signal peptide. Lines 2–6 and 8 show the sequence changes obtained by selecting for nontoxicity of preMAL E–galactosidase fusions. Line 7 shows an additional mutation introduced into the DNA coding for the preMAL E signal peptide by site-directed mutagenesis. Line 6a shows the complete sequence toxicity selection procedure [(315), Fig. III.2, lines 24–30]. Once again, all of the mutations introduced charged or turn-inducing residues within the α-helical hydrophobic core of the signal peptide (Fig. III.2). Even point mutations affecting the center of the core of the shortened signal peptide of pre*MAL E almost completely prevented signal peptide function. This was, in fact, the first time that such strongly negative phenotypes had been recorded for MAL E signal peptide point mutations, although such mutations had been reported in studies of signal peptides of other exported proteins which have shorter hydrophobic cores (509).

Interestingly, two of the mutations identified by Fikes *et al.* (315) affected residues in the −3/−1 region preceding the signal peptide cleavage site (see Section III.A.1; Fig. III.2, lines 29–30), implying that the hydrophobic core stretched into the signal peptidase recognition site. When recombined into an otherwise wild-type *malE* gene, the Leu → Arg at the −2 position substitution allowed processing of pre*MAL E at reduced rates, whereas the Ala → Asp mutation at position −3 completely blocked pre*MAL E processing (314). These results are in agreement with the −3/−1 rule for signal peptidase cleavage sites (Section III.A.1). However, neither mutation completely abolished maltose transport, implying that MAL E was still able to reach the periplasm. This was particularly surprising in the case of the Ala → Asp mutation at position −3 (line 29) since this signal peptide was not cleaved and would therefore be expected to function as a secretory signal sequence–membrane anchor (see below for other examples of mutations affecting signal peptide processing).

Further studies on the ability of cytoplasmic membrane-anchored preMAL E to function in maltose transport (314) introduce a second

---

of a nonfunctional signal peptide corresponding to the mutation shown in line 6. Lines 9–18 show the sequence changes which were obtained by selecting for export (growth on maltose; Exp^+) of MAL E carrying the defective signal produced by the deletion mutation in line 8, and lines 19–23 show the mutationally induced changes which restore signal peptide function to preMAL E carrying this mutation. Line 21a shows the complete MAL E sequence carrying the mutations shown in lines 8 and 21 (referred to as pre*MAL E in the text), and lines 24–30 show the mutations affecting signal peptide function which were obtained by the β-galactosidase toxicity selection procedure. Ascending arrows indicate amino acid insertions; descending arrows indicate amino acid substitutions. Values for export and signal peptide functionality are empirically scored from 0 (totally defective) through to 5 (wild-type activity) according to data presented in the articles cited below. All values are scored according to the steady-state levels of MAL E in the periplasm and/or ability to use maltose as a carbon source (export proficiency) and according to the kinetics of signal peptide processing (processing proficiency). Results for mutations represented in lines 10, 11, 13, 14, 16, and 17 are not available; they are therefore scored as >1 because they restore signal peptide function to an unknown extent. Data taken from Refs. 43, 64, 314, 315, and 978.

approach to the study of prokaryotic signal peptides, namely site-directed and localized chemical mutagenesis. A specifically introduced Ala → Asp mutation at position −3 in the gene for wild-type preMAL E resulted in the production of an unprocessed but functional preMAL E polypeptide. This precursor was almost completely stable but was not accessible to protease when the outer membrane was permeabilized. Fikes and Bassford (314) suggested that the signal peptide was not cleaved and that the precursor remained anchored to the cytoplasmic membrane with the mature part of the polypeptide exposed in the periplasm but somehow protected from proteolytic attack. Thus, a single amino acid substitution at the signal peptidase cleavage site converts a signal peptide into a signal sequence and consequently, a normally soluble protein into a type II membrane protein. Similar mutations have been reported for other types of exported polypeptide (see below).

The major drawback to β-galactosidase-based selection procedures is that certain kinds of mutations cannot be selected (e.g., mutations affecting the basic N-terminus, mutations having very subtle effects, or mutations which improve already functional signal peptides). The only way to determine the importance of each residue, or combination of residues, in a signal peptide is to change every amino acid systematically. This has only been done in the case of the precursor to the *E. coli* outer membrane major lipoprotein (LPP). In some respects, this choice is unfortunate because LPP, unlike most bacterial secretory proteins, is fatty-acylated at the N-terminal Cys^{+1} residue. As discussed in Section III.F.4.a, the signal peptide may be involved in targeting acyl transferases to the precursor lipoprotein, and the processing enzyme, lipoprotein signal peptidase, is different from the signal peptidase I which cleaves most bacterial secretory protein precursors. Nevertheless, Inouye and his colleagues have found that mutations which disrupt the helical structure of the central, hydrophobic core of the LPP signal peptide destroy or reduce its function, whereas "conservative" changes have relatively little effect (883), confirming the results obtained, for example, in studies on the MAL E signal peptide. Kendall *et al.* (568) attempted to improve the efficiency of *E. coli* periplasmic alkaline phosphatase (PHO A) signal peptide by progressively replacing neutral core amino acids by hydrophobic Leu. Signal peptides with nine consecutive Leu residues were more rapidly processed than was wild-type signal peptide, but other changes (e.g., five Leu residues in series) reduced signal peptide efficiency, although not to the extent displayed by most MAL E signal peptide mutations.

Studies on the LPP signal peptide also addressed the importance of the basically charged N-terminus, Met-Lys-Ala-Thr-Lys. The deletion of one or both of the Lys residues had little effect on signal peptide function, whereas signal peptides with the sequence Met-Glu-Asp-Thr-Lys (net charge −1) or Met-Glu-Asp-The-Asn (net charge −2) at the N-terminus were only slowly processed (518,1150). Similarly, changing the net charge of the staphylokinase signal peptide N-terminus from +2 (Met-Leu-Lys-Arg-Ser) to −1 (e.g., Met-Leu-Glu-Gly-Ser) or −2 (e.g., Met-Leu-Glu-Gln-Asp) considerably delayed processing and caused prestaphylokinase to accumulate in the *E. coli* cytoplasm instead of being exported to the periplasm (510,513). Neither LPP nor staphylokinase have an acidic residue at position +2. According to von Heijne's prediction [(435) and Section III.A.1], a similar effect might result from increasing the charge of the N-terminus of the mature sequence (especially in position +2). Thus far, studies on the importance of the N-terminus of mature LPP suggest that its ability to act in consort with the C-terminus of the signal peptide to form a β turn may be crucially important (519). Studies on the effects of small deletions around the cleavage site of pre-β-lactamase also indicated that sequences on each side of the cleavage site exhibit cooperativity (879). Some mutations which replaced the Glu residue at position +2 of *E. coli* bacteriophage M13 coat protein affected processing by signal peptidase I both *in vivo,* which could indicate defective signal peptide insertion into the cytoplasmic membrane, and *in vitro,* which indicates that the cleavage site is not recognized by signal peptidase (94,977). No attempt was made to predict the structural changes caused by these substitutions, but they did not significantly alter the charge balance between the N-termini of the signal peptide and the mature sequence, which was only reduced from +3 to +2. Recent studies by Li *et al.* (633a) show that increasing the incidence of positive charges at the N-terminus of mature alkaline phosphatase (PHO A) reduced export and signal peptide processing.

The effects of substituting residues on the signal peptide side of the cleavage site in M13 precoat protein were studied by Kuhn and Wickner (597). In line with results on preMAL E processing (see above), processing was inhibited by Ala → Thr at position −1 or Ser → Phe at position −3 substitutions (see Section III.A.1). Furthermore, the replacement of the turn-inducing Pro by Ser at the −6 position also abolished processing. None of these mutations affected insertion of mutated procoat into the membrane, suggesting that the cleavage site *per se,* or its accessibility to signal peptidase, had been affected. Similar results were reported by Koshland and Botstein (579) in their studies on pre-β-lactamase signal peptide and by Inouye's group in studies on preLPP and preOMP A signal peptides (263a,519). Plückthun and Knowles (879) recently found that signal peptidase can cleave at sites in mutated pre-β-lactamase which were quite different from the consensus cleavage site sequence (e.g., Val-

Phe-Pro ↓ Glu). However, it was not proven that cleavage was actually due to signal peptidase rather than to another protease or peptidase.

These and other studies confirm the notion that similarities in bacterial signal peptide profiles detected by sequence comparisons reflect constraints on signal peptide variation. However, even relatively small changes in signal peptide sequence may bring about subtle but important changes in signal peptide efficiency (568) which may only be detected in careful kinetic experiments. Further evaluation of the effects of these sequence changes on signal peptide secondary structure may in the future provide valuable information on how signal peptides function. Furthermore, experiments showing that sites within the mature sequence which are several residues away from the cleavage site may affect signal peptide activity or processing indicate that the mature part of a bacterial secretory protein may play a significant role in signal peptide function [e.g., the Asp → Asn mutation at position +12 in M13 coat protein, which prevents signal peptide processing (94), or the Gly → Cys at position +19 and Asp → Tyr at position +283 substitutions in MAL E protein, which partially suppress signal peptide mutations (196,978)]. This view is reinforced by studies on the effects of a single mutation in a signal peptide genetically fused to different mature sequences (622,644,664).

b. Eukaryotic signal peptides

Studies on the effects of mutations altering eukaryotic signal peptide sequences are less extensive than those described in the preceding section. However, they are sufficiently numerous and detailed to allow us to define features important for signal peptide function. Among the most extensively characterized is the signal peptide of invertase, a secreted (periplasmic) protein of the yeast *S. cerevisiae*, which has the following amino acid sequence: Met-Leu-Leu-Gln-Ala-Phe-Leu-Phe-Leu-Leu-Ala-Gly-Phe-Ala-Ala-Lys-Ile-Ser-Ala ↓ Ser-Met. It differs markedly from bacterial signal peptides in that there are no basic residues at the extreme N-terminus and in the presence of a Lys residue at position −4. The hydrophobic core region therefore stretches from positions −19 through −5. The importance of the signal peptide is illustrated by the fact that yeasts produce two forms of invertase; the second, cytoplasmic form is the product of translation initiated at the codon for Met at position +2 in secreted invertase (145,146,862).

Many small insertion and/or deletion mutations which alter the first five residues of the signal peptide do not affect invertase secretion (130,550). However, a significant reduction in invertase secretion was observed when the core region was reduced to seven amino acids by introducing Pro or Arg at position −12. Not surprisingly, longer deletions which significantly reduced the length of the hydrophobic core also drastically reduced invertase secretion. All of these mutations caused unprocessed, nonglycosylated, enzymatically active invertase to accumulate in the cytoplasm (550,863). Schauer *et al.* (1002) identified a mutation causing an Ala → Val conversion at position +1 which severely reduced processing by signal peptidase. This precursor protein was slowly processed between positions +1 and +2 and was only very slowly transported from the ER to the Golgi apparatus, probably because the unprocessed precursor remained firmly anchored in the membrane (see Sections III.B and III.F.1).

Saccharomyces cerevisiae can only use sucrose as a sole carbon source if it secretes invertase. Clones producing invertase with a defective signal peptide are therefore Suc$^-$ phenotypically. This feature was used to screen strains carrying a bank of gene fusion clones in which small fragments of human genomic DNA had been inserted 5′ to a signal-peptide-defective invertase structural gene on a plasmid. Plasmids from Suc$^+$ clones were extracted and the 5′ inserts were sequenced. Peptides with very different sequences were found to function as pseudo-signal peptides or secretory signal sequences. Examples of these sequences are shown in Fig. III.3, together with some examples of peptides which did not function as secretory signal sequences (i.e., invertase remained cytoplasmic and was not glycosylated). Four features are worth noting.

(i) Of the 14 peptides tested, one or possibly two were processed, probably by signal peptidase, indicating that they functioned as true signal peptides.
(ii) Plots of charge distribution (number of charges/unit length) versus overall hydrophobicity indicated that only peptides with high hydrophobicity : charge ratios functioned as signal sequences.
(iii) Most of the peptides which directed invertase out of the cytoplasm included a stretch of 10 or more uninterrupted hydrophobic or neutral residues, although there were some notable exceptions (Fig. III.3).
(iv) It was calculated that approximately 20% of DNA fragments in their library of random, roughly 200-bp fragments encoded functional secretory routing signals. This is much higher than would be expected if signal peptide sequences had evolved for the specific function of targeting proteins into the secretory pathway, indicating that the signal peptide is degenerate (551).

It is not clear whether the degeneracy of the invertase signal peptide is a special case, or whether this is true of signal peptides of other yeast secretory proteins. Recent studies show that minor changes similar to those which inactivate bacterial signal peptides significantly reduce yeast

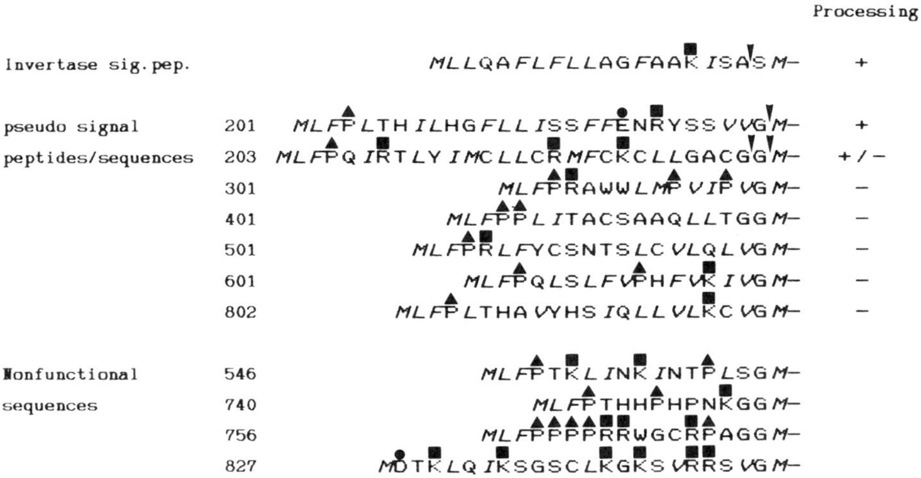

Fig. III.3. Pseudosecretory signal peptides and signal sequences obtained by cloning small fragments of human genomic DNA in frame with a signal peptide-deficient invertase gene (*SUC2*) of *S. cerevisiae*. Representative functional and nonfunctional sequences are presented. Functionality was determined by the ability of the clones carrying the gene fusions to utilize sucrose as a carbon source, by the proportion of the invertase polypeptide pool which was glycosylated (>80%), and by the proportion of pool invertase which was extracellular or in the periplasm between the cell wall and the plasma membrane (>30%). Amino acids are indicated in the conventional single-letter code. Hydrophobic amino acids are in italics, proline residues are indicated by ▲, and charged residues are indicated by ■ (Arg, Lys) or ● (Asp, Glu). All other residues are considered as apolar and neutral. Arrowheads indicate known (invertase) or probable cleavage sites for signal peptidase, and numbers correspond to *SUC2* allele numbers. Data taken from Ref. 551.

prepro-α-factor signal peptide activity (13). It will be interesting to see whether this higher stringency in the case of prepro-α-factor signal peptide is related to the ability of this protein to cross yeast microsomal membranes posttranslationally (Section III.C.3).

Another well-characterized eukaryotic signal peptide is that of bovine preproparathyroid hormone signal peptide. The 12-amino-acid-long hydrophobic core of this signal peptide is flanked on the N-terminal side by two basic amino acids and one acidic amino acid, on the C-terminal side by one acidic and one basic residue (positions −2 and −4, respectively), and by a further four basic residues in the mature part (proparathyroid hormone) of the polypeptide (positions +1, +4, +5, and +6). *In vitro* and *in vivo* studies showed that the shortest deletion into the 5′ end of the gene which affected precursor translocation into the RER removed all of the charged residues and one hydrophobic residue from the signal peptide N-terminus. The authors suggest that the 11-residue-long hydrophobic core remaining in this mutant peptide should normally be sufficient for signal peptide function and that their data indicate that a positively charged N-terminus is essential for signal peptide–RERM interaction [note that another function of this signal peptide, interaction with the signal recognition particle, SRP (Section III.C.2.b), was unaffected by this mutation] (1095). A caveat is required, however, because many natural eukaryotic signal peptides lack basic residues at their N-termini. Mutated bacterial signal peptides lacking positive charges from their N-termini do not function in bacteria but function in yeasts (877) and interact with mammalian SRP to cross dog pancreatic microsomes *in vitro* (347), which again suggests that the positive charges are a unique feature of bacterial signal peptides.

In conclusion, the limits on sequence variability noted by comparing natural eukaryotic signal peptide sequences (Section III.A.1) are also reflected in the effects of mutations which alter signal peptide sequences; i.e., a central hydrophobic core of nine or more amino acids and a signal peptidase cleavage site loosely conforming to the sequence Ala-X-Ala constitutes a functional eukaryotic signal peptide. Eukaryotic signal peptides therefore seem to be subject to fewer structural constraints than their prokaryotic counterparts, as was suggested by the comparative studies discussed in Section III.A.1. Indeed, there are reports of eukaryotic secretory proteins which are correctly targeted, albeit with reduced efficiency, when completely deprived of their signal peptides (86,1045), probably because these truncated protein sequences include a secretory signal sequencelike segment close to their N-termini. Overall, studies on yeast signal peptides suggest that they function even when they contain very few hydrophobic residues and that they have the lowest level of sequence constraints of all signal peptides. Interestingly, the signal peptide of yeast carboxypeptidase Y, a vacuolar protein (Section V.G.5), does not function in mammalian cells, but it becomes functional if its hydrophobicity is increased (84).

5. Are signal peptides obligatorily N-terminal?

Signal peptides are proteolytically removed shortly after translocation of the polypeptide through the RERM or BCM is initiated (Section III.D). In addition, signal peptides must be exposed on the surface of the nascent or completed precursor polypeptide chains in order to be recognized by their cognate receptors. Surface exposure may be favored by their N-terminal location. The option of placing signal peptides at the C-terminus of precursor polypeptides was probably rejected during evolution because precursor synthesis would be completed before routing could commence. As discussed in Section III.C.3, this might exacerbate the problem of maintaining the precursor in a translocation-competent state.

There are examples of secretory signal sequences, the noncleavable but functional equivalent of signal peptides, which are not located at the extreme N-terminus of a secretory protein (Section III.B). These sequences are probably also exposed on the surface of the nascent or completed polypeptide chain so that they can be recognized by their cognate receptors. Results from several laboratories show that the N-terminus of a signal peptide can be lengthened without affecting signal peptide function, although processing efficiency may decline with increasing length, and some extensions may completely abolish signal peptide activity (432,880,974,1095). Simon *et al.* (1049) fused the sequence of the normally cytoplasmic polypeptide globin to the N-terminus of *E. coli* pre-β-lactamase or bovine preprolactin. When mRNA coding for these hybrids was injected into oocytes or was expressed an *in vitro* coupled translation–translocation system with pancreatic microsomes, the precursor polypeptides were processed, presumably by signal peptidase. Even more surprising was the fact that both the prolactin and globin were translocated into the ER lumen when mRNA coding for globin–preprolactin was used. This contrasts markedly with results reported by Kuhn (596) showing that a 154-residue N-terminal segment of a phage M13 precoat–ribulokinase hybrid remained in the cytoplasm when the precursor was cleaved by *E. coli* signal peptidase *in vivo*. Thus, signal peptides may remain functional when extended at their N-termini, depending on the nature of the additional segment.

B. Secretory signal sequences

Sequence analyses of integral plasma membrane proteins consistently reveal the presence of one or more transmembrane segments containing long stretches (approximately 20) of hydrophobic amino acids resembling the core region of signal peptides (368) (Section III.A.1). Studies discussed below indicate that these sequences are functionally equivalent to signal peptides except that they anchor the polypeptide in the membrane [the only known exceptions are ovalbumin, a soluble secreted protein with an uncleaved N-terminal secretory signal sequence (705,1099), and plasminogen activator inhibitor 2 (1220)]. The similarity between signal peptides and signal sequences is underlined by studies showing that inhibition of signal peptide processing causes normally soluble secretory proteins to remain membrane-anchored (209,315,879a), and the observation that even minor sequence changes can convert signal sequences or membrane anchors into (cleavable) signal peptides (640,1032a,1096). Signal sequences may be located almost anywhere along the length of the polypeptide, although they are often close to the N-terminus (see below). Molecular genetic techniques have been used to study structure–function relationships in secretory signal sequences; results from two different approaches are discussed in the following sections.

1. Signal sequence hybrid proteins

As with signal peptides (Section III.A.3), a simple test for signal sequence function is to fuse segments of genes coding for integral membrane proteins to gene coding for reporter proteins to see whether the latter can be rerouted to the RERM or BCM (see Section II.D.4.a). For example, the N-terminus of human asialoglycoprotein receptor H1n, a type II membrane protein, contains a 20-amino acid hydrophobic domain which is capable of directing a cytoplasmic protein (α-tubulin) into microsomes (1071). Similar results were obtained in studies on type II membrane invariant chain (Iγ) of the Class I histocompatibility antigen (640). When mRNA coding for the N-terminus of Iγ fused to RNA coding for the bacterial cytoplasmic protein chloramphenicol acetyltransferase (CAT) was translated *in vitro* in the presence of rough microsomes, the CAT part of the hybrid polypeptide was transported into the lumen of the microsomes, where it was glycosylated (see Fig. II.8). This indicated that the N-terminus of Iγ acted as a signal sequence. The hybrid remained anchored to the membrane with its extreme N-terminus exposed in the cytoplasm (i.e., a type II membrane protein orientation), and was not processed by signal peptidase. The N-terminus of Iγ comprises a stretch of hydrophobic amino acids preceded by a hydrophilic tail. When this hydrophilic tail was drastically shortened, the signal sequence still functioned normally to target the Iγ–CAT hybrid into the RER, but surprisingly, the signal sequence was cleaved off, i.e., it behaved as a signal peptide (640). Inspection of the amino acid sequence of the Iγ–CAT hybrid revealed a potential signal peptidase cleavage site just after the hydrophobic domain of the signal sequence. Lipp and Dobberstein (640) proposed that by deleting the N-terminal tail of Iγ–CAT they had altered the conformation of the signal sequence-anchor domain in the membrane such that the cryptic cleavage site came closer to the catalytic site of signal peptidase.

Proteins in the *E. coli* cytoplasmic membrane have also been studied by gene fusion techniques. Two examples illustrate the utility of this approach for identifying their secretory signal sequences. Manoil and Beckwith (689) fused the gene for the polytopic *E. coli* cytoplasmic membrane protein TSR, a chemotaxis signal receptor, and found that the first transmembrane domain (N-terminus cytoplasmic) was sufficient to lead mature-length bacterial alkaline phosphatase into the periplasm and to anchor it in the BCM. Likewise, the N-terminus of the *E. coli* penicillin-binding protein IB (a type II cytoplasmic membrane protein), which contains the only hydrophobic stretch in the whole length of the polypeptide, also acted as a signal sequence when fused to the mature part of bacterial β-lactamase (272). These studies show that signal sequences in *E. coli* cytoplasmic membrane proteins can replace signal peptides in normally periplasmic proteins (alkaline phosphatase and β-lactamase) but, since signal sequences do not terminate at a potential signal peptidase processing site, the (unprocessed) hybrids remain anchored in the cytoplasmic membrane (see Section III.F.1).

2. Deletion and point mutations

Deletions which result in the production of truncated bi- or polyptopic plasma membrane proteins can also be used to locate secretory signal sequences. Signal peptidase I, a polytopic (type III) membrane protein of *E. coli,* has been extensively studied by this technique to determine which of the three potential transmembrane hydrophobic segments functions as the signal sequence. Proteolysis of the membrane-integrated protein indicated that the third apolar region is entirely periplasmic, as is the N-terminus. The central apolar domain functioned as a signal sequence in hybrid proteins (734) and was therefore the most likely candidate for the signal sequence (see model in Fig. III.9). This was confirmed by site-directed mutagenesis of those parts of the *lep* gene coding for the three apolar regions; only changes in the central domain affected signal sequence function (211,212). From their studies on deletion mutations the *E. coli* gene for the polytopic plasma membrane LAC Y (lactose permease) protein, Stochaj *et al.* (1085) concluded that a maximum of 50 N-terminal residues containing the first long hydrophobic segment could function as a secretory signal sequence.

The most extensively characterized signal sequences of a eukaryotic polytopic plasma membrane protein are those of bovine opsin. This protein has seven transmembrane hydrophobic domains (332). According to current models for the biogenesis of polytopic membrane proteins, every odd-numbered hydrophobic segment should act as a signal sequence–membrane anchor, with every even-numbered hydrophobic segment acting as a stop transfer–membrane anchor signal (88) (see Section III.F.1). To test this hypothesis, Friedlander and Blobel (332) and Audigier *et al.* (28) used molecular genetic techniques to place six of the seven hydrophobic segments in turn at the N-terminus of a truncated opsin molecule lacking all other potential signal sequences. Five of the six segments functioned as secretory signal sequences *in vitro,* although not with the same efficiency. This implies that even hydrophobic segments of a polytopic membrane protein which do not normally function as secretory signal sequences may do so when suitably positioned (see Section III.F.1). The signal sequence of influenza virus neuraminidase, a type II membrane protein, can replace the signal peptide of influenza virus hemagglutinin, a type I membrane protein (104). Markoff *et al.* (691) found that 22 of the 30 mainly hydrophobic N-terminal amino acids were sufficient for signal sequence and membrane anchor activity, whereas the signal sequence became nonfunctional when reduced to only 10 or 7 residues. Charged residues could be introduced into some positions in this signal sequence without affecting RER targeting, but insertions into other positions abolished activity (1040). This signal sequence is unusually long, however, and hence the introduction of charged residues therefore need not necessarily reduce the length of the hydrophobic sequence below the threshold level for signal sequence activity. Studies on the human transferrin receptor led Zerial *et al.* (1233) to conclude that the overall hydrophobicity, rather than any particular sequence, is the key feature of a signal sequence–membrane anchor, but they also suggested that flanking regions may play an important, albeit passive, role.

C. How do secretory routing signals work?

Secretory routing signals may be important in protein export across the RERM or bacterial CM in two ways. First, routing signals may confer upon the precursor polypeptide a specific conformation which is the only one which can interact with the target membrane and be translocated across it. This concept forms the basis of Wickner's "membrane trigger hypothesis" (1182,1183), which proposed that signal peptides affect the folding pattern of secretory polypeptide chains as they emerge from the ribosome. This idea is attractive because proteins fold within a matter of milliseconds whereas polypeptide synthesis may require several minutes from initiation to completion. Signal peptides do indeed affect polypeptide conformation, as judged by protease sensitivity, amphiphilicity, or differences in specific catalytic activity (238,830,946), and retard the folding of the mature part of the polypeptide (846). Sequences within the mature part of the polypeptide were perceived by Wickner to contain some of the information required for its translocation across the membrane, but the signal peptide itself contained the information for targeting the precursor to the appropriate membrane (i.e., there must be a signal peptide receptor). Other features of the trigger hypothesis are that signal peptides insert into the lipid part of the bilayer membrane, not into protein-lined, water-filled channels, and that translocation occurs either co- or posttranslationally.

In the alternative theory, largely developed from the original signal hypothesis by Blobel and Dobberstein (90,91), interactions between signal peptides or sequences and integral BCM or RERM proteins play a much more active and crucial role in protein targeting and translocation, and proteins are translocated across the membranes through water-filled channels (Section III.E.2). Data from experiments designed to test these two hypotheses will be discussed in the following sections.

1. Spontaneous interaction and insertion into membranes

The fact that the sequences of secretory routing signals are so poorly conserved led to the idea that they might not have cognate receptors on the surface of the RERM or BCM. Instead, signal peptides were proposed to be routed the surface of these membranes by electrostatic interactions between positively charged amino acids in the N-terminal part of the signal peptide and the head groups of negatively charged phospholipids in the RERM or BCM. Although charge interactions may be important in bacteria, in which the cytoplasmic membrane is the only one accessible to routing signals, they would be inadequate in eukaryotic cells, which have numerous membranes facing the cytoplasm (see Fig. I.1).

Early studies with bacteria indicated that changes in lipid composition and, more particularly, in cytoplasmic membrane fluidity induced by local anesthetics or by altered unsaturated fatty acid levels resulted in reduced processing of secretory protein precursors (243,836). Similar results were recently obtained in *in vitro* tests using *E. coli* cytoplasmic membrane vesicles and precursors of exported proteins (160). Although these results might reflect the sensitivity of cytoplasmic membrane proteins involved in protein translocation to changes in the lipid environment, they were interpreted as indicating a direct contact between membrane lipids and precursor proteins during protein export. Furthermore, Geller and Wickner (359) found that the precursor of bacteriophage M13 coat protein could be imported into liposomes devoid of membrane proteins.

Chemically synthesized signal peptides undergo dramatic conformational changes when moved from aqueous to apolar (membrane lipidlike) solvents. One of the best-characterized examples is the change from random to ordered conformation of the signal peptide of *E. coli* outer membrane LAM B protein (125). Subsequent analyses confirmed that wild-type LAM B signal peptide could insert into the lipid phase of the monolayer (and therefore probably that of natural bilayers as well) and underwent a transition from random structure to an essentially α-helical structure as it did so. Furthermore, an intermediate transition state involving an essentially β structure was detected after initial contact between the signal peptide and the lipid surface but before insertion had occurred. Sequences corresponding to defective signal peptides with a truncated hydrophobic core did not insert into the lipid phase (127,128). Similar results were obtained with the signal peptide from the outer membrane PHO E protein and the major coat protein of bacteriophage M13 (52), except that the PHO E signal peptide was found to have a β structure in aqueous solvents. These latter studies also revealed a distinct preference for negatively charged lipids during PHO E signal peptide insertion at high monolayer surface pressures. Similar but less extensive studies with a synthetic eukaryotic signal peptide showed that it too binds to liposomes (765).

In models for signal peptide–membrane interactions, N-terminal basic charges interact with negatively charged phospholipid head groups on the cis face of the membrane, triggering conformational changes (random $\rightarrow \beta \rightarrow \alpha$ or $\beta \rightarrow \alpha$) resulting in the insertion of the hydrophobic core into the membrane. The importance of acidic phospholipids (phosphatidyl glycerol or cardiolipin) is demonstrated by the effects of reducing their levels in the *E. coli* envelope or in *E. coli*-derived vesicles on the *in vivo* and *in vitro* processing of secretory protein precursors (1155a). In the model proposed by Batenburg *et al.* (52,52a), signal peptides insert as bent α helices (Fig. III.4B), the bend being induced by a glycine residue, which is often found in the core region of bacterial signal peptides. The C-terminus of the signal peptide is then "flipped" through the membrane to face the periplasm, where it can be cleaved by signal peptidase (Section III.D), while the basic residues anchor the N-terminus to the cis face of the membrane and prevent "headfirst" insertion (see Section III.F.2).

These models are based on studies with isolated signal peptides and ascribe no specific role to the mature part of the polypeptide, which is perceived as being pulled into the membrane by the momentum created by the insertion of the signal peptide. However, residues in the N-terminus of the mature sequence often influence signal peptide function (Sections III.A.1 and III.A.3). The "helical hairpin" and "insertion loop" models (288,517) take this into consideration by proposing that the signal peptide forms one side of a loop which inserts into the lipid bilayer (Fig. III.4C and D), a configuration which Engelman and Steitz (288) considered necessary to neutralize charged residues in the N-terminus of the mature sequence in order to make its insertion into the lipid environment energetically favorable. Studies with hybrid proteins show that the signal peptide N-terminus does indeed remain cytoplasmic (1032a) (Section

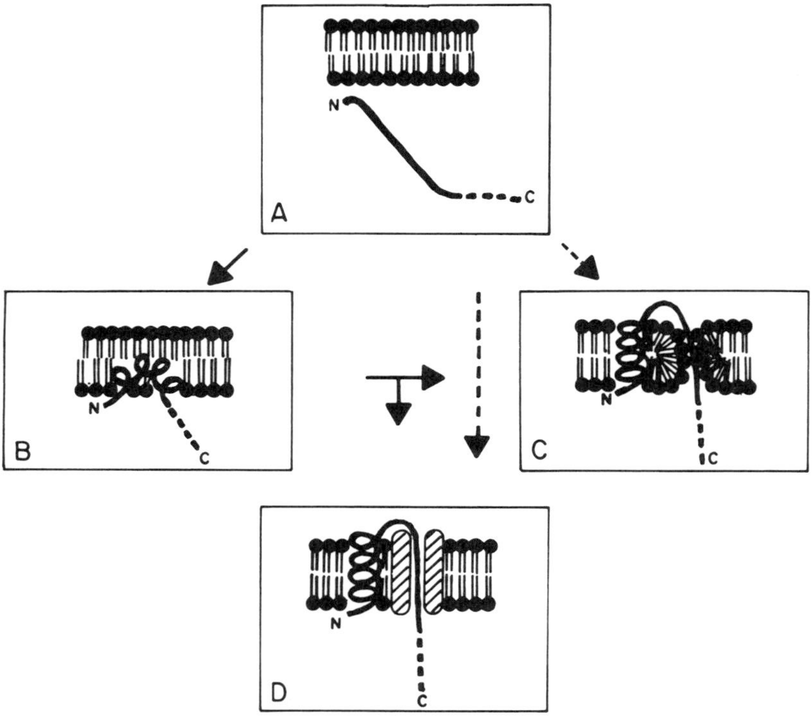

Fig. III.4. Simple models for the insertion of secretory routing signals into lipid bilayers. The N-terminus of a secretory protein (solid line) contains a secretory routing signal (signal peptide or signal sequence). This sequence binds to the surface of the lipid bilayer (A) and subsequently inserts as a looped α helix in which the entire hydrophobic domain of the routing signal is within the lipid part of the membrane (B). Insertion continues as the backbone of the α helix straightens and pulls the following length of the polypeptide into the membrane (C or D). This segment of the polypeptide is often hydrophilic and must therefore be accommodated in a water-filled channel, which may be formed either by lipids in nonbilayer configuration (C) or by one or more proteins (hatched region in D).

III.A.5), as predicted by the loop model. Secretory signal sequences are also proposed to insert as loops or hairpins with adjacent C-terminal segments of bi- or polytopic membrane proteins (28).

The high energy expenditure necessary to translocate charged residues through a hydrophobic lipid environment devoid of water molecules is the strongest theoretical argument for the movement of at least the mature part of secretory proteins through protein-lined, water-filled channels [Fig. III.4.C (1053), Section III.E.2.b]. However, the model proposed by Batenburg *et al.* [Fig. III.4B (52,52a)] sidesteps this controversial issue by invoking the formation of localized nonbilayer lipid structures in the vicinity of the site at which translocation occurs. Polar phospholipid head groups are envisioned as lining a water-filled channel in the membrane through which the polypeptide is threaded. Certain phospholipids can adopt such structures (205), and Batenburg *et al.* have demonstrated that signal peptides induce rearrangements in phospholipid packing (52,52a), which may be a less dramatic manifestation of signal peptide-induced vesicle fusion and aggregation observed by other groups [reviewed in (126)]. Clearly, these models for the early stages of signal peptide insertion into membranes merit very serious consideration.

2. Receptor-mediated interactions between secretory polypeptides and target membranes

There is overwhelming evidence of ancillary factors necessary for routing signal-mediated targeting to the RERM or BCM and for protein translocation. These studies stem from the initial observation that eukaryotic cells contain two classes of polysomes, one synthesizing soluble proteins and the other, membrane-associated class synthesizing secretory proteins (83,450,914). Electron microscopic and other studies by Palade and his colleagues showed that the latter were attached to distinct regions of the ER membrane. [In the yeast *S. cerevisiae,* which has a very limited ER, polysomes are attached to the nuclear envelope and secretory proteins are presumably extruded into the intermembranous space, which is probably contiguous with the lumen of the ER (Section VII.A)]. Furthermore, the polysomes were arranged in regular arrays, which is surprising in view of the high level of fluidity of ER membranes (840,980). This suggests that polysome–RERM interactions result from specific receptor–ligand interactions rather than from electrostatic interactions and that the receptor(s) are firmly anchored in the RERM. Studies by Palade's group and others suggested that the nascent chain of the secretory polypeptide was possibly directed to the RERM by the signal peptide and that further interactions between the ribosome itself and the RERM strengthened this interaction (90,91,423,980), leading to two identifiable classes of membrane-associated polysomes (loosely and tightly bound, respectively) (620,950). The significance of this membrane–polysome association will be discussed in Sections III.C.2 and 3. The following sections discuss the nature of proteins (receptors) involved in targeting nascent chains to the RERM.

a. The membrane-bound routing signal receptor

Rapoport and his colleagues have characterized an RERM protein which binds precursors of several secretory proteins posttranslationally, i.e., in the absence of ribosomes (890,891). The routing signal receptor remains membrane-associated and active when the RERM-derived microsomes are washed with a high-salt solution, indicating that it is probably an integral membrane protein and different from the ribosome-dependent signal recognition particle (SRP) which is only loosely membrane-associated (Section III.C.2.b). Other groups also obtained similar preliminary evidence for a signal peptide receptor in the RERM (715,716,1164). Furthermore, experiments showing that isolated signal peptides bind to microsomes (414,936) and prevent transport of secretory protein precursors *in vitro* and *in vivo* (575,682) also suggest the existence of a membrane-bound signal peptide receptor. Recent cross-linking studies have shown that signal peptides do indeed bind specifically to an ~35 kDa protein in isolated microsomal membranes. The same receptor binds signal peptides directly or via SRP (1159,1189). Two intriguing possibilities are that (i) the receptor is part of the signal peptidase complex which has been partially purified from the RERM or (ii) it is the integral membrane part of the SRP receptor (see Sections III.C.2.c and III.D.3).

The internal signal sequence of cytochrome b_5 does not recognize a routing signal receptor. Bendzko *et al.* (71) suggest that this might be due to the fact that N-terminal signal peptides or sequences are more compact and have a hydrophilic face which interacts with the receptor, whereas the uniformly hydrophobic cytochrome b_5 signal sequence (insertion sequence) does not. Since insertion of cytochrome b_5 into the RERM is also SRP-independent (see Section III.C.2.b), it seems probable that the signal sequence of this protein partitions spontaneously into membranes, although this cannot explain its specific insertion into the RERM (16).

Most bacterial secretory proteins, and probably most integral plasma membrane proteins, are also synthesized on membrane-bound polysomes (908,1066). In contrast to the situation in eukaryotes, inhibition of new rounds of translation initiation with the antibiotic puromycin causes all bacterial polysomes to dissociate from the membrane, indicating that only the nascent polypeptide chain, not the ribosome, is required to anchor polysomes to the BCM (1067). Additional evidence that the signal peptide directs polysomes to the BCM was provided by Rasmussen and Basford (912), who showed that mutations which created nonfunctional signal peptides in preMAL E protein of *E. coli* (see Section III.A.4.a) also caused the precursors to be synthesized preferentially on cytoplasmic ribosomes. These mutations altered the hydrophobic core of the signal peptide, rather than N-terminal basic amino acids, indicating that electrostatic interactions are not the sole basis for membrane–signal peptide interaction (1110a). Furthermore, synthetic signal peptides prevent *in vitro* translocation of bacterial secretory protein precursors into *E. coli* cytoplasmic membrane vesicles in a way which suggests competition for receptor sites (162). The bacterial signal peptide receptor has not been characterized (161,1155).

Two candidates for the *E. coli* signal peptide receptor have been identified genetically. The first, SEC Y (also called PRL A), is a 49-kDa protein which is predicted to span the cytoplasmic membrane 10 times (10,152). Mutations in *secY*, which is in an operon encoding ribosomal proteins, variously prevent export and processing of secretory proteins (1033), suppress the effects of mutations in *secA* (124) and of defective signal peptides in secretory protein precursors (42,277,278,644) (see Section II.F.1), and reestablish membrane association of polysomes synthesizing secretory protein precursors with defective signal peptides (912). SEC Y-deficient inner membrane vesicles are unable to translocate secretory protein precursors (32,297).

The second possible signal peptide receptor polypeptide is the product of the recently identified *secE* gene, an 18-kDa cytoplasmic membrane protein (928). Little is known about *secE* except that a cold-sensitive *secE* mutation blocks protein export and induces increased *secA* expression (see Table III.1 and Section III.E.2). It would be interesting to see whether SEC E or SEC Y proteins are located in specific domains of the BCM using cell fractionation or immunocytochemical techniques.

The *Bacillus subtilis* cytoplasmic membrane can be fractionated into two components with distinct protein contents by sucrose gradient centrifugation (694). One of these fractions is substantially enriched in ribosomes, suggesting that it might be enriched in vesicles derived from cytoplasmic membrane domains specifically involved in the synthesis and export of secretory proteins, i.e., the functional equivalent of the RERM in complex eukaryotes. These membrane fragments presumably contain the signal peptide receptor, as well as other components of the protein translocation apparatus (Section III.C.3).

Although the data described above provide strong evidence for the existence of a routing signal receptor, especially in eukaryotic cells, it is not clear whether or how the receptor–signal complex dissociates to allow the routing signal to penetrate into the membrane. One possibility is that the receptor–signal complexes undergo significant conformational changes which allow the signal to insert into the membrane and the receptor to dissociate from it and to reassociate with other membrane proteins to form a translocation channel around the segment of the polypeptide immediately behind the routing signal (Section III.E.2.b and Fig. III.4).

b. Signal recognition particle (SRP)

The signal recognition particle is a ribonucleoprotein complex. It is found both loosely associated with the RERM and ribosomes, and free in the cytosol. Since its discovery in 1978, SRP has been shown to be essential for targeting most secreted, soluble lysosomal, and integral membrane proteins into mammalian rough microsomes, irrespective of whether the proteins have cleavable or noncleavable routing signals. SRP was initially identified by the fact that salt-washed microsomes were unable to import secretory protein precursors *in vitro,* whereas import was restored when the salt extracts were added back to the coupled translation–translocation mixture (1167). It soon became clear that SRP in the salt extract reduced the rate of elongation of secretory protein nascent chains when translation was carried out in wheat germ lysates in the absence of microsomes. Translation was arrested only when about 70 residues of the nascent chains had been polymerized and was resumed when microsomes were added to the translation reaction (16,641,717,1161). These observations are significant because translation arrest occurs only when the signal peptide of secretory protein precursors has emerged from the ribosome [40–50 residues of nascent polypeptides are assumed to be buried within the ribosome (92)] and continues until such time as SRP binds to its cognate receptor on the microsome membrane surface (Section III.C.2.c). This was perceived as ensuring that translation and translocation are tightly coupled, avoiding the need to translocate full-length precursor proteins, which may adopt tightly folded structures (see Section III.C.3).

Other data are in full agreement with the idea that SRP binds to signal peptides as they emerge from the ribosome and that the combination of SRP plus ribosome causes translation arrest. For example, Ibrahami *et al.* (506) found that the prelysozyme gene could be truncated at its 3′ end, thereby reducing the length of the precursor to 74 amino acids, without affecting SRP-mediated translation arrest. However, synthesis of a 51-residue-long peptide produced by an even shorter gene was not arrested by SRP, and the peptide was not translocated into microsomal vesicles. The latter observation is rather puzzling, since it implies that the signal peptide of the shorter gene product did not bind to the signal peptide receptor. Synthesis of prepromelitin, a natural 70-amino acid-long secretory precursor protein, is also not subject to SRP arrest, probably because its synthesis is completed before its signal peptide protrudes from the ribosome (1237). These results are also in question, however, because Ibrahami (503) has recently reported SRP-mediated translation arrest of prepromelitin synthesis. The difference between these two sets of data may be that Ibrahami's prepromelitin was a peptidyl-tRNA rather than the completed polypeptide chain (748) (see Section III.C.3).

Translation arrest is most obvious in the heterologous wheat germ–pancreas translation–translocation system supplemented with canine SRP; it is much less apparent when the wheat germ lysate is replaced by reticulocyte or tissue culture cell lysates, even when purified SRP is present in large excess (714). This casts doubt on the significance of SRP-mediated translation arrest. Furthermore, Lipp *et al.* (642) have recently shown that SRP does not completely block secretory protein mRNA translation in wheat germ lysates, but rather slows elongation rates to various extents depending on the mRNA. SRP may therefore simply exaggerate natural translation pause sites in mRNA coding for secretory proteins (642). Anderson *et al.* (16) and Perera *et al.* (860) have also reported that integral membrane secretory proteins synthesized in wheat germ lysates are translocated into dog pancreatic microsomes in a SRP-dependent manner but that SRP-mediated translation arrest does not occur. Prehn *et al.* (892) have recently shown that wheat germ extracts contain an SRP which is very similar to mammalian SRP. Furthermore, wheat germ membrane preparations contain an SRP receptor. Secretory precursor proteins were synthesized without SRP-dependent translation arrest in an homologous system comprising wheat germ SRP, its membrane-associated receptor, and a wheat germ lysate. Wheat germ SRP cannot replace mammalian SRP in an otherwise mammalian translation–translocation test system (892).

Thus, the incompatible element in the wheat germ lysate–pancreatic microsome system seems to be the mammalian SRP, which might have slightly different properties from wheat germ SRP. Gilmore and Blobel (373) demonstrated that solubilized, pure SRP receptor relieved SRP-arrested translation in the heterologous wheat germ–pancreas system by displacing SRP from the ribosome. Thus, mammalian SRP may bind more tightly to wheat germ ribosomes than to mammalian ribosomes, thereby preventing their movement along the mRNA until the conformation of the SRP is altered by binding to the SRP receptor. Interestingly, studies discussed below indicate that SRP–signal peptide recognition depends on the presence of the ribosome, suggesting that ribosomes may assist SRP in binding to signal peptides. SRP can bind to signal peptides at almost any moment while the ribosome is still translating the message, although the efficiency of SRP–signal peptide interaction declines as the ribosome approaches the end of the message (8). This again suggests a role for the ribosome in signal peptide–SRP interaction.

The requirement for SRP and its receptor in protein translocation is more clearly understood. Almost all secretory proteins tested require SRP for targeting to the RERM, irrespective of whether or not they are made as precursors. Even bacterial secretory protein precursors interact with mammalian SRP (347,992a). There are, however, a number of exceptions. Foremost among these are short secretory protein precursors which can be imported into microsomes in the absence of SRP or its receptor (1007,1176,1186,1237). Import of these precursors requires a microsomal membrane protein, possibly the signal peptide receptor (1007). Although these results show that SRP is not essential for the transport of these precursor proteins across the RERM, they do not exclude the possibility that SRP–signal peptide interaction does occur as part of the normal routing process *in vivo* (992a). Indeed, studies with hybrid proteins comprising the signal peptide of prepromelitin and various lengths of the promelitin sequence and the cytoplasmic protein dihydrofolate reductase (see Table II.2) show that SRP-independent translocation of small secretory proteins into microsomes is due to their small size; precursors longer than 8–9 kDa required SRP and its receptor for transport, whereas shorter precursors did not (748,749). These and other data (992a) are most compatible with the idea that longer secretory polypeptides need SRP to maintain translocation competence (Section III.C.3).

The second class of secretory proteins whose insertion into rough microsomes does not depend on SRP is made up of certain integral plasma membrane proteins typified by cytochrome b_5 (see preceding section).

SRP is composed of a 7S RNA and six polypeptides (72, 68, 54, 19, 14, and 9 kDa) (1160,1162) which together form a rod-shaped 11S complex which is 5–6 nm wide and 23–24 nm long (20,21). The extremities of the ~300-nucleotide-long 7S RNA are about 80% homologous to a family of repeated DNA sequences called Alu (634,1132,1162). Walter and Blobel (1163) dissociated the RNA and protein constituents of SRP by EDTA treatment followed by ion-exchange chromatography and then reassociated them in active form. The 19-kDa and 54-kDa polypeptides were fully separated by column chromatography, but the 9- and 14-kDa proteins and the 68- and 72-kDa proteins remained associated as heterodimers (Fig. III.5). Walter and Blobel (1163) determined the order in which the different SRP components reassociated on the 7S RNA core. Up to five of the six proteins in the complex seemed to bind directly to the RNA, whereas the sixth protein (54 kDa) only did so indirectly, through the 19-kDa protein (Fig. III.5). This was confirmed by nuclease footprinting assays, which identified two independent RNA–protein interactions (Fig. III.5) (1044). Siegel and Walter (1042,1043) used this unique feature, together with the selective chemical inactivation of SRP components, to separate the three properties of SRP (translation arrest, signal peptide recognition, and receptor-mediated RERM targeting). They found that the 9-kDa and 14-kDa polypeptides were not necessary for RERM targeting of preprolactin but that their absence caused a complete inability to arrest preprolactin synthesis in microsome-free, canine SRP-supplemented wheat germ translation mixtures. Similar results were reported by Scoulica *et al.* (1027), who reduced the size of the complex required for RERM targeting

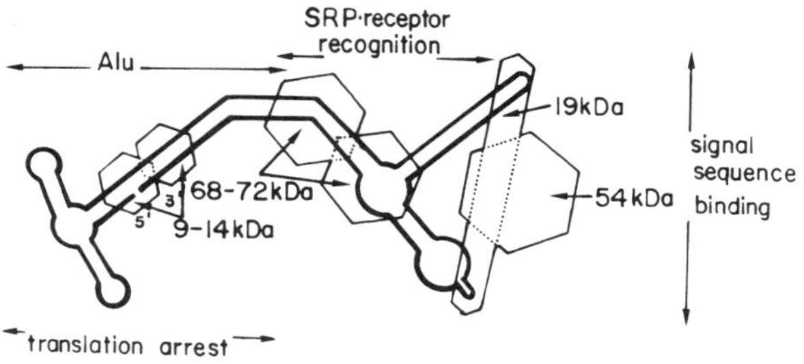

Fig. III.5. Model for mammalian SRP as determined by nuclease accessibility and footprinting and by reconstitution assays. Proteins are assembled on the 7S RNA core as shown. Only the 54-kDa polypeptide is not in direct contact with the RNA. For reasons of clarity, some regions of unpaired structure in the 7S RNA are not shown. Functional domains and the Alu-like region of the 7S RNA are indicated.

by micrococcal nuclease digestion of the terminal Alu sequences from the 7S RNA, leaving only the central RNA region and four polypeptides (72, 68, 54, and 19 kDa). Thus this central region recognizes the nascent polypeptides (see below) and binds to the SRP receptor in the RERM (see below), whereas the hydrogen-bonded flanking Alu regions with the 9- and 14-kDa proteins are required for translation arrest. The mechanisms of translation arrest are far from clear; one possibility is that Alu sequences base-pair with other nucleic acids to slow the progress of the ribosome along the mRNA (1162).

Alkylation of the 54-kDa component reduced SRP–signal peptide interaction, whereas alkylation of the 68–72 kDa complex prevented SRP–SRP receptor interactions (1043). The interaction between the 54-kDa subunit of SRP and the signal peptide was confirmed by elegant cross-linking studies (591,603,1188). Cross-linking also caused microsome-reversible elongation arrest, although to a lesser extent than the natural SRP–signal peptide interaction (1188). In all of these studies, a lysine residue close to the N-terminus of preprolactin was specifically engineered to participate in the cross-linking reaction. However, this does not indicate that basic charges at the N-terminus of the signal peptide are normally directly involved in SRP binding. Many eukaryotic signal peptides lack such charged residues (Section III.A.1). Furthermore, site-directed mutagenic removal of charged N-terminal residues of pre-bovine parathyroid hormone did not prevent SRP-mediated translation arrest of the precursor mRNA (1095), whereas complete loss of SRP arrest was observed when the hydrophobic core of another signal peptide, that of a bacterial enterotoxin subunit, was disrupted by the introduction of a proline residue (505). Thus, the 54-kDa component of SRP probably binds to the signal peptide hydrophobic core.

A requirement for SRP in secretory protein routing in yeast cells has not been demonstrated, although integrity of the *Schizosaccharomyces pombei* gene for 7S RNA is essential for viability (922). This 7S RNA shows no significant sequence homology with mammalian 7S RNA, but certain structural motifs are present in both RNAs, and mammalian SRP proteins bind to this RNA and also to the 7S RNA from another yeast (*Yarrovia* sp.) under stringent conditions (883a). Genetic studies in the yeast *S. cerevisiae* (233) reveal the existence of a number of genes coding for proteins which are apparently important in the early stages of secretory protein routing. Some of these genes may code for components of the hypothetical yeast SRP.

Bacteria do not have a 7S RNA; the nearest equivalent is a 6S RNA, the *E. coli* chromosomal gene for which has been rendered nonfunctional without deleterious effects on protein export or cell viability (617). This seems to rule out the possibility that bacteria have the formal equivalent of mammalian SRP, although one or more bacterial cytoplasmic proteins or 4.5S RNA (883b) may perform at least part of the role of SRP in these organisms (Section III.C.3). Of particular interest are studies on membrane–ribosome complexes of *B. subtilis* and *Staphylococcus aureus*, which revealed the existence of up to four proteins which might be equivalent to eukaryotic SRP (5,470,471,694). The 64-kDa protein component of this *S complex* is protected against proteolytic attack by trypsin by the ribosome and can be cross-linked to other components of the complex (148,149). The 64-kDa protein is unique in that it is only loosely associated with the ribosome; it is also found free in the cytoplasm (470,471). Low Mg^{2+} or prolonged incubation through sucrose gradients releases the S complex from ribosomes, and it is recovered as 76S particles which appear as regularly arranged cages when examined by electron microscopy (148). Whether these particles are laboratory artifacts remains to be determined, but they could be involved in secretory protein–membrane interaction or in maintaining secretory proteins in a translocation-competent state (148) (see Section III.C.3).

c. *The SRP receptor*

Studies on the nature of the SRP receptor were initiated following the observation by Walter *et al.* (1164) that mild proteolysis and high-salt treatment of rough microsomes led to the loss of protein translocation competence. Translocation competence was restored by adding back the released material. The component of the protease-released material which could bind to SRP and to the RERM, and then the RER membrane protein from which the fragment was derived were subsequently identified (375,376,715–717). Meyer and his colleagues named the receptor *docking protein* and believed it to be an integral RER membrane protein comprising an 18-kDa anchor (originally estimated to be 13 kDa) and the normally cytoplasmic domain of 52 kDa (originally estimated to be ~59 kDa), which was released by elastase treatment. The larger fragment was also claimed to release SRP-induced translation arrest (477,717), although Gilmore *et al.* (376) could not reproduce this result. We now know that docking protein is the α subunit of the heterodimeric SRP receptor. The β subunit, an ~30 kDa integral membrane protein which copurifies in equimolar amounts upon SRP affinity column chromatography, can be cross-linked to the α subunit (1101). The α subunit is a peripheral membrane protein which may be anchored to the cytoplasmic face of the RERM via interactions with the β subunit (480,613,1101). Sequence analysis of a cDNA clone encoding the SRP α subunit indicated that the large cytoplasmic domain includes sequences similar to those in nucleic acid binding proteins (613), leading to speculation that the α subunit might bind to the 7S RNA component of SRP.

It is not clear whether SRP receptor is specifically associated with the RERM. Hortsch and Meyer (475) found that the α subunit, like the ribophorins (see below), was specifically associated with rough microsomes when they were separated from smooth microsomes by sucrose gradient centrifugation, whereas other ER marker proteins were equally distributed between rough and smooth microsomes. Immunoelectron microscopic studies also showed the receptor to be specifically located in the RER (478). Tajima *et al.* (1101), however, found that SRP receptor was only enriched 2- to 3-fold in rough microsomes, whereas ribophorins (Section III.C.2.d) and ribosomes were enriched 4.5-fold. These differences may reflect difficulties in separating the two types of microsomes, although Tajima *et al.* suggested that SRP receptor might be located in regions of the smooth ER to which newly created polysomes synthesizing secretory proteins are directed [see also (82)]. SRP receptor is much less abundant than ribosomes on the RERM surface (376), suggesting that it is not involved in any subsequent stage in signal peptide or sequence insertion into the membrane or in protein translocation.

d. The ribosome receptor

The RER is by definition that part of the continuous endoplasmic reticulum which appears rough because large numbers of ribosomes are attached (1133). This attachment seems to be rather tight because it persists during fragmentation of the ER and sucrose gradient centrifugation. The nascent polypeptide chain threading through the membrane may anchor the polysome to the membrane. Inhibition of new rounds of translation initiation by the antibiotic puromycin causes bacterial polysomes to detach from the BCM (Section III.C.2.a) and hence, the BCM does not have a ribosome receptor. This also seems to be the case for smooth ER membranes, which can nevertheless recognize and process secretory protein precursors (82). Early studies suggested that there was a specific interaction between ribosomes and RER microsomes. Binding of ribosomes to these vesicles was unaffected by puromycin alone but was disrupted by puromycin plus high-salt treatment (423,595). Furthermore, salt-released ribosomes could reassociate with the microsomes (588). The major RERM binding site was located on the larger, 60S ribosomal subunit. Polysomes released from microsomal membranes by detergent treatment can also reassociate with salt-stripped membranes. In this case, however, proteins produced by the reassociated polysomes remain on the outer surface of the microsome, rather than being translocated into the lumen, indicating that perhaps some other microsomal component (e.g., signal peptide receptor or protein translocase) has also been extracted by the detergent (103).

The ribosome receptor is inactivated by protease treatment (103) and was tentatively identified as a 105–110-Å, membrane-embedded particle by electron microscopic examination of freeze-fractured RER membranes (370,805). Hortsch and Meyer (476) calculated that ribosome receptors account for ~1% of the total membrane protein content of the RER. Attention has been focused on two proteins (*ribophorins*) which are particularly concentrated in rough microsomes and are as abundant as ribosomes on the RER surface (585,586,690). Three lines of evidence supported the idea that ribophorins are ribosome receptors:

(i) Ribosomal proteins can be cross-linked to ribophorins in the absence of nascent secretory polypeptides.
(ii) Ribophorins are exposed on the microsome surface [as shown by radioiodination with lactoperoxidase and, in the case of ribophorin I, by accessibility to exogenous trypsin (586)].
(iii) Ribophorins associate with polysomes when RER membranes are dissolved in nonionic detergents (585) or when ribosome-enriched inverted microsomal vesicles are formed by mild detergent treatment (587).

However, proteolysis of microsomes was subsequently shown to destroy ribosome binding activity without visibly affecting the amounts or molecular weights of the two ribophorins (479). Furthermore, sequencing analyses of cDNA clones coding for the two ribophorins indicate that they are both type I membrane proteins whose C-terminal cytoplasmic domains are probably too short to have receptor activity (198,422). Thus, the identity of the ribosome receptor remains elusive. A number of other proteins besides ribophorins are enriched in rough microsomes, and ribophorins may play a catalytic role in protein translocation across the RERM or may maintain the characteristic flattened shape of the RER (588).

3. Translocation competence and the involvement of other cytosolic and membrane components

As discussed previously, a number of cytosolic and membrane factors have been shown to facilitate the routing of eukaryotic secretory proteins to the RER. Little attention was devoted to their roles in subsequent stages in the translocation process or to proteins which might play a similar role in bacterial systems. In particular, we must now examine the role of segments of the secretory polypeptide other than the routing signal, especially the idea that, although not containing sequences directly involved in targeting and translocation, these segments must be maintained in a translocation-competent state. This is usually interpreted to mean that secretory polypeptides must not adopt strong secondary structure before they are translocated, thereby allowing the polypeptide to be "threaded" through the membrane. Three different strategies may be envisaged:

(i) Translocation is tightly coupled to translation, so that only very short lengths of the secretory polypeptide span the distance between the ribosome and the RERM or BCM.
(ii) Translocation can occur posttranslationally because the secretory polypeptide does not adopt any strong secondary structure (example A in Fig. III.6).
(iii) Translocation can occur posttranslationally either because cytosolic components prevent secretory polypeptides from adopting strong secondary structures or because membrane proteins denature folded polypeptides immediately prior to translocation (example B in Fig. III.6).

The following sections will examine the extents to which these strategies are adopted in the prokaryotic and eukaryotic secretory pathways.

a. *Cotranslational translocation*

Most eukaryotic secretory protein precursors cross the RERM as nascent polypeptide chains while translation continues on membrane-associated polysomes. Translocation cannot occur posttranslationally *in vitro* except in some very rare cases in which the secretory polypeptide is very short or the ribosome remains attached to the nascent polypeptide (see Section III.C.2). The observation that translation of mammalian secretory protein

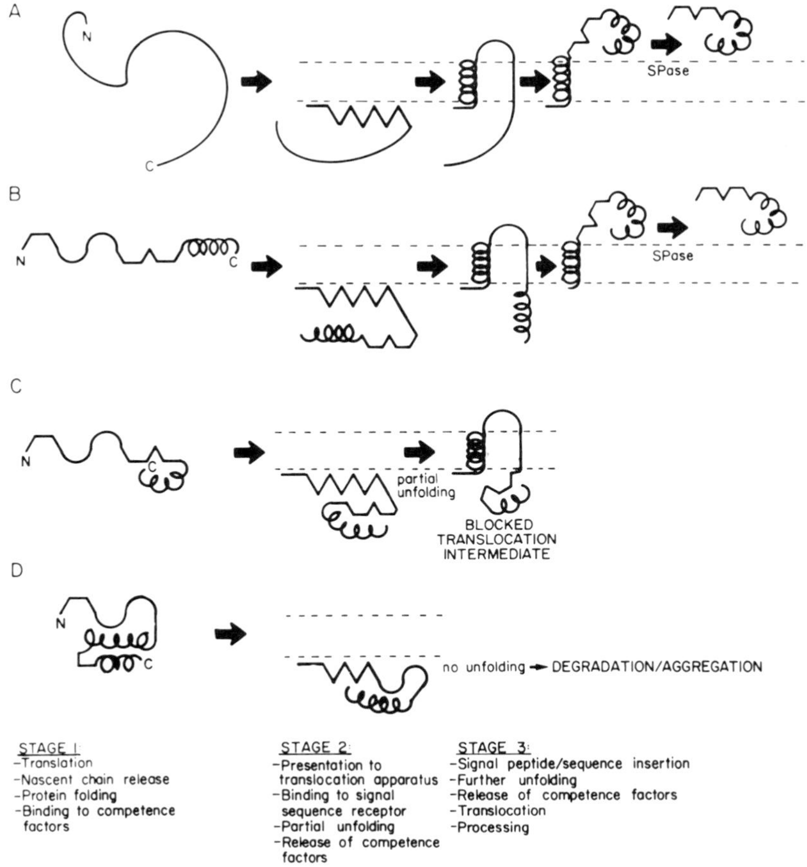

Fig. III.6. Models depicting the effects of secretory protein conformation on its ability to cross the RERM or bacterial CM. In (A) the secretory protein is in an unfolded state with the routing signal exposed on the surface of the polypeptide and therefore able to bind to its receptor and initiate the translocation of the rest of the polypeptide through the membrane. Cytoplasmic components (competence factors) or unfolding are not required. In (B) a moderate degree of secondary structure is tolerated because the routing signal remains exposed on the surface of the polypeptide, which can be unfolded prior to or during translocation. Alternatively, secondary structure formation may be inhibited by cytoplasmic competence factors. In (C) the extent of secondary structure is such that although the routing signal binds to its receptor and inserts into the membrane, the remaining part of the polypeptide cannot be fully unfolded, causing translocation to be aborted at an intermediate stage. This situation may arise either because competence factors do not bind to the secretory protein and therefore cannot prevent folding or because the secondary structure cannot be dissolved by unfoldase. In the extreme case (D), the extent of secondary structure is such that the routing signal is not exposed on the surface of the polypeptide and therefore cannot bind to its receptor. The protein therefore remains in the cytosol and may aggregate or be degraded.

mRNA in wheat germ lysates was arrested by mammalian SRP unless the cognate SRP receptor protein was also present (Section III.C.2) encouraged the view that translation and translocation were very tightly coupled in eukaryotic cells. We now know that this result is at least partially artifactual due to the incompatibility of plant and mammalian translation and secretory protein translocation systems (Section III.C.2.b). Therefore, we have no clear idea of the extent to which the two processes are coupled and we do not know whether relatively large tracts of the polypeptide are polymerized before they are threaded through the membrane. However, it seems likely that SRP and the ribosome are essential to maintain the transport competence of cotranslationally translocated proteins.

In addition, we also have a poor understanding of the extent to which mRNA translation and secretory protein translocation are coupled in bacteria. The rate of translation of mRNA for certain secretory proteins may be significantly reduced in cells producing precursors with altered (inefficient) signal peptides (77) or certain secretory protein–cytoplasmic protein hybrids (446), or when secretory proteins are overproduced (197). Although these results may indicate that cells are able to modulate the amounts of certain secretory proteins produced, they do not indicate that chain elongation is coupled to protein translocation.

Elegant studies by Randall's group have helped to clarify the extent of translation–translocation coupling in bacteria. Their studies showed that although secretory proteins are mostly synthesized on membrane-bound polysomes (Section III.C.2), relatively long stretches of polypeptide could be synthesized before translocation occurred (as determined by signal peptide processing and protease accessibility at the trans side of the membrane) (907). In some cases, full-length secretory polypeptides could be detected before translocation occurred (542,543). Efficient signal peptides may rapidly engage the protein export pathway as they emerge from the ribosome, whereas others may only do so just before or even after the nascent chain has been released from the ribosome. This could explain why even minor changes in signal peptide sequence can switch a bacterial secretory protein precursor polypeptide between co- and posttranslational export modes and alter its dependence on export competence factors (331) (see next section).

b. Posttranslational translocation and competence factors

A small number of eukaryotic secretory polypeptides, and probably the majority of prokaryotic secretory polypeptides, can be posttranslationally translocated across their respective RER or cytoplasmic membranes *in vitro*. Although this may not reflect the situation *in vivo*, these studies show that certain full-length secretory polypeptides maintain translocation competence, implying that they do not adopt a strong secondary structure.

Most of the proteins which can be posttranslationally translocated across microsomal membranes *in vitro* are relatively short (less than 100 residues) (see Section III.C.2.b) and may not adopt strong secondary structures prior to translocation (749). However, even though completed polypeptides in this class of proteins are thought to retain translocation competence, at least one of them, yeast prepro-α factor, is transported into isolated yeast microsomes *in vitro* much faster if denatured by urea (172). Elegant studies by Zimmermann's and Meyer's groups have confirmed that chain length is a crucial feature for SRP-independent translocation of these polypeptides (748,963,1007). Furthermore, Zimmermann's studies of gene fusions demonstrated that the distribution of charged residues along the length of these secretory polypeptides also plays a crucial role, since imbalance of clusters of positive and negative charges leads to the loss of SRP-independent translocation competence (748), as does the formation of intrachain disulfide bridges (749). Another interesting feature of these and other studies is that translocation of such small polypeptides requires ATP and a soluble cytosolic protein or protein complex (303,419,961–963,1007,1172,1186) (see below). Processing of M13 bacteriophage precoat protein by signal peptidase in liposomes also requires ATP, even though ATP is not required for detergent-solubilized signal peptidase I activity. [Precoat is also posttranslationally translocated into microsomes (1186).] ATP and cytoplasmic factors may be required to maintain this class of secretory protein precursors in a translocation-competent state; i.e., they play the same role as that played by SRP and the ribosome in cotranslational translocation (see below). Curiously, the ATP requirement has been shown to vary according to the positions of the charged residues in the mature part of one of this class of polypeptides, suggesting that its consumption may be linked to "unfoldase" activity (749). It is difficult to determine whether ATP is required for cotranslational translocation *in vitro* because it is also needed for translation.

The second group of eukaryotic secretory proteins which can be translocated into and across the RERM after the cessation of chain elongation *in vitro* includes certain integral membrane proteins such as a fragment of glucose transporter (745,746), a truncated β-lactamase–globin hybrid (860), and human placental lactogen (150) and yeast prepro-α factor in mammalian microsomes (346). These proteins retain translocation competence only while ribosomes remain attached to the nascent chain, implying that the ribosome and SRP prevent the polypeptide from adopting a tightly folded conformation prior to translocation (992a). GTP is also required for the translocation of these proteins into rough microsomes, probably after SRP has bound to its receptor (190,460). A GTP-binding protein may participate in this reaction since nonhydrolyzable GTP analogs are also effective (460,1193a). GTP and/or ATP hydrolysis may be involved in another step in the import of these proteins (4a,860) (see below). The term "ribosome-coupled" translocation has been proposed to distinguish it from authentic posttranslational translocation of completed polypeptides (348).

One of the most intriguing recent developments in the study of protein import into the RER is the demonstration that a family of 70-kDa heat shock proteins (HSP70) are required for efficient posttranslational transport. *Heat shock proteins* are produced in larger amounts at high temperatures and are thought to bind to and prevent heat denaturation of cytosolic proteins (857). Dissolution of heat shock protein complexes requires ATP (857). Thus, transitory involvement of heat shock proteins like HSP70 as competence factors which prevent secretory proteins from folding tightly was proposed as a possible source of the ATP requirement in secretory protein export.

Two lines of evidence confirm this hypothesis. First, yeast mutants with only one functional gene for HSP70-type proteins (*SSA1* under *GAL* promoter control; other HSP70 genes were mutated) become Sec^- and accumulate unprocessed prepro-α factor when *GAL* promoter expression was turned off (234). Precursors of at least two other secretory proteins, invertase and carboxypeptidase Y, also accumulated when the level of HSP70–SSA1 protein was reduced. Full-length preprocarboxypeptidase Y may be inefficiently translocated into yeast microsomes, but preinvertase translocation is obligatorily cotranslational *in vitro* (420). Thus HSP70–SSA1 protein, which differs from most heat shock proteins in that it is synthesized constitutively (and accounts for up to 2% of total cell protein), seems to be required for maintaining secretory protein competence in yeasts. It is probably significant that prepro-α factor, which can be imported posttranslationally (and presumably without SRP) was the most affected of the three secretory proteins studied; export competence of invertase and carboxypeptidase Y may be maintained in the absence of HSP70 by SRP and tighter translation–translocation coupling (see below).

These *in vivo* studies were complemented by others showing that the same HSP70 protein dramatically improved *in vitro* import of prepro-α factor into yeast microsomes (172). In this case, HSP70 was purified by ATP affinity chromatography, which again suggests that ATP could be involved in its action, and was separated from another competence factor present in yeast cytosolic extracts. It is significant that import of urea-denatured prepro-α factor into yeast microsomes does not require HSP70 (172).

The only detailed studies on the role of HSP70-type proteins which have been published so far are those described above. However, Zimmermann *et al.* have recently found that HSP70 can partially replace the cytosolic factors which are required for posttranslational, SRP-independent import of proteins into dog pancreatic microsomes (1240a). Thus, HSP70, which is highly conserved in all eukaryotes tested, may play a central role in posttranslational protein transport across the RERM. Its role in cotranslational (SRP-dependent) transport is less clear, but it may be that relatively long stretches of polypeptide are polymerized before translocation starts; folding of the nascent chain may be prevented (or the folded chain unfolded) by HSP70-type proteins. The signal recognized by HSP70 remains to be determined; one possibility is that it binds to surface-exposed hydrophobic sequences (e.g., signal peptides) (857). However, HSP70 also seems to be involved in the import of mitochondrial protein precursors, which often do not have significant stretches of hydrophobic amino acids (Section VI.B.5).

Can other eukaryotic secretory proteins be maintained in a translocation-competent state once their synthesis is complete? Maher and Singer (680) suggested that reducing disulfide bridges in secretory polypeptides allowed them to retain translocation competence, but Ibrahami (504) showed that this was an artifact caused by the reduced translation of the secretory protein mRNA in the presence of the reducing agent. Thus, large secretory protein precursors probably lose translocation competence very quickly if they do not immediately enter into the translocation machinery in the RERM.

Tight translation–translocation coupling in eukaryotic systems might explain why hybrid proteins comprising secretory routing signals and normally cytoplasmic polypeptides can be translocated into the lumen of the RER *in vitro* (see Section III.A.3). While similar hybrid polypeptides have not been extensively studied in bacterial *in vitro* systems, *in vivo* studies suggest that only hybrid polypeptides which do not adopt a strong secondary structure can be translocated across the *E. coli* cytoplasmic membrane (Section III.A.3). *In vivo* studies on periplasmic proteins suggested that they are translocated across the *E. coli* cytoplasmic membrane domain by domain, even when translation is complete, and that translation pauses may accentuate this effect (907). Furthermore, *in vitro* studies indicate that full-length precursors of periplasmic and outer membrane proteins can be translocated across the cytoplasmic membrane (161,360,750,1155,1216). These studies suggest that bacterial secretory proteins must remain translocation-competent and that cytoplasmic proteins fused to secretory targeting signals either cannot do so or are not recognized by factors which maintain the translocation competence of normal secretory proteins.

Kinetic studies *in vivo* suggest that *E. coli* secretory protein precursors rapidly lose translocation competence (909). *In vitro* assay systems provide an excellent way of studying the maintenance or loss of translocation competence. It has consistently been shown that only a small proportion (5–25%) of secretory protein precursors are translocated and processed in these systems, suggesting that the remaining polypeptides were not translocation (export)-competent. Translocation competence is progressively lost during prolonged incubation (161,200).

The requirement for additional cytosolic components and ATP for efficient *in vitro* translocation of bacterial secretory proteins was recognized at an early stage (157,360,751,1216). The role of ATP in the bacterial translocation assay remains obscure, especially because *in vivo* studies suggest that the cytoplasmic membrane energy potential provides the main source of energy for protein translocation (37,831), and the *in vitro* assay is not totally independent of membrane energy potential (159,1155,1216). One interpretation is that ATP hydrolysis (158) is required for an "unfoldase" enzyme which acts prior to translocation (303). If this is the case, then it is required for both co- and posttranslational translocation in the bacterial *in vitro* system (159).

Recent studies are beginning to reveal the nature of soluble proteins or protein complexes which are required for protein translocation in the bacterial *in vitro* assay system (Table III.1). Crooke and Wickner (200) found that a cytosolic extract could restore translocation competence of a denatured form of preOMP A, an *E. coli* outer membrane protein. Their studies suggested that this factor (an ~65-kDa protein called *trigger factor*) bound to preOMP A to maintain the initial translocation-competent state (200a). Trigger factor was found to be unnecessary, however, if preOMP A protein was purified in 8 *M* urea and then rapidly renatured, although competence was very rapidly lost unless trigger factor was added (200a). Trigger factor is an abundant ribosome-associated protein which appears to have a membrane receptor and which binds in stoichiometric amounts to preOMP A protein (636a). Indeed, purified trigger factor is the only cytosolic protein required for preOMP A transport into *E. coli* membrane vesicles *in vitro,* although other competence factors such as SEC A (see below) may have been present in the membrane preparations used. The competence factor identified by Weng *et al.* (1180) also contained an ~60-kDa polypeptide. Fandl and Tai (297) found that increasing the amount of soluble extract overcame the effects of a translo-

Table III.1. Possible functions for *E. coli sec* gene products

Protein	Size (kDa)	Location	Effect of absence	Possible function(s)	References
SEC A	92	Cytoplasm (membrane)	Precursor accumulation	Routing signal receptor	36,810,811
SEC B	12	Cytoplasm	Prevents export of some precursors	Competence factor	185
SEC D	64	Cytoplasm	Precursor accumulation; induction of *secA*	Competence factor	349
Trigger factor	65	Cytoplasm	Not tested[a]	Competence factor	200
SEC E	18	Cytoplasmic membrane	Precursor accumulation; induction of *secA*	SEC D or trigger factor receptor; translocase; energy coupler	928
SEC Y	49	Cytoplasmic membrane	Precursor accumulation; induction of *secA*	SEC A receptor; routing signal receptor; translocase	10,36,1033

[a] Gene not identified.

cation defect resulting from the use of vesicles devoid of SEC Y activity. Their results were interpreted as showing that increasing the amount of a trigger factor-like substance might lengthen the period during which the precursor polypeptide remained translocation-competent, thereby increasing the time span during which productive interaction between the precursor polypeptide and the membrane could occur.

The trigger factor structural gene has not yet been identified, but one likely candidate is *secD* (349). Mutations in *secD*, which codes for a soluble 64-kDa protein, prevent the export and processing of a large number of secretory proteins in *E. coli*. However, difficulties in purifying secretory protein precursors have so far restricted studies on the role of trigger factor to studies of its effect on preOMP A protein, and it is therefore not known whether the absence of trigger factor would produce the pleiotropic phenotype exhibited by SecD$^-$ mutants. Another possible candidate as the trigger factor structural gene is *groEL*, which codes for a 57-kDa protein (originally estimated to be 65 kDa) known to be required for bacteriophage assembly (445). GRO EL protein is highly homologous to a chloroplast protein which binds to nuclear-encoded and chloroplast-encoded subunits of the enzyme ribulose-1,5-biphosphate carboxylase-oxygensae (RUBISCO). As discussed in Section VI.C, this binding protein (BP or *chaperonin*) facilitates the formation of RUBISCO heterooligomers prior to their transport into thylakoids, probably by preventing folding prior to assembly. GRO EL protein is not known to be required for protein export in bacteria, but it is essential for cell viability.

Two other potential export competence factors have been identified genetically in *E. coli*. The first of these, the *secA* gene product (Table III.1), is required for the export and processing of a large number of secretory proteins (809,811). SEC A is also required for signal sequence-dependent insertion of some but not all cytoplasmic membrane proteins (1201). The *secA* gene is essential for cell viability. Interestingly, temperature-sensitive *secA* mutations affect the export of different precursors to different extents, suggesting that different precursors have different levels of SEC A dependence (643), which may be related to their co- or post-translational modes of export. Recent studies show that SEC A protein is required in very small amounts for *in vitro* import of preOMP A protein into *E. coli* membrane vesicles, implying that SEC A protein acts catalytically rather than by binding to the precursor to prevent it from folding prior to translocation (139a). One intriguing possibility is that SEC A protein binds to signal peptide–trigger factor complexes and delivers them to the membrane-associated signal peptide receptor. SEC A protein may thus "select" precursors from a pool of completed and nascent polypeptide chains in the cytoplasm. Most of the pool of SEC A protein seems to be free in the cytoplasm, although small amounts may be membrane- or ribosome-associated (636). It remains to be seen whether these minor pools of SEC A protein are important for protein export or whether SEC A protein has other functions besides protein export.

One interesting feature of SEC A is that translation of *secA* mRNA is increased when export is blocked either by other *sec* mutations (928, 949a) or by hybrid proteins which become jammed at translocation sites in the membrane (810, 949a). This suggests that cells can detect an abnormally high level of secretory protein precursor accumulation in the cytoplasm and can react accordingly. It is not known whether synthesis of other SEC proteins is similarly regulated.

Mutations in *secA* are suppressed by *prlA* mutations in the *secY* gene (124), which codes for a cytoplasmic membrane protein (Table III.1). Other mutations in *secY* prevent protein export and cause precursor accumulation (524) and can suppress the effects of mutations which reduce signal peptide efficiency (42). Furthermore, both SEC A and SEC Y are required for the association of polysomes with the *E. coli* cytoplasmic membrane (912). Thus SEC Y protein is a good candidate for the hypothetical SEC A and signal peptide receptor (Table III.1; see also Sections III.C.2.a and III.E.2).

The other export competence factor which was identified through the effects of *sec* mutations is SEC B, which is unusual in that it is required for the export of only a subset of secretory protein precursors (599), most of which are apparently exported posttranslationally *in vivo* (185). One of these precursors, preMAL E protein, has recently been shown to require SEC B protein for import into *E. coli* vesicles *in vitro* (185). Furthermore, results obtained in the same study suggest that SEC B protein might bind to a stretch of fewer than 37 amino acids in the mature part of MAL E to prevent the precursor from folding into a protease-resistant conformation (185). Alternatively, this region of MAL E might provoke rapid folding of the precursor except when SEC B protein is bound to it. In this case, SEC B could bind to the signal peptide (1171a) or elsewhere in the polypeptide. Mutations in the *malE* gene which reduce the extent of secondary structure in preMAL E protein reduced its dependence on SEC B for export. Some types of mutant MAL E precursors, however, could not be exported and prevented the export of other precursor proteins, presumably because they formed irreversible complexes with SEC B protein. Export of normal precursor proteins was restored when SEC B protein was overproduced (185). Together, these results suggest that SEC B binds to preMAL E protein to prevent it from folding prior to its interaction with the translocation machinery in the cytoplasmic membrane. There are obvious parallels between SEC B and trigger factor (see above), although they are different proteins (Table III.1).

It is an interesting observation that several groups found that *E. coli* export competence factors fractionated as complexes during sucrose gradient centrifugation (751,910,1180). Future studies may reveal the extent to which several proteins may act together to preserve export competence and to deliver secretory protein precursors to the translocation machinery in the cytoplasmic membrane.

Recent results suggest that there may be parallels between eukaryotic and *E. coli* secretory competence factors. Bacterial cell extracts can substitute for yeast cytosolic extracts in the *in vitro* import of preproα factor into yeast microsomes (303). It is not clear whether the bacterial extract replaced the yeast HSP70 proteins (see above), but it may be significant that the heat shock response (as well as *secA*) is induced in *E. coli* cells when export is blocked by secretory protein–β-galactosidase hybrids (525) and that protein import into BCM vesicles requires ATP (see above). The bacterial equivalent of the HSP70 class of proteins is DNA K protein (857), which is not known to be required for protein export. GRO EL is also a heat shock protein (see above). Preliminary evidence suggests that trigger factor can substitute for SRP and vice versa in some *in vitro* systems (200b,992a).

The so-called S complex of proteins, which binds to *Bacillus subtilis* ribosomes synthesizing secretory polypeptides (149) (Section III.C.2.b), may also be involved in maintaining translocation competence. The 64-kDa protein component of the S complex, which is both ribosome-associated and free in the cytoplasm (149), could be the *Bacillus* equivalent of trigger factor or SEC D protein. Genetic studies of the components of the *B. subtilis* secretory pathway have only recently been initiated (5,574a).

c. *Proteins which cannot cross the RERM or BCM*

Of the three possible scenarios presented at the beginning of this section for the delivery of translocation-competent secretory proteins to the translocation apparatus in the RERM or BCM, most of the evidence points to the critical role played by competence factors which prevent proteins from folding during their brief stay in the cytoplasm. Therefore, proteins which are not recognized by these competence factors, or which are only partially protected, may fold prior to the initiation of translocation and may not be able to unfold upon contact with the translocation machinery; this may be the case, for example, with certain secretory protein–cytoplasmic protein hybrids. This may also happen when competence factors are saturated due to the overproduction of normal secretory proteins. In these cases, polypeptides may become jammed in the translocation pathway or may not even be recognized by the translocation machinery (e.g., signal peptide receptor) because the signal peptide is buried within the folded polypeptide (examples C and D in Fig. III.6). In the latter case, the precursor proteins may aggregate in the cytoplasm or be degraded (361).

D. Signal peptidases

Signal peptidases are a special class of endopeptidases with relatively well-defined substrate specificities, including a conserved tripeptide on the N-terminal side of the cleavage site (Section III.A.1). Prokaryotic secretory precursor proteins are cleaved by eukaryotic signal peptidase(s) and vice versa (638,753,1105,1173,1176). Signal peptidases of the bacterium *E. coli* have been more extensively characterized than the eukaryotic enzymes and will accordingly be dealt with first.

1. Bacterial signal peptidase I

E. coli signal (leader) *peptidase I* was first described by Mandel and Wickner (687), who observed specific processing of the precursor of M13 phage major coat protein (precoat) by an enzyme located in the cell envelope. The enzyme was subsequently overproduced and purified and was found to be mainly inner membrane protein and to have a molecular weight of ~36,000 (1199,1200). Wolfe *et al.* (1200) estimated that there are ~500 copies of signal peptidase per cell, which compares with ~10^6 secretory protein precursors processed per generation. Precoat was shown to be processed by signal peptidase I either in detergent-solubilized form or in proteoliposomes (359,1175,1184), and the enzyme also correctly processed precursors of other authentic *E. coli* secretory proteins *in vitro* (1199). A further indication that most *E. coli* secretory protein precursors are processed by signal peptidase I is that purified precoat signal peptide inhibits its action on precursors of other signal peptides *in vitro* (1184).

Attempts to replace the wild-type chromosomal signal peptidase I (*lep*) gene by an inactivated gene were unsuccessful, indicating that *lep* is an essential gene (215). Progressive depletion of signal peptidase I levels by reducing expression of the *lep* gene (under the control of a tightly regulated promoter) caused the cells to accumulate precursors of many secretory protein precursors (209). This again confirms that signal peptidase I is the major *E. coli* signal peptidase.

Signal peptidase I has two transmembrane domains; the C-terminus, which includes the catalytic domain, is exposed in the periplasm (210,222,734). This corresponds with the observation that signal peptidase cleavage only occurs when precursors have penetrated the cytoplasmic membrane *in vitro* (1046,1175) and with *in vivo* kinetic experiments showing that processing occurs as the nascent or completed precursor polypeptides are translocated through the cytoplasmic membrane (542,543).

2. Bacterial lipoprotein signal peptidase

A limited number of secretory protein precursors produced by both Gram-positive and Gram-negative bacteria are unique in that they have a Cys residue at position +1 which must be fatty-acylated before processing by signal peptidase can occur. Furthermore, the sequence to the N-terminal side of the cleavage site, although similar to that recognized by signal peptidase I, is more highly conserved, and stretches to four rather than three residues (details of the modification process are discussed in Section III.F.4.a). The enzyme which recognizes these sequences has been called *signal peptidase II* or, more usually, *lipoprotein signal peptidase*. Tokunaga *et al.* (1113,1114) showed that antibodies against signal peptidase I did not inhibit lipoprotein signal peptidase action and that fatty-acylated precursors were the only substrates recognized by the enzyme. The latter feature was confirmed by Dev and Ray (236), who purified the enzyme and showed that it acted on fatty-acylated precursors but not on unmodified forms of these precursors or on a precursor normally cleaved by signal peptidase I. The two signal peptidases are therefore quite distinct and do not have overlapping substrate specificities.

The cyclic peptide antibiotic globomycin probably kills *E. coli* cells as an indirect consequence of its ability to inhibit lipoprotein signal peptidase noncompetitively (236a,499,507,520). Incubation in the presence of globomycin, a structural analog of the lipoprotein signal peptidase cleavage site (236a), causes cells to accumulate large amounts of preLPP, the precursor of the major *E. coli* outer membrane lipoprotein (531). Processing of other major outer membrane proteins is not affected by globomycin (507). Globomycin sensitivity is considerably reduced both by mutations which prevent either preLPP synthesis or its fatty acylation (606,618) and by overproduction of lipoprotein signal peptidase (1115). Reducing the levels of lipoprotein signal peptidase also causes lipoprotein precursors to accumulate and leads to cell death (1213). Although globomycin also inhibits the processing of Gram-positive lipoprotein precursors, it does not kill these bacteria, possibly because they do not produce large amounts of any lipoprotein (432).

E. coli lipoprotein signal peptidase is an 18-kD cytoplasmic membrane polypeptide (236,1116,1117). Sequence analysis of the cloned structural gene (*lsp*) indicates that the polypeptide probably has four transmembrane domains (516,1228). The membrane topography of lipoprotein signal peptidase has not been determined, but the catalytic site is presumably located in the periplasm. Lipoprotein signal peptidases are unique to bacteria.

3. Other bacterial signal peptidases

Although the combined activities of the peptidases described above account for all of the signal peptidase requirements in *E. coli,* a third enzyme with the same substrate specificity as signal peptidase I has been purified from this bacterium. The specific activity of this enzyme, tentatively named *signal peptidase III,* is 500-times higher than that of signal peptidase I. It is also larger than signal peptidase I (58–60 kDa compared to 36 kDa) (913). The role of signal peptidase III in protein export remains to be determined, but it is unlikely to function alone, because processing of precursors which are also substrates for signal peptidase III *in vitro* is prevented *in vivo* by reducing the level of signal peptidase I.

A unique signal peptidase may process certain types of fimbrial subunit precursors. Fimbriae of certain Gram-negative bacteria including *Neisseria* sp., *Pseudomonas* sp. and *Vibrio cholerae* (692,718,997) are cell surface appendages assembled from polypeptide subunits which are synthesized as precursors with apparently normal signal peptides. The signal peptides, however, are processed at a site just after the positively charged N-terminus, rather than at the normal processing site after the hydrophobic core. Other secretory proteins are processed normally in these bacteria, indicating that they have signal peptidase I-type activity (389,394).

4. Eukaryotic signal peptidase

Eukaryotic signal peptidases have mainly been characterized in relatively crude detergent extracts of microsomal membranes. The enzyme is inactivated by ionic detergents such as sodium deoxycholate but not by nonionic detergents, an observation which forms the basis of the *in vitro* assay for signal peptidase activity in detergent-solubilized microsomes (531,532). Many secretory protein precursors are not processed by the extracted enzyme, probably because, once released from the ribosome, they fold into conformations which cannot be recognized by the enzyme or which do not interact correctly with it in the detergent micelles. Active enzyme has been partially purified from microsomes by (i) salt extraction (to remove loosely associated proteins), detergent solubilization, ion-exchange chromatography, and gel filtration (339) or by (ii) deoxycholate extraction, sucrose gradient centrifugation, and reconstitution into liposomes (686). A more recent report (294) describes a five-stage purification procedure leading to a 42-fold enrichment of canine signal peptidase activity from detergent extracts to salt-washed microsomes. When analyzed by SDS-PAGE, the purified material was found to contain stoichiometric amounts of six polypeptidase (25, 23, 22, 21, 18, and 12 kDa). Only two proteins were present in signal peptidase purified from hen oviduct microsomes (36a). Both ribophorins and SRP receptor were separated from signal peptidase during early stages in the purification procedures. Mumford *et al.* (755) purified a pancreatic zinc metalloendopeptidase with similar cleavage specificity to signal peptidase, but its identity with signal peptidase has never been formally proven.

The gene for one subunit of canine signal peptidase has been sequenced (1032b), as has the *S. cerevisiae* gene *SEC11* coding for one component of the yeast signal peptidase (96). The latter gene is essential for signal peptide processing and for cell growth, but neither gene shows any sequence homology with the *E. coli lep* gene although it is by no means obvious that the *E. coli* and eukaryotic enzymes should be related.

The catalytic site of eukaryotic signal peptidase is located on the trans, noncytoplasmic face of the membrane. Thus, signal peptidase is cryptic in the absence of detergents or protein translocation through the microsomal membrane, and protease treatment of rough microsomes does not inactivate the enzyme. Processing by signal peptidase is usually considered as evidence that protein translocation has occurred (Section II.F.3), but at least one protein, hepatitis B virus precore protein, can be processed by RERM signal peptidase and then released into the cytoplasm (348). This case of aborted translocation is probably caused by the failure of the mature part of the protein to engage correctly in the translocation channel. Although signal peptidase is usually assumed to be located exclusively in the RERM, smooth microsomes can also process secretory protein precursors (92). It is not known whether signal peptidases present in the smooth and rough microsomes are the same enzyme or whether plant, animal, and simple eukaryotic endoplasmic reticula have the same signal peptidase (1100).

5. Signal peptide peptidase

What happens to the signal peptide once it is cleaved from the secretory protein precursor? It is difficult to imagine the retention of liberated signal peptide in the membrane for two reasons. First, free signal peptides might prevent productive interaction between signal peptides on other secretory proteins and the RERM or BCM (162,575,682,1184). Interestingly, studies by Simon *et al.* (1049) suggest that signal peptides may be cleared much more slowly in *in vitro* systems with dog pancreatic microsomes than *in vivo*, an observation which may explain certain discrepancies regarding the efficiency of protein translocation across the RERM *in vitro* and *in vivo*. Furthermore, free signal peptides might also affect the functioning of later stages in the eukaryotic secretory pathway, as suggested by Koren *et al.* (575) and Austen *et al.* (30). Second, signal peptides have been shown to have deleterious effects on lipid monolayers and bilayers *in vitro* (127,1035) and might therefore exert similar effects *in vivo* were they to accumulate in large amounts.

Bacterial and eukaryotic signal peptides are rapidly degraded after they are cleaved from the precursor polypeptide (415,500,532,850,1243). Thus, signal peptide clearance might be initiated by a carboxypeptidase, with subsequent stages in signal peptide breakdown resulting from the action of one or more other peptidases. According to Novak *et al.* (792), several soluble and membrane-associated peptidases are responsible for the clearance of the preLPP signal peptide in *E. coli*. One of these enzymes, *cytoplasmic membrane peptidase IV* (833), has been identified (500) and purified (792), and its structural gene has been cloned, inactivated by mutation, and substituted for the chromosomal wild-type gene (508,1092). Signal peptide clearance occurs more slowly in this strain, but the bacteria do not suffer any ill effects due to the slightly increased accumulation of signal peptide. A mutation altering the hydrophobic core of the preLPP signal peptide was reported by Pollitt and Inouye (882) to cause the depolarization of the *E. coli* cytoplasmic membrane. They proposed that the mutant signal peptide could not interact correctly with the translocation machinery and consequently perturbed membrane structure. However, another possibility is that the signal peptide was processed normally but was not degraded by signal peptide peptidases and that depolarization resulted from the continued presence of the signal peptide or its association with the translocation machinery. It is not known whether peptidase IV degrades other *E. coli* signal peptidases.

E. Protein translocation

We now turn our attention to the ways in which polypeptide chains of secretory proteins cross the RERM or BCM. Several models explain how the N-terminus of secretory proteins inserts into the membrane (Section III.C.1). In all of these models, the N-terminus inserts as a loop, with the C-terminal part of the polypeptide remaining in the cytoplasm. In the following sections, we shall consider how translocation might continue.

1. Simple models for protein translocation

In the simplest of the models presented in Section III.C.1, the mature part of the polypeptide is "pulled" into a nonbilayer part of the membrane.

The mature polypeptide might be threaded through this "channel" until the C-terminus appears on the trans side of the membrane. If the routing signal is cleaved, the mature sequence will be released in soluble form (Fig. III.7A), corresponding to extracellular or periplasmic proteins in Gram-positive and Gram-negative bacteria, respectively, or to a protein

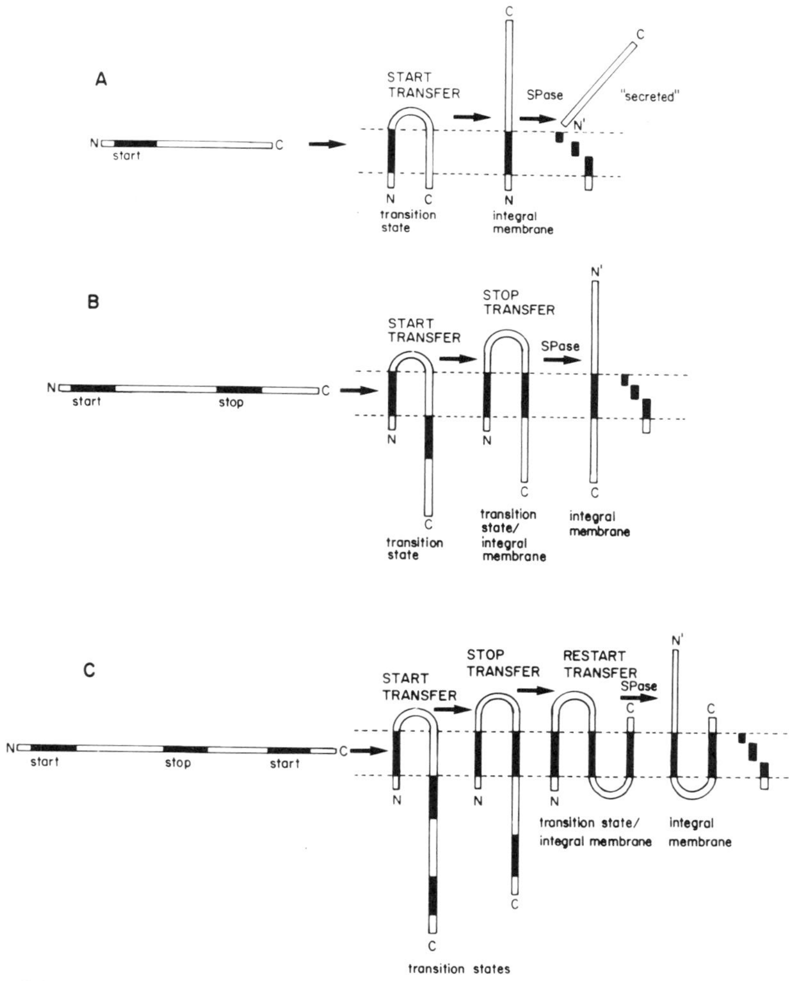

Fig. III.7. Models for the insertion and topogenesis of membrane secretory proteins. Shaded areas are hydrophobic segments of the polypeptides. (A) Soluble secreted polypeptide (processed by signal peptidase) or type II membrane protein; (B) type I membrane protein (processed) or simple polytopic membrane protein; (C) polytopic membrane protein. "Start" corresponds to a secretory routing signal (signal peptide or signal sequence–membrane anchor) and "stop" to a stop transfer–membrane anchor sequence.

destined to reside temporarily or permanently in the lumen of the ER. Although this model by no means fully accounts for all known aspects of the translocation process, it serves as a basis from which to discuss the nature of ancillary factors required for efficient translocation to occur.

2. Role of the membrane in protein translocation

Do proteins in the RER or bacterial cytoplasmic membranes play active or passive roles in protein translocation? We have already seen that various membrane components such as signal peptidases and ribosomes, SRP, competence factors, and signal peptide receptors are directly involved in early stages of the translocation process. Other RERM proteins including ribophorins (479) (Section III.C.2.d) and a protein which is loosely associated with the cytoplasmic face of the yeast RERM (975) have also been implicated in protein translocation. The SEC E (928) and SEC Y (525) proteins of the *E. coli* cytoplasmic membrane are also required for protein export. Although it has been proposed that both may function as receptors, they may play other roles in the protein translocation through the membrane. Specific regions of the cytoplasmic membrane of the bacterium *B. subtilis* contain a subset of proteins which might be involved in protein translocation (694), and it remains to be seen whether the heterogeneity of *E. coli* membrane vesicle content, such as that shown in Fig. II.1, reflects a similar specialization of certain cytoplasmic membrane domains for protein export (1110a). Overproduction of nonexportable signal peptide–β-galactosidase hybrid proteins saturates export sites in the *E. coli* cytoplasmic membrane. Because these sites are quite abundant [~20,000 per cell (523)], it may be possible to detect them by immunocytochemistry or cell fractionation, which would greatly facilitate the identification of components of the bacterial protein export machinery. The potential roles played by membrane proteins in protein translocation are discussed in the following sections.

a. *Unfoldase*

Membrane-associated proteins may unfold secretory proteins as they cross the membrane. This activity would be in addition to the cytosolic heat shock protein HSP70 and other cytosolic competence factors. Moreover, ATP hydrolysis associated with HSP70 release from protein complexes almost certainly occurs at the membrane surface and may involve additional proteins present on the cytoplasmic face of the RERM.

b. *Translocation channel*

There are strong theoretical arguments and experimental evidence for the translocation of secretory polypeptides through a water-filled channel rather than through the lipid environment of the bilayer membrane, for which the energy requirement would be prohibitively high (688,1053). Evidence that protein translocation occurs through a water-filled channel was reported by Gilmore and Blobel (374), who showed that nascent translocation intermediates are accessible to aqueous reagents. The possibility that signal peptides could form the pore was discounted by Singer *et al.* (1053), who favored the idea that the channel was formed by a specific "translocase" protein, which folds around the secretory polypeptide as it penetrates the membrane (Fig. III.4). Hydrophobic (stop transfer–anchor) sequences in secretory polypeptides were proposed to cause the dissociation or lateral opening of the channel, releasing the anchor segment into the hydrophobic environment of the lipid bilayer (1054). The diameter of the pore must be a minimum of 1 nm in order to accommodate at least one of the two segments of the loop structure which insert into the membrane (see Fig. III.4).

The putative translocase protein(s) have not yet been identified. Wiedmann *et al.* (1189) have suggested that signal sequence receptor polypeptide could form a channel or at least a channel gate. Trypsin-treated rough microsomes process prepromelitin, but the protein does not reach the RER lumen (1237), suggesting that the protease destroys a component of the translocation machinery. The channel may be formed by one or more of the polypeptides which are present in the signal peptidase complex which has been purified from RER membranes. The membrane proteins SEC E or SEC Y could form part of the protein translocation channel in *E. coli* (Table III.1). Alternatively, the channels could be formed from lipids in nonbilayer conformation with hydrophilic head groups lining the walls (see Fig. III.4). It would be interesting to see whether domains of bacterial cytoplasmic membranes or the RERM involved in protein export contain unusually high amounts of lipids with a strong tendency to form such nonbilayer structures. The RERM is thought not to contain unusually large amounts of nonbilayer configuration (706).

c. *Energy coupling*

We have already seen that the posttranslational translocation of secretory proteins across the RERM *in vitro* requires ATP and have considered the possibility that ATP hydrolysis is required to maintain or achieve translocation competence (Section III.C.3.b). ATP hydrolysis is almost certainly involved in the action of one set of competence factors (HSP70) common to the eukaryotic secretory and mitochondrial protein routing pathways (Section III.C.3). ATP hydrolysis may also fulfill the energy requirement for translocating secretory polypeptides through the RERM, in which case a membrane-associated ATPase may be required for energy coupling (4a). Cotranslational translocation may also require ATP hydrolysis, but it has not yet been possible to test this directly.

Protein translocation across the *E. coli* cytoplasmic membrane *in vitro* also requires ATP hydrolysis (157,360,1155,1216), which may again be partly utilized to maintain translocation competence. An OMP F–LPP hybrid protein with a mutated (unprocessed) signal peptide cleavage site undergoes ATP-dependent, competence factor-independent translocation into *E. coli* proteoliposomes, implying that ATP hydrolysis is required for translocation (1217). Rather surprisingly, the membrane potential fulfills all of the energy requirements for protein translocation in bacteria *in vivo* (37,214,216,284,760). A membrane potential is also required for protein translocation in bacterial *in vitro* systems (754,1216), implying that two separate stages in the translocation process require two different energy sources. Either component of the transmembrane potential (proton or chemical gradient) may be used to drive protein export *in vivo* (37). Both are presumably used either to energize a membrane-associated protein unfoldase or to drive protein translocation.

3. One or two secretory pathways?

In this chapter, we have seen that signals within secretory proteins target them to the bacterial cytoplasmic membrane or to the membrane of the rough (and possibly the smooth) endoplasmic reticulum in eukaryotic cells. In all cases, soluble cytoplasmic and integral proteins in the target membranes are responsible for efficient targeting and possibly also for translocation of the secretory protein. Signals within the secretory polypeptide determine whether it is released on the trans side of the membrane or remains inserted in it (see next sections). Since routing signals of prokaryotic and eukaryotic secretory proteins are structurally similar and often interchangeable (Section III.A.2), one might have predicted that components of the signal recognition, processing, and translocation machinery would be very similar in the two groups of organisms. Figure III.8 summarizes our present knowledge of the events involved in translocating targeted secretory proteins to and across the eukarytoic RERM. Two pathways are envisaged, one which is independent of SRP and its receptor and one which is SRP-dependent. Our knowledge of the corresponding pathway leading secretory proteins across the bacterial cytoplasmic membrane is somewhat less extensive, but it is clear from the comparison shown in Table III.2 that the two pathways appear to be quite different. It is by no means obvious that the early stages in the eukarytoic secretory pathway are more complex than the corresponding stages in the prokaryotic pathway, as is often supposed. It may be that the signal peptide

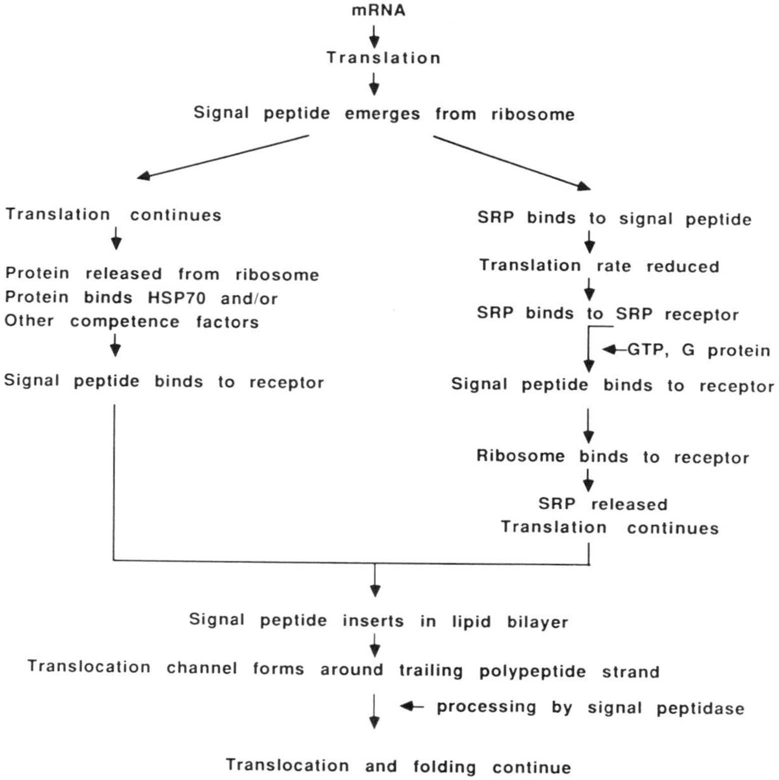

Fig. III.8. Schema showing the early stages in the translocation of a secretory protein across the membrane of the endoplasmic reticulum.

or sequence is the only major feature common to both pathways and that differences in all other features indicate the independent evolution of the two pathways. It will be interesting to determine the extent to which components of the two systems are interchangeable.

F. Membrane protein topogenesis

As illustrated in Fig. III.7A, failure to process a soluble secretory protein precursor causes the protein to remain anchored in the membrane with the C-terminus on the trans side of the membrane and the N-terminus on the cis, cytoplasmic side; i.e., it would become a type II bitopic membrane protein (104,640,879a).

Table III.2. Principal factors involved in the initial stages of protein export and secretion in eukaryotes and prokaryotes

Feature	Eukaryotes	Prokaryotes
Signal peptide or sequence	+	+
Kinetics	Mainly cotranslational	Co- or posttranslational
ATP dependence		
Co-translational	Unknown	+
Posttranslational	+	+
Membrane potential dependence	−	+
Competence factors	SRP ribonuclear protein; heat shock protein 70	SEC A protein; SEC B protein; SEC D protein; trigger factor; 4S,8S,12S,16S complexes
Competence factor receptor	SRP receptor	SEC E or SEC Y protein (?)
Signal peptide receptor	35-kDa protein	SEC E or SEC Y protein (?)
Ribosome receptor	+ (not identified)	−
Translocases	Not identified	Not identified
Signal peptidase	Protein complex	Two types; simple polypeptides

1. "Start" and "stop" signals

What happens if an additional stretch of hydrophobic residues is present on the C-terminal side of the signal peptide? As shown in Fig. III.7B, translocation of the polypeptide continues until these hydrophobic residues are fully within the membrane. At this time, the lipids at the translocation site may revert to a bilayer structure, or the translocation channel may dissociate, and the hydrophobic residues interact with the fatty acid chains of the membrane lipids to prevent further translocation. Cleavage of the signal peptide thereby produces a type I bitopic membrane protein, whereas the absence of processing by signal peptidase produces a simple polytopic membrane protein with both the N- and C-termini in the cytoplasm.

How does this model, and in particular the idea that long stretches of hydrophobic amino acids cannot pass through the RERM or BCM, correspond to the known configuration of typical type I membrane proteins? Many type I membrane proteins have two stretches of hydrophobic amino acids, one in the N-terminal signal peptide and one within the segment of the mature polypeptide. Several experiments show that the hydrophobic nature of these "stop transfer–membrane anchor" sequences is of paramount importance. Among the simplest and most elegant of these experiments are those of Davis and his collaborators on bacteriophage f1-encoded gpVIII protein, which is inserted into the cytoplasmic membrane of infected *E. coli* cells. Deletions which removed the membrane anchor caused the protein to be exported to the periplasm (227). *In vitro* linker mutagenesis was used to replace the hydrophobic domain by other hydrophobic sequences in order to determine the minimum length and overall hydrophobicity of potential membrane-anchoring sequences. Results obtained by Davis and Model (226) indicated that a minimum of 16 contiguous hydrophobic or neutral amino acids were required and that membrane anchorage was independent of the precise sequence of the residues in the anchor segment. Similar though less extensive results were obtained, for example, in studies on the membrane-anchoring domains of vesicular stomatitis virus (VSV) G protein (4) and of influenza virus hemagglutinin (259).

Figure III.7C shows what might happen if there are three long stretches of hydrophobic amino acids in the sequence of a transmembrane protein. The first two segments behave as routing signal and stop transfer–membrane anchor sequences, respectively, whereas the third hydrophobic segment reinitiates translocation of sequences located on its C-terminal side; i.e., it functions as a *restart* signal sequence. It will not, of course, be processed by signal peptidase. An alternating pattern of restart and stop transfer signals will automatically lead to a complex membrane topology of alternating hydrophobic (transmembrane) and hydrophilic (extramembrane) sequences, as proposed by Blobel (88).

In order to test this model, Audigier *et al.* (28) dissected the gene for bovine opsin and fused DNA coding for six of the seven proposed transmembrane stretches of the polypeptide in turn to DNA encoding the extreme N-terminus of the protein. The hybrids were thus devoid of all other long stretches of hydrophobic amino acids. All four potential secretory signal sequences (one start signal and three restart signals) were able to translocate the N-terminus of the polypeptide into the lumen of microsomes while themselves remaining anchored in the microsomal membrane. This usual orientation (the N-terminus of the routing signal usually remains on the cis side of the membrane) might arise because the polypeptides "flip" in the membrane after insertion has occurred in the normal manner (see next section). One of the two stop transfer signals was also found to function as a signal sequence in the truncated opsin polypeptide, whereas another stop transfer signal was clearly devoid of signal sequence activity (see also Ref. 1180a).

These observations suggest that some but not all stop transfer sequences can function as routing signals, and vice versa, when placed in a different context [see also (972a,1225,1233)]. The inability of some sequences to perform both functions may be related to differences in their overall hydrophobicity and "membrane-seeking" potential, both of which seem to be higher in routing signals than in stop transfer sequences (276). The interaction between restart signal sequences and the RERM is now known to be SRP-independent (1180a).

This model for membrane protein topogenesis is based on the idea that signal sequence C-terminus penetrates into the membrane to reach the trans side while the N-terminus remains on the cis side. Almost all of the well-characterized polytopic membrane proteins seem to be organized in this way. Furthermore, several groups have noted that the cis borders of the transmembrane segments are often marked by one or more positively charged amino acids (441a), as is the case for bacterial signal peptides (Section III.A.1). This distribution of positive charges is particularly striking in the case of cytoplasmic membrane proteins of *E. coli,* which, as noted by von Heijne (437), have most of their basic residues in their cytoplasmic (cis) loops and most of their acidic residues in their periplasmic (trans) loops. These residues may function in consort with the hydrophobic segments to determine membrane topology. A simple way of testing this idea is to study the effect of deleting the codons for these basic residues from a bi- or polytopic membrane protein. Results from several studies on eukaryotic secretory membrane proteins suggest that basic charges on the cis side of transmembrane segments are not needed for membrane topogenesis (206,1242). However, such mutations may produce subtle changes in membrane topology by weakening membrane anchor function or by allowing normally ''excluded'' sequences from the cis side of the membrane to penetrate into the lipid bilayer (207).

2. Exceptions

A small number of unprocessed integral membrane secretory proteins have their N-termini on the trans side of the membrane. Examples include truncated derivatives of bovine opsin with one signal sequence–stop transfer sequence close to the N-terminus (see last section) and a number of ''natural'' integral membrane proteins including *E. coli* signal peptidase I (211–213) hepatitis B virus surface antigen (271), influenza B glycoprotein (1193), and the ER cytochrome P_{450} (628) (Fig. III.9). Space does not allow these exceptions to be discussed individually, but the main question to be answered is how the N-terminus of the polypeptide is translocated through the membrane. Two possible explanations have been suggested: either the first transmembrane segment is an unusual class of signal sequence, which naturally inserts ''head first'' into the membrane (628,1096), or it inserts in the normal ''tail first'' manner and then flips around (28,438). In the case of signal peptidase I and other polytopic

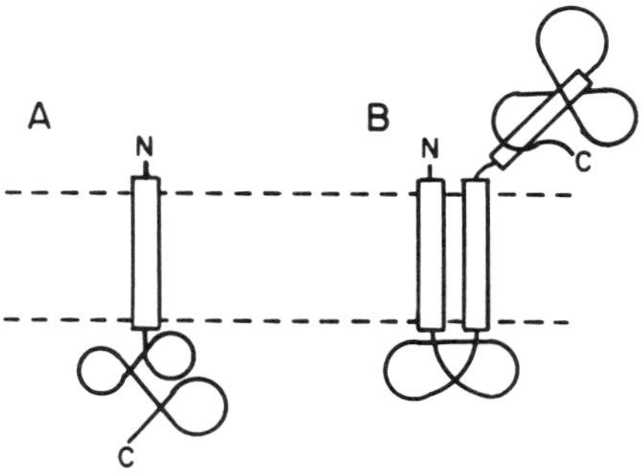

Fig. III.9. Schematic representation of the organization of two classes of membrane proteins [cytochrome P_{450} (A) and *E. coli* signal peptidase I (B)] which have N-terminal hydrophobic segments (boxed regions) inserted "headfirst" into the RER and bacterial cytoplasmic membranes respectively.

membrane proteins, this would allow a second signal sequence located immediately behind the first to function in the conventional "tail first" way. Sequences located immediately after the first "headfirst" signal sequence may prevent it from penetrating into the membrane in the conventional way (442). Neither model has been rigorously tested, and it may be that there are two distinct classes of insertion modes for these unusual signal sequences. The N-terminus of cytochrome P_{450} (type A in Fig. III.9) is endowed with signal sequence-type function, as illustrated by its SRP-dependent insertion into the RERM (984), but the C-terminus remains cytoplasmic irrespective of whether it is the authentic P_{450} sequence (628) or foreign sequences derived from other secretory proteins (984). However, positive charges introduced at the extreme N-terminus of P_{450} convert it into a normal, processed signal peptide (1096). Thus, the absence of positive charges from the P_{450} signal sequence may allow it to insert headfirst into the RERM. Insertion of *E. coli* signal peptidase I (type B in Fig. III.9) into the cytoplasmic membrane, however, depends upon an internal signal sequence which acts in consort with either of the two flanking hydrophobic domains (213).

3. Unusual peptide membrane anchors

There is no *a priori* reason why a segment of a fully translocated protein should not reinsert into the membrane it has crossed and anchor it there. These membrane anchors may differ from stop transfer signals and yet retain the ability to penetrate into the lipid bilayer. One anchor sequence which may function in this way is the C-terminus of the *E. coli* cytoplasmic protein DAP A (penicillin-binding protein 5) (529). The unusual amphiphilic helix of this anchor has a high hydrophobic moment, which might allow it to partition into cytoplasmic membrane lipids from the periplasmic (trans) side of the membrane after signal peptide processing and translocation of the entire mature DAP A sequence through the membrane.

4. Membrane proteins with lipid anchors

Covalent modifications by lipids or fatty acids are among the many posttranslational changes which occur during protein transit through the secretory pathway. Unlike many of the other cases of posttranslational modification, there is good evidence that fatty acyl and lipid groups influence membrane protein topology. Indeed, there are numerous examples of membrane proteins whose sole anchor is a fatty acid or lipid group. This group of posttranslational modifications is therefore considered here because of their possible influence on membrane protein topogenesis.

a. Bacterial lipoproteins

One of the most extensively characterized groups of fatty-acylated proteins are the so-called bacterial lipoproteins. Over 12 different types of lipoproteins have been identified in bacteria. They include

(i) the major outer membrane protein of *E. coli* and related bacteria (LPP) (417,512)
(ii) a β-lactamase produced by several strains of Gram-positive bacteria (785)
(iii) lysis proteins involved in the release of *colicins,* a special class of toxins produced by *E. coli* (183)
(iv) an *E. coli* enzyme (FTS I) involved in cell wall synthesis (766)
(v) the TRA T proteins encoded by conjugative plasmids and involved in mating pair formation and plasmid exclusion phenomena (864)
(vi) cell surface and extracellular enzymes of *Klebsiella* sp. (904) and *Vibrio* sp. (535)
(vii) a cytochrome subunit from *Rhodopseudomonas* sp. (1181)

The unifying feature of all of these proteins is that they are synthesized as precursors with signal peptides which are processed by lipoprotein signal peptidase (Section III.D.2). Radioactive fatty acids (usually palmitate) and glycerol can be specifically incorporated into them, and processing is inhibited by globomycin (Section III.D.2). These features indicate the sequence of events leading to the production of the mature lipoprotein (Fig. III.10). Full details of these modification reactions have recently been reviewed by Wu and Tokunaga (1208); the following salient details

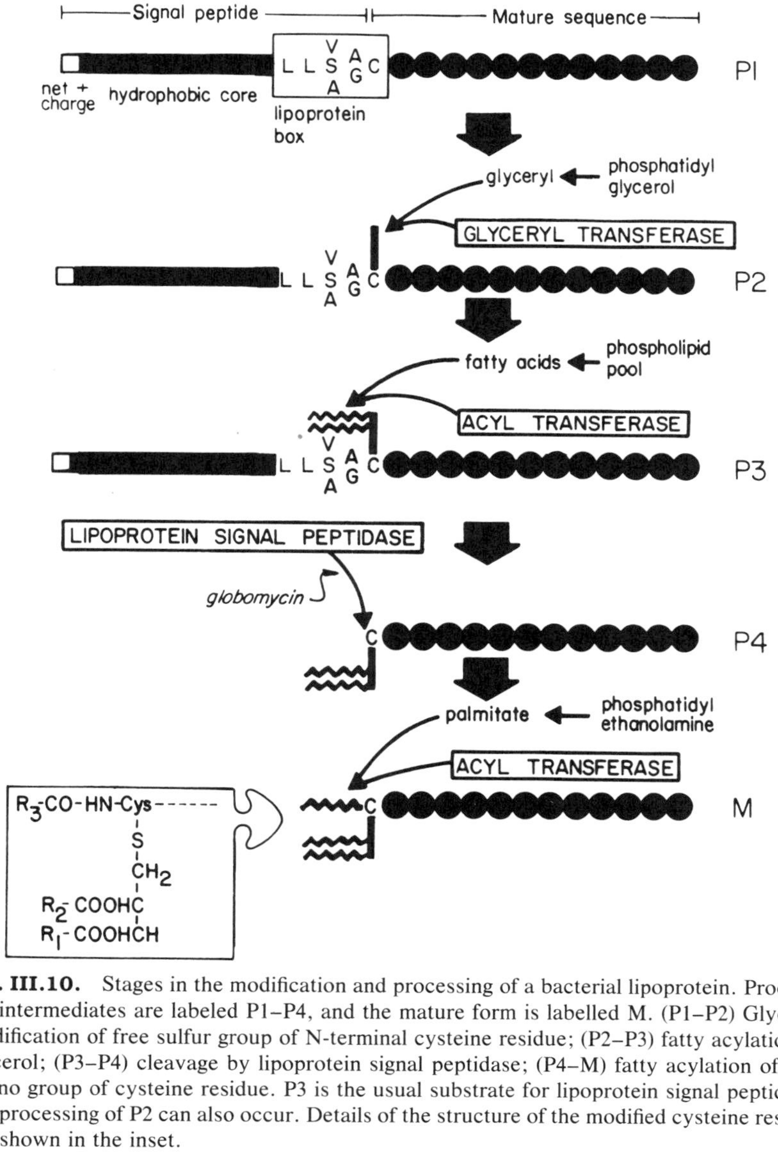

Fig. III.10. Stages in the modification and processing of a bacterial lipoprotein. Processing intermediates are labeled P1–P4, and the mature form is labelled M. (P1–P2) Glycerol modification of free sulfur group of N-terminal cysteine residue; (P2–P3) fatty acylation of glycerol; (P3–P4) cleavage by lipoprotein signal peptidase; (P4–M) fatty acylation of free amino group of cysteine residue. P3 is the usual substrate for lipoprotein signal peptidase, but processing of P2 can also occur. Details of the structure of the modified cysteine residue are shown in the inset.

concern the timing of the events, the sources of the covalently attached groups, and the sites at which modification occurs.

The glyceryl group, derived from phosphatidyl glycerol, is first attached to the free sulfhydryl group of the cysteine residue at position +1 of precursor lipoprotein by prelipoprotein glyceryl transferase. This is then acylated at two sites by one or more acyltransferases to give a diglyceride lipoprotein, the substrate for lipoprotein signal peptidase. Cleavage of the signal peptide liberates a free amino terminus, which is the site for a further acyl transferase (*N*-acylase)-mediated fatty acylation reaction. Both ester- and amide-linked fatty acids are derived from the large phospholipid pool; recent evidence suggests that ester-linked fatty acids are selected at random, whereas the amide-linked group is derived exclusively from the 1-acyl moiety of phosphatidyl ethanolamine (528). All of these reactions occur on the trans side of the cytoplasmic membrane, and the enzymes concerned are presumably integral cytoplasmic membrane proteins. The diglyceride structure also forms part of the recognition site for lipoprotein signal peptidase and presumably, the conserved sequence at the C-terminus of the signal peptide forms part of the recognition sequence of the specific glyceryl and acyl transferases (the lipoprotein box) (Section III.D.2).

Several lines of evidence suggest that the covalently modified N-terminal cysteine residue of bacterial lipoproteins is a primary membrane anchor:

- (i) Normally soluble secretory proteins can be converted into membrane-anchored proteins by replacing their signal peptides by the signal peptide and N-terminal mature region of a known lipoprotein (285,369).
- (ii) Mutations which change the codon for Cys at position +1 prevent fatty acylation and processing by lipoprotein signal peptidase and sometimes allow processing by signal peptidase I. The nonacylated mature proteins are released from the membrane (431).
- (iii) Some lipoproteins, notably the β-lactamases, are apparently released from the membrane by proteolysis at a site near the N-terminus. Cell-associated (lipo)-β-lactamases have a relatively long half-life (~3 hr). The processing step is catalyzed by one or more endogenous proteases (526,784).

The fatty acyl chains of the (lipo-)β-lactamases anchor the polypeptide in the outer face of the cytoplasmic membrane (783). There is confusion, however, regarding the site of membrane insertion of the fatty acyl chains of lipoproteins produced by Gram-negative bacteria. When the (lipo-)β-lactamase gene from *Bacillus licheniformis* is expressed in *E. coli*, the enzyme is apparently localized to the outer membrane but is not exposed on the cell surface (i.e., it is probably facing the periplasm). Endogenous lipoproteins of *E. coli* and related bacteria are apparently located in either the outer or inner (cytoplasmic) membranes (507). The sequences of several of these lipoproteins have been deduced from the nucleotide sequences of their structural genes. One lipoprotein which seems to be located in the cytoplasmic membrane, NLP A, has a potential membrane-spanning peptide segment in addition to its fatty acids (248,1229). This segment of the polypeptide may function as a stop transfer signal, anchoring the protein in the cytoplasmic membrane. The same may also be true of the FTS I protein (766). However, recent evidence suggests that amino acids immediately after the fatty-acylated cysteine residue may determine the lipid bilayer into which the fatty acids will insert; the presence of an acidic residue in the region +2–+5 appears to correlate with insertion in the cytoplasmic membrane, both in natural lipoproteins and in hybrid proteins (1214). Thus, a single amino acid seems to determine the final location of the class of secretory proteins in Gram-negative bacteria.

All other sequenced lipoproteins seem to be located in the outer membrane and are devoid of potential membrane-anchoring sequences other than the fatty acyl groups. One particular lipoprotein, the enzyme pullulanase of the Gram-negative bacterium *Klebsiella pneumoniae,* is known to be located on the cell surface, but cell fractionation experiments (Section II.A.1) indicated that it is located in both inner and outer membranes (286). We have suggested that this may be due to the release of pullulanase-enriched micelles from the outer membrane during fractionation. We encountered similar difficulties in attempts to localize the low-molecular-weight lysis proteins by cell fractionation techniques (183,899), which raises doubts concerning the use of such techniques to localize membrane-bound lipoproteins. However, it seems reasonable to assume that lipoproteins are normally anchored to the inner face of the outer membrane by their fatty acyl groups. Cell-surface lipoproteins such as pullulanase require additional secretion functions for correct localization (286,287) (Section IV.C.1.b). Wu and Tokunaga (1208) have suggested that lipoproteins transit through nonbilayer regions of the cell envelope where the cytoplasmic and outer membranes are in contact. There is some evidence that such structures exit, but their role in protein export and secretion is far from clear (Section IV.B.3).

b. Phosphatidylinositol lipoproteins

A significant number of eukaryotic cell surface membrane proteins (currently about 20) of diverse origins and functions [e.g., the surface VS-G protein of trypanosomes (310), the acetylcholine receptor (342), the Thy-1 major surface glycoprotein of rodent thymocytes and neurons (191), a yeast cell surface protein (191a), and murine histoincompatibility proteins H-2, Qa, and Tla (1086)] are anchored in the membrane by glycosyl–phosphatidylinositol (PI) attached to their C-termini [reviewed in (201) and (658)]. There are no known examples of plant or bacterial proteins with PI anchors. The structure of the lipid anchor (Fig. III.11) has been deduced mainly through work on the VS-G protein. This and all other PI proteins are characteristically released from the membrane by PI-specific phospholipase C, which cleaves between the diacylglycerol anchor and the phosphate group (Fig. III.11). The glucosamine–inositol linkage is also characteristically sensitive to nitrous acid. The ethanolamine is amide-linked to the free α-carboxyl group of the C-terminal amino acid. All members of this group of lipoproteins are antigenically related, probably because the inositol–glucan region is highly immunogenic.

Not all members of this group of lipoproteins have identical modifications; VS-G protein, for example, has one ethanolamine residue whereas mammalian proteins have two or three, and different sugar residues and linkages have been reported (465). The fatty acid groups may also vary in different proteins: the fatty acid in VS-G is myristic acid, whereas Thy-1

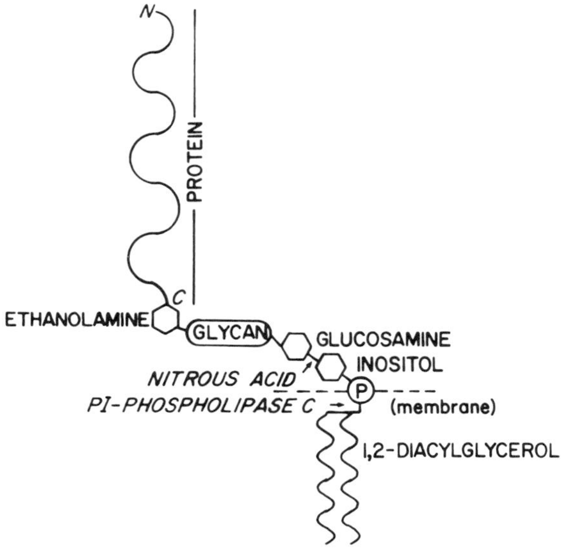

Fig. III.11. Model for the structure of the membrane anchor segment of a typical phosphatidylinositol lipoprotein. The cleavage sites for nitrous acid and phospholipase C are indicated.

and acetylcholine esterase contain mainly palmitic and stearic acids. Different C-terminal amino acids can provide the free carboxyl group (e.g., Ser or Asp in variants of VS-G protein, Cys in Thy-1, and Gly in acetylcholine esterase). Even closely related PI lipoproteins may have structurally different glycolipid anchors (935).

Kinetic evidence suggests that PI lipoproteins are modified in the ER during or very shortly after synthesis (311). They are all synthesized with signal peptides, so processing presumably occurs in the RER. A precursor glycosyl–PI anchor has been identified in trypanosomes, suggesting that the PI anchor is fully assembled before it is attached to the protein (584), although other studies (41a) suggest that the lipid may be galactosylated at a later stage. An unusual feature of PI lipoproteins is that the polypeptides initially have a C-terminal hydrophobic membrane anchor sequence (stop transfer signal), which is cleaved off before or as modification occurs (657). There does not appear to be any consensus sequence around the cleavage site, although only a few PI lipoprotein genes have been sequenced.

It is not clear why the transition from peptide to lipid membrane anchor should be necessary. It could allow endogenous phospholipase C-catalyzed release of PI lipoproteins from the cell surface, as recently demonstrated for the surface protease of the parasite *Plasmodium falciparum* (120). Other possibilities are that the lipid allows greater lateral mobility in the membrane than the original peptide anchor or that the lipid acts as a sorting signal to target PI lipoproteins to particular domains on the cell surface (573). Clear evidence that PI acts as an anchor comes from studies by Caras *et al.* (143), who genetically fused the last 37 residues of the PI lipoprotein decay-accelerating factor to the C-terminus of glycoprotein D of herpes simplex virus, thereby converting the normally constitutively secreted glycoprotein into a membrane-anchored form.

c. *Palmitoylated eukaryotic secretory proteins*

Different eukaryotic cells have been shown to produce a significant number (up to 50) of minor, surface-associated proteins which can be specifically labeled by radioactive palmitic acid (e.g., mammalian transferrin receptor, influenza virus hemagglutinin, vesicular stomatitis virus (VSV) G protein, apolipoprotein A-1, and ankyrin, a cytoskeletal protein). Most palmitoylated proteins are also glycosylated, but glycosylation is not a prerequisite for palmitoylation, as shown by the failure of the glycosylation inhibitor tunicamycin to block palmitoylation (679) (Section II.C). The palmitate is characteristically released by hydroxylamine treatment, indicating ester linkages to internal Ser (O-ester) or Cys (S-ester) residues, the former being cleaved only by high concentrations of hydroxylamine (679,1016,1028,1197). There are no obvious similarities around the modification sites of the small number of palmitoylated proteins for which the corresponding cDNA sequences are known.

VSV G protein, the most extensively characterized palmitoylated protein, has a thioester-linked palmitate residue. Site-directed mutagenesis has been used to identify the modified cysteine residue (951). The palmitate group could anchor this and similar proteins in the membrane, although all of them also seem to have additional transmembrane peptide segments which could also perform this function. The nonacylated variant of VSV G protein constructed by Rose *et al.* (951) still reaches the cell surface, indicating that the palmitate is not the sole membrane anchor or cell surface targeting signal. Other functions, including membrane fusogenic activity (607), have been proposed to reside in the fatty acid moiety.

The acyl transferase(s) performing the modification reaction(s) has a relatively high preference for palmitic acid, which is supplied as palmitoyl-coenzyme A. Most kinetic studies indicate that palmitoylation occurs about 20 min after protein synthesis is complete, probably in the Golgi (679,812). Palmitoylation occurs just before complex glycosylation, which also occurs in the Golgi (249) (Sections V.D,E). Furthermore, a mutant form of VSV G protein which is not transported beyond the Golgi apparatus is not palmitoylated (1236). Other proteins, such as the spike protein of Semliki Forest virus, are palmitoylated within 1 min of synthesis, suggesting that the reaction occurs in the RER (1017). Results from a cell-free acylation assay also indicate that palmitoyl-CoA transferase is located in a "late" ER compartment, with little activity being detected in the Golgi (78,1015). These results were confirmed by the purification of palmitate-specific acyl-CoA transferase from ER membranes, together with a fatty-acylesterase which may modulate biological functions associated with the presence of covalently linked fatty acids (78). The acylesterase is only active on certain types of palmitoylated protein, raising the possibility that the attachment of different ester-linked palmitate groups is catalyzed by different acyltransferases located in the membranes of either the RER or the Golgi. Studies with secretion-defective mutants of the yeast *S. cerevisiae* (Section II.F.1.b) also indicate that palmitoylation occurs in the ER (1179).

An entirely different group of eukaryotic proteins has a second type of fatty acid modification, involving mainly myristic acid linked by a hydroxylamine-resistant amide bond (679,1197). Although myristate may anchor these proteins in membranes, they are not true secretory proteins (Section V.H).

Further reading

Briggs, M. S., and Gierasch, L. M. (1986). Molecular mechanisms of protein secretion: The role of the signal sequence. *Adv. Protein Res.* **32,** 109–180.

Hortsch, M., and Meyer, D. I. (1986). Transfer of secretory proteins through the membrane of the endoplasmic reticulum. *Int. Rev. Cytol.* **102,** 215–242.

Randall, L. L., Hardy, S. J. S., and Thom, J. R. (1987). Export of protein: A biochemical view. *Annu. Rev. Microbiol.* **41,** 507–541.

Walter, P., and Lingappa, V. R. (1986). Mechanism of protein translocation across the endoplasmic reticulum membrane. *Annu. Rev. Cell Biol.* **2,** 499–516.

Wu, H., and Tai, P. C., eds. (1987). "Protein Secretion and Export in Bacteria," Curr. Top. Microbiol. Immunol., Vol. 125. Springer-Verlag, Berlin.

CHAPTER IV

Later stages in the prokaryotic secretory pathway

In the previous chapter, we saw that secretory routing signals are the only features of secretory proteins which are directly involved in their targeting to and their translocation across the bacterial cytoplasmic membrane. This chapter deals with features of the same proteins which determine whether the protein will remain in the periplasmic space or be inserted into the outer membrane or secreted into the medium. Prokaryotes have only very limited capacity for posttranslational covalent modification of secretory proteins. Indeed, the only authenticated examples of such covalent modifications are those in bacterial lipoproteins (Section III.F.4.a), the asparagine-linked rhamnose in a surface layer protein (712a), and the apparently extensive glycosylation of a myxococcal surface protein (678a). Disulfide bridges in bacterial secretory proteins probably form spontaneously, which is again in contrast to the situation in eukaryotes (see Section V.B.2).

A. Secretory proteins without sorting signals

The cell wall of Gram-positive bacteria represents only a minor impediment to the release of soluble secretory proteins into the medium (392,1138). Some Gram-positive bacteria produce a proteinaceous coat perforated by pores, which might also restrict the free passage of secreted proteins to a limited extent (993). In contrast, the outer membrane of Gram-negative bacteria prevents the release of periplasmic proteins and can only be breached by specific types of authentic secreted proteins (Section IV.C), by treatment with membrane perturbants, or by mutations which increase the permeability of the outer membrane by distorting its lipid, lipopolysaccharide, or protein content (897,903).

The medium and the peirplasmic spaces are, therefore, the final destinations for the "default" branch of the secretory pathways in Gram-positive and Gram-negative bacteria, respectively. Consequently, most secreted proteins of Gram-positive bacteria are localized to the periplasm when their structural genes are expressed in Gram-negative bacteria (897). There are some minor differences, however, as described in the following sections.

1. Secondary proteolytic processing

Secreted proteins of Gram-positive bacteria are frequently processed at the N-termini after signal peptide removal (378,582,719,834,843). Only two periplasmic proteins are known to be similarly processed: *E. coli* penicillin G acylase α and β subunits are derived from a single polypeptide, apparently during translocation through the cytoplasmic membrane (1023), and a single amino acid is proteolytically cleaved from the mature sequence of *E. coli* alkaline phosphatase (768). Proteolytic processing of the secreted proteins may be necessary either for their release from the membrane or for their activation. Alternatively, the proteins may simply be partially degraded by extracellular or cell-associated proteases which are of no consequence for secretion or activation (554,940,1104a). In some cases, longer "precursor" forms have been stabilized by growing cells in medium containing protease inhibitors (562,1104a), but the inability to select mutants totally devoid of extracellular protease activity has restricted studies on this aspect of protein secretion.

N-terminal proteolytic processing is thought to be important in one unusual aspect of protein secretion by Gram-positive bacteria, namely, the release of certain β-lactamases (formerly called penicillinases). These β-lactamases have a fatty-acylated cysteine at the N-termini of their mature sequences and belong to the "lipoprotein" group of bacterial secretory proteins (Section III.F.4.a). The fatty acyl residues are thought to anchor β-lactamase at the surface of the bacteria producing them (*Bacillus licheniformis, Bacillus cereus,* and *Staphylococcus aureus*), and the enzymes appear to be released by proteolytic cleavage close to the N-terminus, presumably by a secreted or cell surface protease. At least two extracellular versions of nonacylated β-lactamase have been detected (526,772,1052), suggesting that further proteolytic processing occurs after the enzyme has been released. The cell-associated form of β-lactamase, which persists for up to 3 hr, can be released by trypsin treatment, as can other *B. licheniformis* cell-surface lipoproteins which are apparently not normally released into the medium (783). Results of our own studies on the secretion of the lipoprotein pullulanase by *Klebsiella pneumoniae* (Section IV.C.1.b) suggest that lipoproteins can be released from cell surfaces as micelles, without prior proteolytic cleavage. This raises the possibility that fatty-acylated β-lactamases are released as micelles and that subsequent proteolysis releases the monomeric forms of the enzyme. However, prior clustering of β-lactamase on the cell surface, which might be necessary to reach the critical concentration required for micellation (capping), could not be demonstrated by immunocytochemistry (407).

Certain secreted proteases (e.g., subtilisin and elastase) are produced by Gram-positive and Gram-negative bacteria as precursors with unusually long (up to 227 residues) N-terminal extensions (80a,1073,1102,1219). The extreme N-termini of these precursors resemble classical signal peptides and have consensus cleavage sites in the expected positions (Section III.A.1). The regions between these cleavage sites and the N-termini of the mature sequences (the propeptides) are hydrophilic and highly charged. It has been proposed that they could anchor the protease at the cell surface, assure correct folding of the polypeptide chain, or maintain the protease in a zymogen (inactive) form until it reaches the outside of the cell.

Power *et al.* (885) observed that mutations which inactivated *B. subtilis* subtilisin protease caused the polypeptide to accumulate at approximately 1000 sites at the cell surface, suggesting that the propeptide anchors the protease in the cell wall or cytoplasmic membrane and that propeptide processing is autocatalytic. Production of a second, catalytically active subtilisin resulted in cleavage and release of the inactive enzyme. Similar results were reported by Vasantha and Thompson (1139), who fused DNA coding for the prepro-region of subtilisin to DNA coding for protein A or β-lactamase. The hybrid genes were expressed in a *B. subtilis* host lacking secreted neutral and alkaline proteases, which failed to process the propeptide and secreted the hybrids as pro-forms. Wong and Doi (1202) confirmed that preprosubtilisin was processed at the predicted signal peptide cleavage site in an *in vitro* assay; they found no evidence for autocatalytic processing. Ikemura *et al.* (512), who reported expression of the gene for preprosubtilisin in *E. coli,* confirmed that processing of the propeptide was autocatalytic. Furthermore, when the prepro-region of subtilisin was replaced by the signal peptide from the *E. coli* outer membrane OMP A protein, mature-length but inactive subtilisin was exported into the *E. coli* periplasm. Active subtilisin was only obtained when the OMP A signal peptide was fused to the prosubtilisin polypeptide, suggesting that the propeptide is necessary for subtilisin to fold in the correct conformation.

Gram-positive bacteria synthesize at least two other secreted proteins as unusually long precursors, nuclease A-B of *Staphylococcus aureus* (721,1038) and lipase of *Staphylococcus lyticus* (378), but it is not known how these proteins are processed or whether the propeptides play a role in secretion or enzyme folding.

2. Secretion of specialized enzyme complexes

Although most soluble secretory proteins are probably released as free monomers by Gram-positive bacteria, some assemble into surface protein layers, whereas others are released as high-molecular-weight complexes comprising several enzymes which act together to degrade insoluble substrates. The most thoroughly studied example of the latter are the *cellulosomes* secreted by several cellulose-degrading bacteria such as *Clostridium thermocellum*. Cellulosomes first appear as protuberances on the cell surface and are subsequently released or sloughed off into the medium. They contain various proteins with different substrate specificities and activities which act together to degrade cellulose polymers (608). The cellulosome also contains an additional protein which causes bacteria or the cellulosome to adhere to the substrate. Mutants devoid of this protein do not adhere to cellulose and do not form cell-surface protuberances, indicating that the protein may be both an adhesin and the backbone on which cellulases are assembled (56). Contact with cellulose causes the normally compact cell-surface cellulosome to form fibrous "corridors" and also causes secreted cellulosomes to coat insoluble cellulose (55).

The precise structure of the cellulosome has not been determined; it is not known, for example, whether it contains membrane-derived lipids. Sequencing of several *C. thermocellum cel* (cellulase) genes indicate that they encode precursor polypeptides with typical signal peptides. Expression of these genes in *E. coli* results in the release of soluble cellulase into the periplasm, indicating that they do not have intrinsic membrane anchor sequences (68,195).

B. Outer membrane proteins

Outer membrane proteins (OMPs) of several Gram-negative bacteria, principally *E. coli,* have been extensively characterized. Studies have focused on how they are organized in the membrane and how they are targeted to the outer membrane.

1. Cell surface organization

Major integral OMPs became popular for the study of membrane protein structure owing to the ease with which they could be purified in large amounts. Spectroscopic analyses indicate the predominance of β-sheet structures in the two classes of proteins which have been analyzed, the porins and the OMP A protein (773,952,1151). This is in contrast to "normal" transmembrane proteins, in which hydrophobic α helices predominate in the transmembrane domains (Section II.D.1), and also to sequence analyses and structure predictions (60,87,179,727,828), which initially suggested a predominance of α helices. Sequence analyses of genes coding for other OMPs from *E. coli* and from closely or distantly related bacteria indicate that the absence of typical transmembrane hydrophobic segments is a general feature of this class of membrane proteins.

The membrane organization of the *E. coli* porins (OMP C, OMP F, PHO E, and LAM B) and OMP A protein has been extensively characterized using molecular techniques including the mapping of antigenic epitopes (631,1004) and protease cleavage sites (1005), the study of the products of hybrid porin genes (182,633,728,791,1121), sequence analyses of mutants affected in the adsorption of bacteriophages (179,578,740) or epitope-specific antibodies (231,632), and insertion (linker) mutagenesis of the structural genes (105,153,329). Although these studies do not reveal the complete structure of any of these proteins, they have been particularly useful in predicting their conformation in the membrane (see example in Fig. IV.1). A striking feature of these proteins is that only very short segments of the polypeptides are usually exposed on the outer membrane surface, the only exception being OMP A, which has a long periplasmic tail (Fig. IV.1).

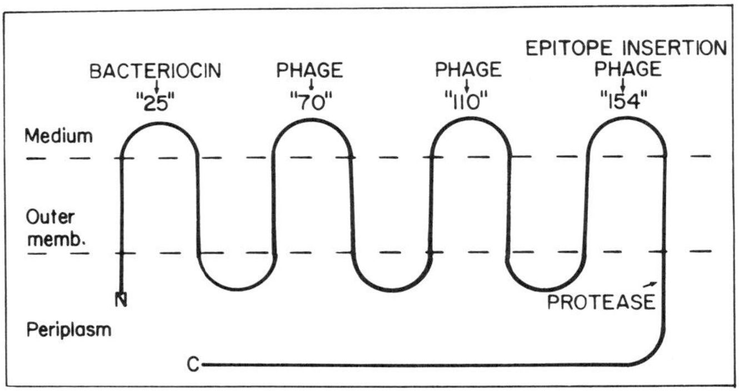

Fig. IV.1. Folding pathway of the OMP A protein in the *E. coli* outer membrane as determined by proteolysis, by the analysis of mutant proteins with defective bacteriophage (phage) receptor sites, by the analysis of hybrid proteins and bacteriocin receptor activities, by comparison with homologous sequences of corresponding proteins from related bacteria, and by epitope insertion mutagenesis. Exposed domains are numbered according to the position of the amino acid residues along the length of the polypeptide. Transmembrane segments are predominantly in a β-sheet configuration (60,182,328,329,740,1151).

The trimeric structure of porins was demonstrated both by cross-linking studies (842) and by analysis of native complexes (952). Trimerization complicates attempts to determine the topology of these polypeptides in the membrane and adds an additional dimension to studies on their export and insertion (see below). Porins also form a regular lattice on the cell surface (1081) and interact with lipopolysaccharide (LPS) and the underlying peptidoglycan (941,952) (see Fig. I.2). Other OMPs may also interact with LPS and peptidoglycan (1026). Several OMPs have now been crystallized and preliminary X-ray crystallographic data are available (345). These and other studies on planar crystals (252,629) should soon provide fine-structure membrane topology maps of these proteins.

2. Outer membrane sorting signals

Do OMPs contain sorting signals? Studies with *lacZ* gene fusions (Section III.A.3) and internal gene deletions suggested that LAM B protein might have an outer membrane sorting signal close to its N-terminus (74,75). This region of LAM B includes an 11-residue sequence which is also present in the putative transmembrane segments of other OMPs (787). It should be noted, however, that at most only a minor fraction of the LAM B–LAC Z hybrid proteins containing this segment was sorted to the outer membrane, the remainder becoming jammed at protein export sites and subsequently accumulating in the cytoplasm (1118,1120,1152). Furthermore, studies with other proteins indicate that overall OMP conformation, rather than specific, linear sorting signals, are required for insertion into the outer membrane (106,328,530). The proteins may fold into an insertion-competent state during transit through the cell envelope and then partition into the outer membrane (see below). The ability to fold into this insertion-competent state would be adversely affected by deletions or insertions which disrupted the folding pattern or which prevented stable oligomerization (728), and the putative OMP sorting signal may in fact be a nucleation site for correct folding or promote contact with the outer membrane (572a,b).

3. The outer membrane sorting pathway

Several studies have attempted to determine whether OMPs migrate through the periplasm in free, monomeric form or in vesicles, whether there are regions of contact (bridges) between the inner and outer membrane, and where and when the oligomeric porins adopt their final conformation.

Early studies showing that LAM B protein inserted preferentially near the septum (979) were later shown to be incorrect. In fact, newly synthesized LAM B and other OMPs are apparently inserted into the outer

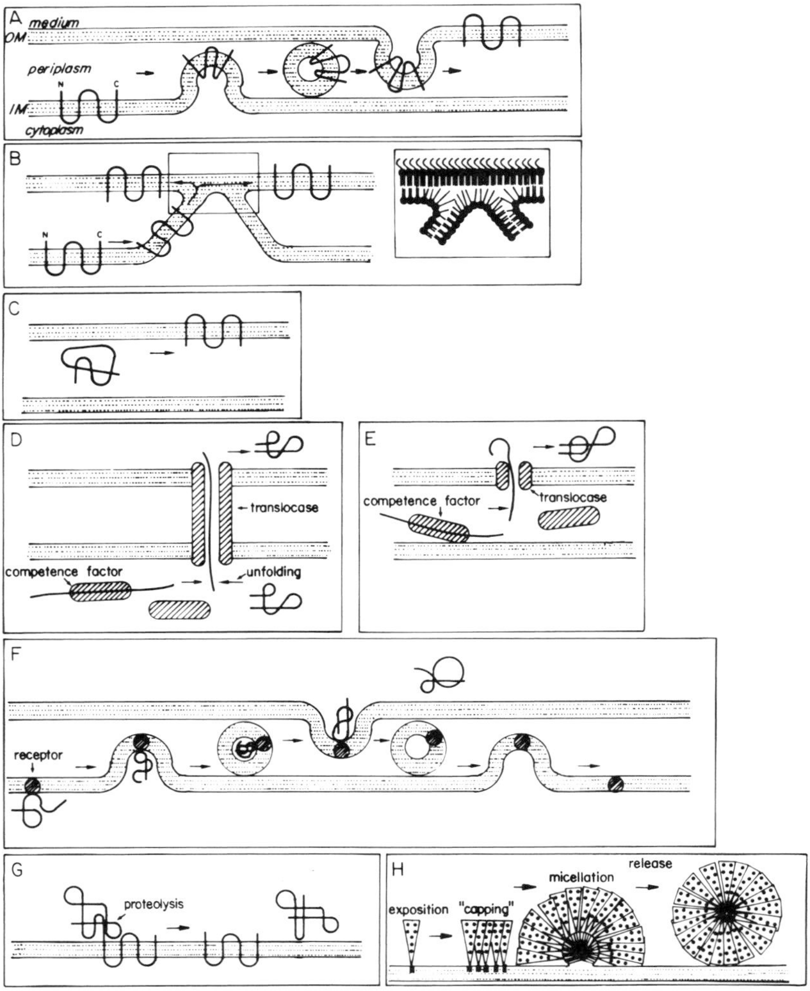

Fig. IV.2. Models for the insertion and transport of proteins into and across the outer membrane of Gram-negative bacteria: (A) Insertion of OMPs via membrane vesicles. The OMP inserts into specific regions of the cytoplasmic, or inner, membrane (IM), possibly at sites where LPS is synthesized. Vesicles then bud from the surface of the inner membrane, diffuse through the periplasm, and fuse with the outer membrane (OM). (B) Insertion of OMPs via membrane fusion sites. The OMP inserts into a specific region of the inner membrane as in (A) but migrates through adjacent regions of the envelope formed by nonbilayer lipids (a possible example is shown in the insert) or by membrane fusogenic proteins and then diffuses into the outer membrane. (C) Insertion of OMPs via soluble periplasmic intermediates. The OMP is released into the periplasm. Its interaction with other outer membrane components such as LPS or other proteins and/or partitioning into the outer membrane cause it to adopt its final conformation. (D) Protein secretion across both inner and outer membranes from a soluble cytoplasmic intermediate. The protein may be maintained in a secretion-competent state by a competence factor, or it may be unfolded as it passes through a channel formed by the specific secretion functions (e.g., α-hemolysin of *E. coli*). (E) Transport of a periplasmic intermediate across the outer membrane through a protein-specific channel. The intermediate may be maintained in an unfolded state by perimembrane at randomly distributed sites (67,1064,1154). Presumed intermediates in the OMP sorting pathway accumulate in the periplasm (106,329,530) or in a membrane-associated, detergent-soluble form (presumed cytoplasmic membrane) (530) when outer membrane insertion is prevented by deleting part of the corresponding structural gene. Intermediates in the sorting pathway have been detected in the cytoplasmic membrane (202) and in a soluble (presumed periplasmic) form (530) by pulse–chase experiments. Furthermore, spheroplasts from which the outer membrane has been peeled away by lysozyme-EDTA treatment release porin polypeptides into the medium (713), which again suggests the existence of a periplasmic intermediate in the sorting pathway.

Freudl *et al.* (330) noted that the cytoplasmic membrane intermediate of OMP A protein had a different conformation from mature, outer membrane OMP A protein, the final conformation being attained only after the polypeptide bound the lipid A portion of LPS. Overproduction of OMP A protein caused this intermediate to appear in the periplasm as well. The intermediate form of OMP A did not fold into its characteristic β-sheet conformation, suggesting that LPS is required for correct folding and outer membrane insertion. Trimerization and insertion of porin polypeptides could also involve LPS (97a). Kinetic studies suggest that the formation of stable, membrane-integrated trimers is a slow process involving metastable dimeric or trimeric intermediates (455,917a,1153) and that the transition from monomer to trimer occurs in the outer membrane (46). It is thus difficult to determine the stage at which OMPs interact with LPS and adopt their final conformation. OMP–LPS complexes forming at an intermediate stage in their assembly into the outer membrane could only exist in either of the two lipid bilayers or in vesicles or micelles between the two membranes.

One possibility is that vesicles budding from the surface of the cytoplasmic membrane pass through the periplasm to fuse with the outer membrane (Fig. IV.2A). Two observations argue against this model. First, the gellike structure of the peptidoglycan in the periplasm would severely restrict vesicle movement (118,458). Second, vesicles budding from the

plasmic competence factors or may be unfolded during translocation through the channel. (F) Secretion of a soluble cytoplasmic protein via vesicles which shuttle between the inner and outer membranes. Specificity is determined by an inner membrane receptor protein to which the secreted protein binds. The protein–receptor complex is packaged into vesicles. Changing conditions alter the conformation of the protein and cause it to be released from its receptor either in the vesicle or at the cell surface. The vesicle then reforms and returns to the inner membrane (e.g., α-hemolysin). (G) Release of an N-terminal segment of an integral membrane protein via autoproteolysis (only the outer membrane is shown) (e.g., IgA proteases). (H) Release of an outer-membrane-anchored lipoprotein by micellation (only the outer membrane is shown) (e.g., pullulanase).

cytoplasmic membrane would inevitably contain cytoplasmic material, which would be released into the medium upon fusion with the outer membrane (Fig. IV.2A). Vesicular transport of proteins from the inner to the outer membrane must be extraordinarily efficient to handle the large amount of protein and LPS assembled in the outer membrane, and yet significant amounts of cytoplasmic proteins are not detected in the medium. Alternatively, if OMPs do indeed pass through a soluble periplasmic phase (Fig. IV.2C), they presumably interact first with phospholipids in the inner face of the outer membrane and only adopt their final configuration as they become fully integrated into the outer membrane. In this case, LPS and OMPs must reach the outer membrane by separate routes. Monoclonal antibodies could be useful tools with which to probe the conformational changes which occur during the insertion of proteins into the outer membrane.

It has also been proposed that OMPs migrate to the outer membrane through contact bridges between the inner and outer membrane (Fig. IV.2B). There is little doubt that such regions of contact do indeed exist (57), that some of them probably play a role in cell division (192), and that they can be released as a special class of vesicles containing unique enzyme functions when cells are physically disrupted (58,59,522) (see also Fig. II.1). Bayer (57) proposed that some of these membrane junctions mark sites of LPS and OMP export, but this attractive hypothesis has not been rigorously proven, and it does not explain the presence of an apparently periplasmic intermediate stage in OMP sorting (see above). It may be that Bayer's fusion sites are transient structures marking the sites of vesicle movement between the inner and outer membranes, possibly in regions of the envelope where the peptidoglycan has a lower degree of cross-linking. In any event, their exact structure remains to be determined; they must surely contain phospholipids in nonbilayer structure, together with LPS and OMPs in their correct orientation (Fig. IV.2B). Clearly, much work remains to be done before we understand how OMPs are exported and assembled.

C. Secreted proteins of Gram-negative bacteria

The outer membrane prevents nonspecific release of proteins into the medium, yet many species of Gram-negative bacteria secrete proteins into the medium (897,903). Recent studies on the mechanisms of protein secretion in Gram-negative bacteria reveal the existence of several different pathways, the best-characterized of which are discussed in the following sections.

1. Secretion via an outer membrane intermediate

OMPs generally have only short segments exposed on the cell surface and could not act as intermediates in a secretion pathway unless they were released as vesicles, which does not appear to be the case. However, two classes of secreted proteins have been shown to have outer membrane intermediates. They differ in the way in which they are inserted in the membrane and released.

a. Proteolytic release of a fragment of an integral OMP

The IgA-specific protease of *Neisseria gonorrhoeae* has a C-terminal segment whose structure is typical of that of OMPs and which inserts into the outer membrane when the protein is produced in *E. coli*. Export and sorting to the outer membrane presumably occur via the normal pathway, which includes processing of a typical signal peptide. Proteolytic activity resides in the N-terminus, which is proposed to cross the outer membrane through a pore formed by the barrel-stave structure of the C-terminus. Autoproteolysis at sites in the center of the polypeptide resembling the normal cleavage site in the IgA substrate result in the release of the N-terminal part of the protein into the medium (881) (Fig. IV.2G). The IgA proteases of *Hemophilus influenzae* (406) and the *Serratia marcescens* serine protease (1218) may be released by similar mechanisms. The fate of the C-terminal peptides which remain in the outer membrane has not been determined; they are presumably degraded, because the continued presence of large, nonspecific pores in the outer membrane might be deleterious to the cell.

b. Release of surface-localized lipoprotein

Pullulanase is a starch-hydrolyzing enzyme produced by *Klebsiella pneumoniae*. The protein is synthesized as a precursor from which the signal peptide is processed by lipoprotein signal peptidase (904) (see Section III.D.2). Pullulanase is initially localized to the cell surface, to which it is anchored by the fatty-acylated N-terminal cysteine. Pullulanase accumulates first in the outer membrane, and is subsequently released into the medium as pullulanase-enriched vesicles or micelles (904) (Fig. IV.2H). *E. coli* cells expressing the pullulanase structural gene alone fail to localize pullulanase to the cell surface or to secrete it into the medium. Cell surface localization and secretion can be achieved in *E. coli*, however, if the strain also carries genes flanking the pullulanase structural gene cloned from the *K. pneumoniae* chromosome. Some of these genes are coregulated with the pullulanase structural gene (286,287). Their inactivation by mutation either in recombinant *E. coli* (287) or in *K. pneumoniae*

(576) blocks surface localization. The complex organization of the pullulanase secretion genes indicates that several functions may be required to translocate the pullulanase polypeptide and attached fatty acyl chains through the outer membrane. The failure to isolate mutations specifically affecting pullulanase release suggests, however, that it may occur spontaneously. It will be interesting to see whether pullulanase secretion functions interact directly with pullulanase, either prior to or during translocation through the outer membrane, and what factors determine the specificity of this unusual secretion pathway.

c. *Outer membrane intermediates of other secreted proteins*

Studies on the effects of membrane perturbants led Lory *et al.* (655) to propose that toxin A was secreted by *Pseudomonas aeruginosa* via an outer membrane intermediate, possibly by passing through regions of cytoplasmic–outer membrane contact. However, more recent studies show that toxin A accumulates in the periplasm when its structural gene is expressed in *E. coli* (254), suggesting that it has a periplasmic intermediate (see below).

2. Secretion via a periplasmic intermediate

Many genes coding for secreted proteins from various Gram-negative bacteria have been expressed in *E. coli*. The gene products are generally localized to the periplasm rather than secreted into the medium (487,565,856). These proteins are invariably synthesized with signal peptides (897) and presumably enter the periplasm by default (i.e., *E. coli* lacks sorting functions which are present in strains from which the genes were derived). Although complete secretory pathways have not yet been reconstituted in *E. coli,* mutations affecting their function in the original producing bacteria lead to the accumulation of the normally secreted protein in the periplasm (22,304,485,486). Of particular interest is the fact that the heat-labile enterotoxin I (cholera toxin) produced by *Vibrio cholerae* accumulates in the periplasm, where the six subunits (1 A plus 5 Bs) associate to form the active halotoxin. The polypeptide thus adopts its final tertiary structure prior to translocation across the outer membrane, which is in contrast to almost all other known examples of the movement of proteins across membranes (453,454). A *V. cholerae* mutant which fails to secrete enterotoxin has been isolated (464), implying the existence of a specific pathway for extracellular sorting (456). B subunits can be secreted in the absence of A subunits, whereas the reverse is not the case. This suggests either that only the B subunits carry a secretion sorting signal or that sorting signals in the A chain are not functional unless the B chains are present.

Overall, these results indicate the existence of several pathways for translocating proteins across the outer membrane. How is specificity maintained? Secreted proteins could be recognized by receptors which package them into specific "secretory" vesicles, which then fuse with the outer membrane, but it is difficult to imagine how vesicles could form around the periplasmic intermediates of heat-labile enterotoxin. Alternatively, receptors could target the polypeptide directly to a specific secretion apparatus in the outer membrane. This secretion apparatus might be a channel through which the unfolded polypeptide is threaded (Fig. IV.2E) or a more complex structure which accommodates folded oligomers and which functions in a completely different way from all other known protein translocation systems. It will be interesting to see whether pools of active periplasmic enzymes accumulating in the absence of a cognate sorting function can be subsequently chased into the medium when the missing sorting gene is expressed.

3. Secretory pathway-independent secretion

Extracellular α-hemolysin of *E. coli* is synthesized without a signal peptide and seems to cross the cytoplasmic and outer membranes without direct involvement of the secretory pathway (308,309). Instead, hemolysin secretion depends on the expression of two genes (*hlyB* and *hlyD*) located immediately downstream from the hemolysin structural gene in the hemolysin operon. The products of these two genes are envelope proteins which are thought to interact with hemolysin to channel it through the envelope, either as an unfolded polypeptide (Fig. IV.2D) or in a specific class of vesicles whose formation and specific entrapment of hemolysin might be determined by HLY B or HLY D (Fig. IV.2F) (377,675,676). A putative ATP-binding site similar to that found in some protein kinases is also present in the HLY B polypeptide. Nonconservative amino acid substitutions in this region dramatically reduce hemolysin secretion, suggesting that ATP, which is required for hemolysin secretion (377), is directly coupled to HLY B-mediated transport of the hemolysin polypeptide, or that HLY B protein is an ATP-dependent inhibitor of premature folding of α-hemolysin (577a). Recent studies by Holland's group led to the identification of a secretion signal within the last 27 amino acids of the hemolysin polypeptide (677). This particularly interesting result should help to define the way in which hemolysin interacts with its cognate secretory functions.

The extracellular proteases of *Erwinia chrysanthemi* are also synthesized without cleavable signal peptides, although a short, N-terminal propeptide is autocatalytically removed, probably once the proteases have been secreted into the medium. Their secretion depends on the expression of *E. chrysanthemi* genes located close to the protease structural genes (P. Delepelaire and C. Wandersman, personal communication). Another group of extracellular proteins which are released by Gram-negative bacteria without direct involvement of the secretory pathway are *bacteriocins*, a group of toxins which act only on bacteria closely related to the producing strain (895). *Colicins* (bacteriocins produced by *E. coli*) do not have signal peptides and accumulate in the cytoplasm. Their release depends on the expression of a gene located downstream from the colicin structural gene in the colicin operon. The product of this gene, an envelope lipoprotein, activates a normally cryptic phospholipase whose action on phospholipids produces lysolipids and free fatty acids. These detergentlike products permeabilize the cell envelope, allowing periplasmic and some cytoplasmic proteins (including colicins) to leak out into the medium (193,900–902). Thus, colicins are not secreted *sensustricto* but rather are released by partial lysis.

4. Cell appendages

Although not actually released into the medium, protein constituents of cell surface appendages (fimbriae, pili, and flagellae) must reach the outside of the cell and are thus considered to be secreted proteins. The subunits of most types of fimbriae from Gram-negative bacteria are synthesized with typical signal peptides (see next section for exceptions), and their translocation across the outer membrane depends on the expression of a coregulated gene coding for a periplasmic protein which probably binds to the fimbrial subunits en route through the cell envelope [unpublished data, quoted in (637)]. Another gene in the same operon codes for an OMP which anchors assembled fimbriae to the cell surface (34). Some types of fimbrial subunits which are synthesized with apparently typical signal peptides and yet are processed at sites close to their N-termini instead of at the normal consensus cleavage sites (718,997) (Section III.D.3). Processing of these fimbrial subunit precursors is presumably carried out by a special signal peptidase (Section III.D.3). The subunits of another class of cell appendages, the flagellae of Gram-negative bacteria, are not made as precursors (1235) and appear to migrate to the tip of the growing flagellum through a central canal. Mutations affecting "capping" at the tip of the flagellum or flagellin polymerization cause flagellin release into the medium (466,511).

5. Activation and secondary processing of secreted proteins

Although most bacterial secretory proteins are not subject to secondary processing, some proteins secreted by Gram-negative bacteria must be processed to become active. Examples include a secreted protease of *Pseudomonas aeruginosa* which acquires elastase activity when the *laxA* gene is expressed [LAX A protein may also be involved in elasatase secretion (381,1000,1001)], secreted aerolysin (hemolysin) of *Aeromonas hydrolytica* (486), and secreted *Erwinia chrysanthemi* proteases (P. Delepelaire and C. Wandersman, personal communication). *E. coli* α-hemolysin is activated prior to its secretion by the product of the *hlyC* gene, which is located in the same operon as the hemolysin structural gene and its secretion functions (781).

Further reading

Mézés, P. S. F., and Lampen, J. O. (1985). Secretion of proteins by bacilli. *In* "The Molecular Biology of the Bacilli" (D. A. Dubnau, ed.), vol. 2, pp. 151–183. Academic Press, Orlando, Florida.

Pugsley, A. P. (1988). Protein secretion across the outer membrane of Gram-negative bacteria. *In* "Protein Transfer and Organelle Biogenesis" (R. A. Das and P. A. Robbins, eds.), pp. 607–651. Academic Press, San Diego, California.

Pugsley, A. P., and Schwartz, M. (1985). Export and secretion of protein by bacteria. *FEMS Microbiol. Lett.* **32,** 3–38.

Tommassen, J. (1988). Biogenesis and membrane topology of outer membrane proteins in *Escherichia coli. In* "Membrane Biogenesis" (J. A. F. Op den Kamp, ed.), pp. 351–373. Springer-Verlag, Berlin.

CHAPTER V

Later stages in the eukaryotic secretory pathway

A. General concepts

The secretory pathway of eukaryotic cells comprises a succession of compartments, the *secretory organelles,* through which proteins pass en route to their final destinations. Although each secretory organelle has its own special characteristics, a number of basic features are common to all of them. Some proteins pass more or less unimpeded through the entire length of the secretory pathway, whereas others are retained in secretory organelles. This raises two of the fundamental questions which will be addressed in this chapter, namely, what determines whether a protein will be taken out of circulation at a particular step in the pathway, and how tight is the separation between secretory organelles?

Ultrastructure and cell fractionation studies indicate that there is little or no physical link between the RER and the Golgi. Furthermore, specific proteins are known to reside in individual compartments of the Golgi apparatus. As we shall see, the physical separation of Golgi enzymes is also indicated by the succession of posttranslational modifications to which secretory proteins are subjected, by *in situ* immunocytochemistry, and by the separation of Golgi-derived vesicles containing different enzymes. How then do secretory proteins move between these compartments, and is the secretory pathway a continuous gradient of secretory organelles or are they functionally and structurally independent?

A feature common to all stages of the secretory route beyond the RER is that proteins move between secretory organelles in specific classes of transport vesicles. Consequently, soluble secretory proteins never come into direct contact with the outer face of the organelle to which they are being targeted and therefore can play no direct part in sorting. Integral membrane proteins usually have segments exposed on the cytoplasmic face of the transport vesicle which could be recognized by receptors on the target organelle. Microinjected antibodies recognizing the C-terminal, cytoplasmic tails of plasma membrane proteins can prevent their transport to the cell surface (25,590), but this is probably due to antibody-induced changes in protein conformation which make the protein incompetent for transport along the secretory pathway, rather than to inhibition of receptor–secretory protein interactions.

In this chapter, we will consider three different ways in which sorting of secretory proteins might occur:

(i) All secretory proteins have signals which target them successively through the secretory organelles and then on to their final target; some proteins remain in different organelles because they lack the signal necessary for targeting to the next organelle in the pathway.

(ii) Secretory proteins only have sorting signals for the last stage in the secretory pathway; some proteins have retention signals which prevent them joining the bulk flow through the secretory pathway, thereby causing them to be retained in specific organelles.

(iii) As (ii), except that secretory organelle proteins pass through the secretory pathway in the bulk flow and are recycled via signal-dependent counterflow.

If either of the last two models are correct, then molecules devoid of retention or sorting signals should travel through the secretory pathway as part of the bulk flow. The rate at which molecules are transported out of the cell will thus be determined by the rate at which they diffuse to sites within the RER and Golgi where transport vesicles are formed and depart en route to the next compartment in the chain. This flow rate has recently been measured with the N-glycosylation acceptor tripeptide Asn-Tyr-Thr (Section V.B.1). If this tripeptide contains a radioiodinated Tyr residue, its progress through the secretory pathway, which it seems to enter by diffusing across the RERM, can be followed by chromatography and autoradiography. Wieland *et al.* (1190) found that the tripeptide was glycosylated in the ER (this prevented it from leaking back into the cytosol), then trimmed by mannosidases located in the Golgi (Section V.D), and finally secreted into the medium. Tripeptide was detected in the medium after 5–10 min, depending on the cell type, which is considerably faster than the time required for the transport of most secretory proteins through the secretory pathway. Thus, bulk flow is very efficient, implying that diffusion through the lumen of the ER and Golgi can occur relatively easily and that there is massive, vectorial movement of vesicles between the organelles and the cell surface. The way in which this result influences our understanding of protein sorting in the secretory pathway is discussed in following sections.

B. Protein modification, proofreading, and retention in the endoplasmic reticulum

Soluble and integral membrane secretory proteins fold once they have crossed the RERM; they are also often covalently modified. The following sections deal with the types of modifications which occur in the ER and their role, if any, in protein sorting. Palmitoylation and phosphatidyl inositol modification of secretory proteins are discussed in Section III.F.4.c; only their role in protein transit through the secretory route will be discussed here.

1. Glycosylation reactions in the RER

Most secretory proteins are nitrogen (N)-glycosylated in the RER. Because we are primarily concerned with the effects of glycosylation on protein targeting, only general details of the glycosylation reactions will be given here.

a. *N-Glycosylation*

The sequence of reactions shown in Fig. V.1, which is common to both yeasts and higher eukaryotes, results in the addition of a $(\text{Glucose})_3$–$(\text{Mannose})_9$–$(N\text{-acetylglucosamine})_2$ complex onto Asn residues in the sequence Asn-X-Ser or Thr (the acceptor peptide, in which X can be any

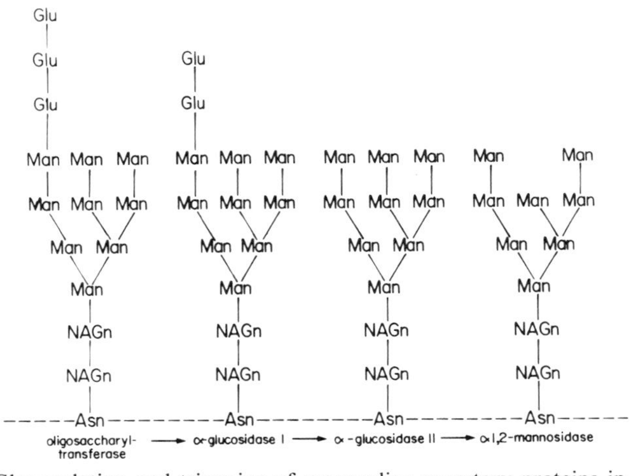

Fig. V.1. Glycosylation and trimming of mammalian secretory proteins in the ER. The complete [$(\text{glucose})_3$–$(\text{mannose})_9$–$(N\text{-acetylglucosamine})_2$]–[$(\text{Glu})_2$–$(\text{Man})_9$–$(\text{NAGn})_2$] complex is transferred to asparagine (Asn) residues in the lumen of the ER from donor dolichol–oligosaccharide complexes. Figure redrawn from Ref. 577.

amino acid except possibly proline or aspartate). Isolated tripeptides can act as acceptors only when the two extremities are blocked. Longer acceptor peptides are more rapidly glycosylated (577). As shown in Fig. V.1, the entire complex is assembled before it is attached to the Asn residue in the lumen of the RER.

At least some sugars cross the RERM as intermediates complexed with the long-chain lipid dolichol, whereas others may cross the RERM as nucleotide 5′-diphosphate–sugar complexes. However, it is by no means certain that the complex is assembled entirely within the lumen of the RER. For example, there is no evidence that GDP–mannose can be transported across the RERM, leading to the proposal that the five mannose residues are added via GDP–mannose donors on the cytoplasmic face of the RERM, whereas later modifications, and possibly earlier modifications, occur in the lumen. This presupposes that the dolichol-PP–$(NAGn)_2$ complex can "flip" from the lumen to the cytoplasmic face and then back again once the five mannose residues have been added [see (452) for further discussion of the topology of glycosylation reactions in the RER]. The preferred donor lipid–carbohydrate complex is the complete complex shown in Fig. V.1 (1129). However, truncated versions lacking glucose and even mannose residues can act as donors *in vitro,* and under-glycosylated complexes can act as donors *in vivo* in yeasts (598) and protozoa (577). Variations in the mannose content of transferred oligosaccharides have also been reported (577).

The transfer of the first *N*-acetylglucosamine (NAGn) residue from the UDP complex to the dolichol, is inhibited by the antibiotic tunicamycin, which therefore blocks all N-glycosylation. This provides a valuable technique for studying the role of N-glycosylation in protein traffic. The yeast (*S. cerevisiae*) gene (*ALG7*) coding for the RERM-associated, tunicamycin-sensitive enzyme (UDP-NAGn : dolichol-P-transferase) was cloned by virtue of its ability to rescue tunicamycin-treated cells when present on multiple copy number plasmids (931). Null mutations in *ALG7* are lethal. Mutants affected in other stages in the yeast glycosylation pathway have been isolated by mannose suicide selection (Section II.F.1.b). Only mutations blocking the earliest stages in the pathway are lethal; incomplete dolichol-linked oligosaccharides containing a minimum of four mannose residues allow normal growth, presumably because they can be attached to secretory proteins if the full-length lipid–oligosaccharide complex is absent (see above) (598).

The complete oligosaccharide chain is transferred onto the Asn acceptor site in the lumen of the RER. N-glycosylation is generally thought to occur as the polypeptide is being threaded through the RERM, while it is still in its unfolded state. However, some acceptor sites are not glycosylated. This may be because they rapidly become inaccessible within the structure of the folded polypeptide, although other explanations are possible (577,968). An unexplained anomaly is that some ER membrane glycoproteins have glycosylated residues exposed on the cytoplasmic face of the ER (1). One explanation could be that additional NAGn transferases are present on the cytoplasmic face of the smooth ER, where these glycoproteins are predominantly located.

b. Glucose and mannose trimming

The initially homogeneous oligosaccharide is processed immediately following its attachment to the polypeptide chain, initially by the removal of glucose residues by one or more glucosidases. Glucose trimming of glycosylated vesicular stomatitis virus (VSV) G protein has been reported to occur cotranslationally (27). Further processing steps are different in yeasts and in animal cells. In *S. cerevisiae* cells, a single α1-2-linked mannose residue is replaced by α1-3 mannose. A mutation in the *GLS1* gene coding for α1-2 glucosidase does not affect removal of the mannose residue or subsequent chain elongation, whereas removal of the ninth α1-2-linked mannose residue may be essential for outer chain elongation, which occurs in the Golgi (Section V.D.1) (598). Further processing of animal cell glycoproteins also occurs in the Golgi, although further mannose trimming of resident ER proteins and reglycosylation may both occur in the ER (577,847).

c. O-Glycosylation

Ser and Thr residues on secretory proteins of yeasts and possibly other fungi can be O-glycosylated in the ER (429). The process is less well characterized than N-glycosylation but seems to involve the direct transfer of mannose residues from dolichol-1-mannose onto acceptor amino acids. Studies with acceptor oligopeptides suggest that no particular sequence is required around the modification site (54,621). Further processing of the mannose residue occurs in the Golgi apparatus (429) (Section V.D.2). Yeast secretory proteins may be both N- and O-glycosylated. Most studies indicate that O-glycosylation occurs exclusively in the Golgi apparatus (Section V.D.2).

2. Protein folding and disulfide bridge formation

Secretory proteins apparently fold spontaneously as they are extruded across the RER membrane (194,1222). However, disulfide bridge formation is probably catalyzed by protein disulfide isomerase (PDI), which is loosely associated with the RERM (325,326). The formation of soluble protein complexes occurs shortly after synthesis (604), but trimerization of VSV G protein and influenza virus hemagglutinin (HA), both of which are integral membrane proteins, occurs only 7–10 min after synthesis (193,194,364,589) and involves the selection of polypeptides from a random pool of prefolded monomers rather than from a restricted pool of monomers synthesized on a single polysome (108). Kreis and Lodish (589) suggest that this delay may be due to the segregation of PDI in a late "compartment" of the ER. Studies by Doms *et al.* (250) demonstrated the existence of a transient monomeric form of a temperature-sensitive, mutant G protein. The monomer aggregated rapidly and accumulated in the ER at the nonpermissive temperature. ATP-dependent trimerization and exit from the ER occurred when the cells were shifted to a permissive temperature. Although these studies were performed with a mutant G protein, and aggregation is probably not an intermediate step in the folding of normal G protein, Doms *et al.* consider their results to reflect the normal requirement for ATP in the trimerization of G protein. HA trimerization is ATP-independent.

According to the papers discussed above, oligomerization of G and HA occurs just before they leave the ER, 5–10 min after synthesis. This view is challenged, however, by Yewdell *et al.* (1222), who found that monoclonal antibodies specific for oligomerized HA reacted only with proteins in the Golgi apparatus, and not with HA in the ER (see next section). A possible explanation for this ambiguity is that HA monomers fold and trimerize in the ER and are then transported to the Golgi apparatus, where further modifications alter the conformation of the HA trimers to produce the antigenic sites recognized by the antibody used by Yewdell *et al.*

3. Competence for transport to the cis Golgi

Different proteins transit through the secretory pathway together, but the rates at which they are secreted may vary considerably (335,343,615,1221). The site at which the secretion lag is most prominent seems to be transit from the RER to the Golgi (646). It has been suggested that this delay might reflect the need for secretory proteins to interact with specific receptors in the RER for transport to the Golgi and that carbohydrates might form part of the recognition signal (320,647,648). Alternatively, proteins may be retained in the ER until they have folded correctly; different folding kinetics may result in different retention times in the ER (589,1192). Indeed, exit from the ER of one of the proteins studied by Fitting and Kabat (320) coincided with a partial proteolysis event, and polypeptides were "selected" for exit from a random pool of "new" and "old" proteins. Furthermore, studies discussed in the preceding section showed that oligomerization of virus-encoded membrane proteins occurs just before they are transported to the Golgi.

If secretory proteins are indeed retained in the ER until they are folded and oligomerized in the correct way, then exit-incompetent proteins should either remain indefinitely in the ER or be secreted at very much reduced rates. This could explain why genetically altered or hybrid proteins are sometimes translocated into the RER without difficulty and yet do not transit further through the secretory pathway (104,281,364), and why incomplete glycosylation or glucose trimming and failure to process signal peptides sometimes affect secretion kinetics (96,577,649,1045). Even minor sequence changes have been reported to affect protein conformation and exit from the ER (1192).

What happens to incorrectly folded proteins in the ER? Results from numerous studies indicate that they stay in the ER (604,681) or are degraded either in the ER (642a) or in the lysosome (723). These proteins do not seem to precipitate in large aggregates. Instead, a specific, major ER protein seems to bind to some incorrectly folded proteins, thereby preventing their exit from the ER. Haas and Wabl (413) first detected this protein (BiP) complexed with immunoglobulin heavy chains synthesized in the absence of light chains (heavy and light chains are only transported to the Golgi as complexes) (97), and it was subsequently found to be complexed with incorrectly oligomerized or monomeric HA (194,364), nonglycosylated invertase (*in vitro* in dog pancreatic microsomes) and incorrectly folded prolactin (557), and recombinant human factor VII in Chinese hamster ovary cells (561). It was not detected in association with aggregated VSV G protein (250) or with incorrectly folded (mutant) class I histocompatibility antigen (1192). The dissociation of secretory protein–BiP complexes requires ATP (758), which might explain part of the ATP requirement for protein movement along the secretory pathway, but because ATP is usually present in cell lysates, some BiP complexes (such as G–BiP) may dissociate during extraction. BiP has also been shown to associate with nascent polypeptides as they enter the lumen of the ER (557).

Two interesting features of BiP are that its synthesis is stimulated by glucose deprivation and that it is structurally related to a heat shock protein (758,857). Glucose starvation is likely to reduce glycosylation, thereby increasing the proportion of incorrectly folded secretory proteins in the ER. Studies have confirmed the idea that the extent of glycosylation can affect the association between secretory proteins and BiP, as well as their rate of secretion (251). The accumulation of misfolded (mutant) proteins in the ER also increases BiP synthesis (583), possibly because the cell senses that its store of BiP has been sequestered into protein complexes. Thus, synthesis of BiP may be increased according to requirements (857), but studies on the "proofreading" or quality control role of BiP are still at an early stage. A different, cytoplasmic quality control protein may be responsible for proofreading cytoplasmic domains in transmembrane proteins to ensure that they too are correctly folded and oligomerized (406a).

Protein conformation, rather than any specific structural feature such as glycosylated residues, is thus likely to determine whether a protein is competent to leave the ER. This feature is illustrated by studies on the effects of glycosylation on VSV G protein. Machamer *et al.* (672) used site-directed mutagenesis to replace the glycosylated Asn residues. Nonglycosylated G protein was transported from the ER to the Golgi but did not reach the cell surface. When only one of the Asn residues was deleted, the protein reached the cell surface; thus G protein with one glycosylated Asn residue can be transported through the entire secretory pathway. G protein export became temperature-sensitive when new glycosylation sites were added (671). Kotwal *et al.* (581) noted that some natural G protein variants have only one oligosaccharide whereas others have none at all and suggested that compensatory changes in primary structure may allow nonglycosylated protein to fold into a secretion-competent conformation. Thus, glycosylation is probably needed for correct protein folding but is not directly implicated in the formation of secretion competence signals. At least one secreted protein, ovalbumin, is not glycosylated.

4. Specific retention of resident ER proteins

The ER contains a large number of proteins, some of which it shares with the contiguous nuclear membrane (117). They include proteins involved in secretory protein translocation through the RERM, signal peptidase, ribophorins, BiP, enzymes involved in lipid synthesis and in protein glycosylation, and cytochrome P_{450}. The characterized proteins are similar to other secretory proteins: They often have signal peptides (198,630,1089), are glycosylated, and may be located in the lumen or in the RERM.

Rothman (965) has argued that it may be physically impossible for the ER to prevent the escape of endogenous proteins, and particularly membrane proteins, to the cis Golgi. He cited two observations in support of the idea that these proteins could leave the RER as part of the bulk flow through the secretory pathway, and then recycle from the Golgi to the ER:

(i) Some ER proteins are detected in significant amounts in vesicles derived from the cis Golgi when cells are fractionated (102).
(ii) Bulk lipid flow out of the ER necessitates efficient recycling, possibly from the cis Golgi, which could provide "carriers" for the recycling of ER proteins.

Most ER proteins are, however, almost completely excluded from the Golgi (663,697). Furthermore, ER proteins are not terminally glycosylated (117,1089), which means that they do not reach the medial or trans Golgi (Section V.E). Mannose residues on some ER proteins are not trimmed beyond the stage catalyzed by ER mannosidase (117,954,1089), but a lysosomal protein carrying an ER retention signal (see below) is phosphorylated by NAGn phosphotransferase, indicating that it reaches the early cis Golgi (858). Significantly, the phosphate groups are not modified by NAGn phosphodiesterase, which may be located in a different, ER-distal Golgi compartment (Table V.1). Warren (1166) has proposed that ER proteins are salvaged from an intermediate compartment, the so-called *transitional element* (Section V.C), located between the ER and the cis Golgi.

Recycling of "escaped" endogenous ER proteins, rather than receptor-mediated retention in the ER, is the currently favored model for the specific accumulation of proteins in the ER (151). It is not clear whether soluble and ER membrane proteins are both subject to the same retention mechanism, but studies on the rates of diffusion of membrane proteins in the ER indicate that RERM proteins are more restricted than BiP and that the mobility of BiP devoid of its ER retention–salvage signal (see below) cannot be distinguished from that of normal BiP (151).

Several studies have sought to determine the nature of the ER retention–salvage signal(s). The rotaviral type II ER membrane protein VP7 was found to be secreted when the two potential N-terminal transmembrane segments were deleted (884), suggesting that retention–salvage depended on a membrane anchor domain. However, subsequent studies showed that the entire N-terminus was absent from mature VP7 (1084) and thus that some other feature must account for VP7 retention in the ER (883c). Although deletion of the C-terminal transmembrane and cytoplasmic tails of the type I ER membrane protein E19 of adenovirus also causes the protein to be secreted, deletion of the last eight residues (FIDEKKMP) of the C-terminal, cytoplasmic tail alone causes the protein to appear on the cell surface, suggesting that the C-terminus contains the ER retention–salvage signal. Furthermore, cell-surface interleukin 2 receptor protein β chain was converted into a resident ER protein following fusion of FIDEKKMP to its cytoplasmic C-terminus (832). The cytoplas-

Table V.1. Location of proteins in the Golgi stack

Enzyme or protein	Location	Evidence	References
NAGn phosphotransferase	Early cis	Fractionation	235,380
		Kinetics	379,614a
NAGn phosphodiesterase	Late cis	Fractionation	235,380
		Kinetics	379,614a
Mannose-6-phosphate receptor (275 kDa)	Cis	Immunocytochemistry	131
58-kDa Antigen	Cis	Immunocytochemistry	996
Palmitoyl CoA transferase	Cis or medial	Monesin	905
		Kinetics	249,540, 1016
		Cell fusion assay	970
Mannosidase I	Medial	Fractionation and endo H-resistance	265
NAGn transferase I	Medial	Immunocytochemistry and endo H-resistance	267
		Fractionation	265
		Cell fusion assay	970
	Medial (or trans)	*In vitro* assay	39,40
Mannosidase II	Disperse	Immunocytochemistry	796
	Medial	Fractionation	267
Fucosyltransferase	Medial	Fractionation	265
45-kDa Antigen	Medial	Immunocytochemistry	169
Galactosyltransferase	Trans	Immunocytochemistry	955
		Lectin (ricin)	399
		Cell fusion assay	970
		In vitro assay	966
Sialotransferase	Trans	Immunocytochemistry	957
		Lectin (wheat germ)	1110
	Late trans	*In vitro* assay	540
		Immunocytochemistry	1098
O-glycotransferases	Trans	Kinetics	540
Uridine diphosphatase	Trans	Enzyme cytochemistry	156
Tyrosine sulfotransferase	Late trans	Kinetics	33
		In vitro assay	33
Acid phosphatase	Late trans	Enzyme cytochemistry	299

mically exposed signal could be recognized by "salvage receptors" in the transitional element.

These studies on the E19 protein are reminiscent of earlier, seminal work on the soluble ER protein BiP which resulted from the observation that three soluble proteins in the lumen of the ER, including BiP, had the same four residues, Lys-Asp-Glu-Leu (KDEL), at their C-termini. Munro and Pelham (759) found that if this sequence was deleted or extended, BiP was secreted into the medium. This suggests that KDEL is the ER retention signal and that it must be at the extreme C-terminus, which is presumably the last segment of the polypeptide to fold and may therefore be exposed on the surface of the protein. In a complementary study, DNA coding for KDEL was fused to the end of a cDNA clone coding for secreted lysozyme. The resulting hybrid protein was retained in a perinuclear region probably corresponding to the ER (759).

Further studies are required to determine whether KDEL is the "universal" ER retention signal. Preliminary studies suggest that some mammalian ER proteins have the sequence RDEL at the C-terminus, whereas yeast ER proteins have the sequence HDEL (858a). HDEL or KDEL is presumably recognized either by an endogenous ER membrane protein, which could anchor soluble proteins in the ER, or, more likely, by the recycling receptor located in the transition element or cis Golgi. Indeed, a putative HDEL receptor gene (*ERD1*) has recently been identified in yeast cells (858,858a). It is not clear what triggers the release of BiP from its receptor once it is recycled to the ER.

β-Glucanase is a typical lysosomal protein; small but significant amounts of it, however, are retained in the ER through a specific interaction with an endogeneous ER protein, the esterase egasyn. Studies by Medda *et al.* (704) indicate that β-glucanase–egasyn interaction is blocked by inhibitors of esterase activity, leading to β-glucanase secretion (rather than sorting to the lysosome). Whether this is a physiologically significant mechanism for retaining proteins in the ER remains to be determined. Only about 10% of egasyn is complexed with β-glucanase (666), so perhaps it also binds to other resident ER proteins. It would be interesting to see whether egasyn itself has a C-terminal KDEL-like sequence.

C. Transport from the endoplasmic reticulum to the Golgi

As discussed above, secretory proteins do not move to the cis Golgi until they attain an exit-competent state, and some proteins are specifically retained in the ER, probably by recycling of escaped proteins. Both normal secretory proteins (1223) and exit-incompetent proteins have been reported to accumulate in a specific region of the smooth ER, the vacuolar transitional element (79,996), from which vesicles either migrate to and fuse with the cis Golgi (41) or coalesce to form new Golgi cisternae (797). Specific membrane proteins [e.g., the product of the *SEC12* gene in yeasts (767a)] may be required to form this specialized domain of the ER and hence directly or indirectly assist vesicle formation and protein transport to the Golgi.

Secretory pathway "shuttle" vesicles are difficult to isolate because they have very short half-lives (567). However, two groups have recently isolated what appear to be ER–Golgi shuttle vesicles. Lodish *et al.* (649) used shallow sucrose gradients with deuterated water in which these shuttle vesicles or fragmented transitional elements were separated from the slightly heavier ER or Golgi-derived vesicles. The secretory protein studied by Lodish *et al.*, α-1 antitrypsin, appeared in the shuttle vesicles about 5 min after synthesis. Deoxynojirimycin, an inhibitor of ER glucosidase, prevented the appearance of antitrypsin in the vesicles. Nowack *et al.* (797) used a similar preparation of shuttle vesicles in an *in vitro* assay for bulk protein transfer from the ER to the cis Golgi. Protein transfer was unidirectional and required a cytosolic macromolecule together with nucleotide triphosphates, as observed *in vivo* (41). Earlier work by Jamieson and Palade (533) also showed that ATP depletion caused secretory proteins to accumulate in transitional elements, but this study did not distinguish between energy requirements for protein folding and for vesicular transport from the transitional element to the Golgi.

A different *in vitro* assay was developed by Haselbeck and Schekman (428) to study protein movement from the ER to the Golgi in yeast cell extracts. Their assay uses donor ER vesicles derived from a strain carrying the *sec18*ts mutation, which blocks protein movement from the ER, and the *mnn1* mutation, which prevents terminal (Golgi) mannosylation of secretory proteins. Invertase accumulated in the ER grown at the restrictive temperature was mannosylated after transfer to recipient Golgi from wild-type cells. The transfer efficiency was low, however, possibly because invertase accumulated at the restrictive temperature did not become exit-competent at the permissive temperature *in vitro,* or because recipient Golgi were saturated. As in the mammalian system, protein transfer was ATP-dependent and required soluble cofactors including SEC18 protein (269b) as well as proteins on the surface of the recipient Golgi. A similar but more efficient reconstitution system using gently lysed yeast-cells has recently been developed (35a). Results obtained with this system indicate the probable requirement for GTP in ER to Golgi traffic.

D. Protein modification in the Golgi

Secretory proteins are subjected to further chemical modification as they transit through the Golgi. These processes are discussed in the following sections (except palmitoylation, which was covered in section III.F.4.c); their significance with regards to compartmentalization of the Golgi apparatus and their role in protein targeting will be discussed in later sections of this chapter.

1. Further processing of N-linked saccharides

In Section V.B.1, we saw that the basic oligosaccharide core on Asn residues is trimmed by glucose- and mannosidases before secretory proteins leave the ER. Following their arrival in the Golgi, mannose residues on secretory proteins may be processed in one of two ways, depending on whether they are targeted to the lysosome.

a. Phosphorylation of lysosomal enzymes

Soluble lysosomal proteins carry phosphorylated mannose residues which function as lysosomal sorting signals (Section V.G.5). They undergo specific mannose phosphorylation catalyzed by one or two *N*-acetylglucosamylphosphotransferases and *N*-acetylglucosamine-1-phosphodiester-α-*N*-acetylglucosaminidase (phosphodiesterase) in different cis Golgi compartments (614a). Sequential action of these enzymes results in the transfer of *N*-acetyl glucosamine-1-phosphate from UDP-NAGn to any one of five mannose residues in the oligosaccharide core, followed shortly afterwards by the removal of the *N*-acetylglucosamine to expose the phosphomannosyl group (577). Goldberg and Kornfeld (379) found three partially phosphorylated peptides in β-glucuronidase, indicating that lysosomal enzymes may not be uniformly phosphorylated.

Lysosomal enzymes presumably contain signals which are recognized by the phosphorylating enzymes (919). Deglycosylated (endoglycosidase H-treated) lysosomal cathepsin D is not a substrate for the phosphorylation reaction *in vitro,* but it inhibits phosphorylation of intact lysosomal enzymes. Proteolytic fragments of glycosylated cathepsin D are also not phosphorylated, and do not inhibit phosphorylation when they are dephosphorylated, indicating that the "phosphorylation signal" is probably not a linear sequence of amino acids (612). Frog oocytes are also able to recognize the phosphorylation signal of human cathepsin D (301). Renin is closely related to cathepsin D and yet is secreted by mammalian cells, presumably because it lacks the phosphorylation signal. However, renin produced in oocytes is phosphorylated, remains intracellular, and is degraded (presumably in the lysosome) (302). This suggests that renin has a phosphorylation signal which is recognized by amphibian but not by mammalian phosphorylating enzymes.

b. Other processing of mannose residues

Outer chains on lysosomal and nonlysosomal secretory proteins of complex eukaryotes may be further trimmed by Golgi mannosidase I to leave five mannose residues on the core oligosaccharide. Further modifications involve the addition of an *N*-acetylglucosamine residue by *N*-acetylglucosaminyltransferase I, further removal of two mannose residues by

Golgi mannosidase II, fucosylation of the innermost *N*-acetylglucosamine by fucosyltransferase, and the addition of galactose, *N*-acetylglucosamine, and sialic acid residues by appropriate transferases (Fig. V.2). Oligosaccharides of both lysosomal and nonlysosomal proteins may be further modified by sulfation of mannose and *N*-acetylglucosamine residues and O-acetylation of sialic acid residues (577).

Outer chain modification in yeasts is markedly different. Golgi mannosyltransferases extend the basic $(man)_8$ core oligosaccharide to produce large mannan structures typical of many yeast mannoproteins, which may carry as many as 150 mannose residues (598).

2. O-Linked oligosaccharides

The single O-linked mannose residue on yeast glycoproteins (Section V.B.1.c) may be further modified by the addition of up to four more mannoses transferred from GDP-mannose (598,1107). The reaction is

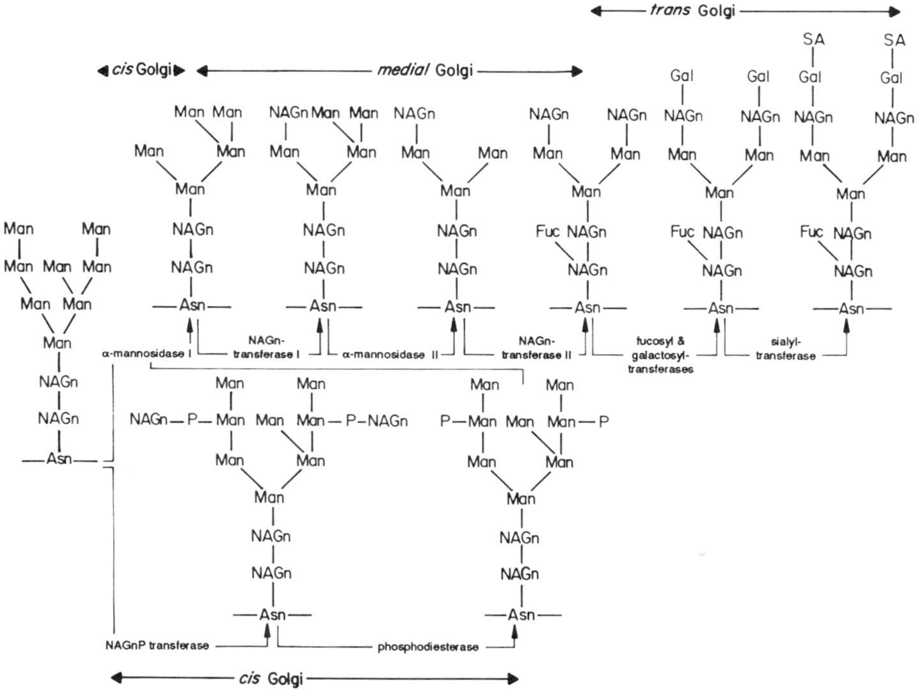

Fig. V.2. Sequence of phosphorylation, trimming, and glycosylation of core-glycosylated secretory proteins in the Golgi apparatus of mammalian cells. Asn, asparagine; Man, mannose; Fuc, fucose; NAGn, *N*-acetylglucosamine; Gal, galactose; P, phosphate; SA, sialic acid. Redrawn from Ref. 577.

blocked in *sec* mutants, which prevent secretory proteins from reaching the Golgi (429).

In complex eukaryotes, hydroxyl groups of Ser or Thr residues are O-glycosylated by enzymes thought to be located exclusively in the Golgi (418,786). Acceptor octapeptides can be O-glycosylated. *N*-acetylgalactosamine is the primary sugar in O-linked oligosaccharides; further residues of galactose, sialic acid, fucose, *N*-acetylgalactosamine, and *N*-acetylglucosamine can then be added. Lipid intermediates are not involved in O-glycosylation, and the reaction is not inhibited by tunicamycin (418), but it is not known whether other sugar transferases are involved in both N- and O-glycosylation.

3. Tyrosine sulfation

Glycosylated and unglycosylated secretory proteins are major substrates for tyrosine sulfation (501). Tyrosynylprotein sulfotransferase is enriched in Golgi-derived membrane fractions and has its active site oriented toward the Golgi lumen (619). The sulfate donor is 3′-phosphoadenosine 5′-phosphosulfate. Almost all tyrosinated proteins have an acidic residue, a glycine or proline residue, and no cysteines, basic residues, extended secondary structure, or N-glycosylation sites close to the modified tyrosine, but there does not appear to be a strict consensus sequence around the modification site (474,619). Inhibition of tyrosine sulfation is reported to retard the exit of a secretary protein from the TGN (331b).

4. Proteolytic processing

Many secretory proteins undergo secondary proteolytic processing following removal of the signal peptide. The best characterized examples of this class of processed secretory proteins are α factor and killer toxin of yeasts. The α-factor peptide is present four times in the pro-α-factor protein which reaches the Golgi apparatus. This probably represents a way of reducing "shipping costs" since α factor itself may be too small to be efficiently transported to the RERM and it would be inefficient to pad the polypeptide out with redundant sequences (602). Julius *et al.* (545) found that pro-α-factor processing was blocked by *sec* mutations, which prevented protein movement through the Golgi, and that mature α factor was normally present in secretory vesicles en route to the cell surface. Thus, processing occurs in the Golgi. Some viral coat proteins such as influenza virus HA are also proteolytically processed in the trans-Golgi network (TGN) or trans Golgi. Processing of mammalian prohormones, which is similar to that of pro-α factor, occurs in secretory granules budding from the trans Golgi [see Section V.G.4.c and (1003)]).

The four 13-residue-long α-factor peptides in pro-α factor were found to be separated by 6–8 residues (Lys-Arg-Glu-Ala-Asp-Ala-Glu-Asp) (602). This suggests that at least one processing protease has trypsin- or chymotrypsin-like activity. Killer toxins are processed at similar sites (245,1074). In fact, the enzyme which performs this initial processing step, the product of the *KEX2* gene, is a Ca^{2+}-thiol protease which cleaves between basic residues. The enzyme is inhibited by anti-α1-trypsin and can correctly process mammalian proalbumin (53). Strains mutated at *KEX2* secrete unprocessed pro-α factor (546). The product of the *STE13* gene, a membrane-associated aminopeptidase (544), processes the N-terminal tetrapeptide, and further processing of the C-terminus is performed by *KEX1*-encoded carboxypeptidase (245). Other secretory proteins produced by different species of yeasts may also be processed by one or more of these enzymes (695b).

E. Organization of the Golgi cisternae

Results from a number of experimental approaches (summarized below) show quite conclusively that individual Golgi cisternae are separate, biochemically and functionally distinct entities organized according to a very strict pattern and that secretory proteins progress in a synchronized wave from one end of the stack of cisternae to the other en route to their final destinations. According to these data, Golgi stacks must contain at least three distinct cisternae (generally referred to as cis, medial, and trans according to their orientation with respect to the ER). Heterogeneity within these "domains" indicates that the actual number of cisternae is probably higher than three (see Table V.1, Fig. V.3) (299). Cells normally have one or a very limited number of Golgi stacks. The stacks break up into clusters of many vesicles during mitosis, apparently to ensure equal partitioning of Golgi components to daughter cells, although the number of Golgi clusters produced is far in excess of that required for this purpose. Golgi breakdown presumably involves membrane fission, so each cluster of vesicles contains cis, medial, and trans components (662). Protein secretion is usually shut down during mitosis or meiosis, but the yeast *S. cerevisiae* continues to process and secrete invertase during mitosis, implying that the Golgi fragments remain active (685).

In general, results from different analyses give a coherent picture of the cis to trans organization of Golgi cisternae (see Table V.1), on the basis of which a simple map can be drawn (Fig. V.3). The segregation of modifying enzymes presumably allows prosthetic groups to be added or removed according to a strict sequence and prevents competition between processing enzymes which could act on the same substrate (266).

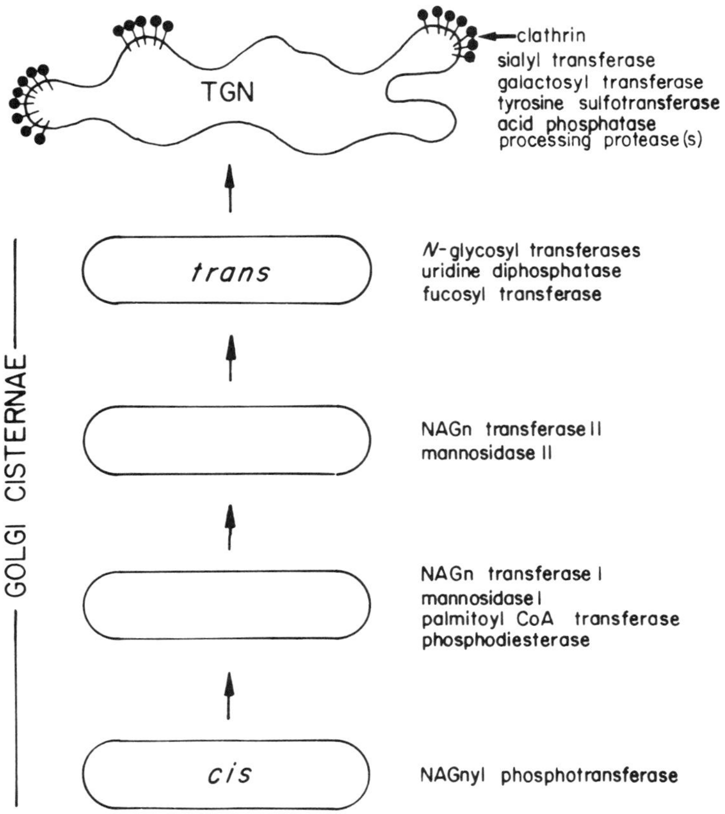

Fig. V.3. Possible topology of mammalian Golgi compartments showing the location of enzymes and the site of clathrin accumulation. The minimum number of compartments compatible with the data listed in Table V.1 is five, but the actual number of Golgi cisternae may be higher than that shown. NAGn, *N*-acetylglucosamine; TGN, trans-Golgi network.

1. Cytochemical studies

Early studies showed that one or two cis (ER-proximal) Golgi cisternae were preferentially stained during prolonged exposure to osmium [presumably due to strong reducing conditions in these cisternae (333)]. Subsequent histological and cytochemical tests showed that many enzymes involved in protein glycosylation and other reactions, as well as some proteins with unknown function, were not evenly distributed through the Golgi stock but were compartmentalized in one or two cisternae (Table V.1). The techniques employed include immunocytological detection of proteins using specific antibodies, enzymatic cytochemical reagents for specific enzymes, and the detection of lectin-specific sugar residues on terminally modified glycoproteins in transit through the Golgi (Table V.1). As noted by Farquhar (299), some studies with different cell lines give conflicting results, suggesting that the organization of the Golgi stack may vary depending on cell type and function.

2. Kinetics of processing

Intermediates in the secretion pathway can be detected by pulse–chase experiments in which oligosaccharides or other prosthetic groups are labeled by the metabolic incorporation of radioactive precursors. The sequence in which processing occurs can thus be determined and correlated with the location of processing enzymes in the Golgi cisternae.

3. Density fractionation of Golgi membranes

Golgi enzymes involved in N- and O-glycosylation, sulfation, and palmitoylation of secretory proteins can be separated by density gradient centrifugation of Golgi-derived vesicles, which apparently differ in density due to differences in cardiolipin content (817). Enzymes identified in cis Golgi by kinetic and histological or cytochemical tests are generally found in heavier Golgi membrane fractions, and density is now widely used to define the approximate location of Golgi proteins in the stack (Table V.1).

4. Inhibition of Golgi functions by monesin

The ionophore monesin slows or arrests intra-Golgi transport and inhibits late (trans) Golgi functions, and can thus be used to distinguish between early and late Golgi processing events. Monesin also causes secretory proteins to accumulate in medial or late Golgi compartments, which become distended and vacuolated, aiding their identification by cytochemical methods (400,905,1108). However, as discussed by Dunphy and Rothman (266), studies on the effects of monesin in different cell types often give conflicting results and do not necessarily indicate the precise site of secretory protein accumulation or processing.

5. Endoglycosidase H-resistance

Oligosaccharides on secretory proteins are only cleaved by endoglycosidase H before they are processed by NAGn transferase I and mannosidase II. These enzymes are probably located in medial Golgi cisternae. Thus, endoglycosidase H-sensitivity provides a simple test for determining whether secretory proteins have reached these cisternae (267).

6. Cell fusion and *in vitro* assays for processing

Rothman *et al.* (969,970) demonstrated that pulse-labeled VSV G protein present in the Golgi of a cell lacking a particular modification enzyme can be modified upon fusion with a cell producing the modifying enzyme. The Golgi cisternae from donor and recipient cells did not fuse, and the G protein, rather than the modifying enzyme, was transferred from one cisterna to the other. These observations have implications for the mechanisms of protein movement between cisternae (Section V.F) but, like *in vivo* kinetic experiments, may also indicate the sequence in which prosthetic groups are added or removed (Table V.1). *In vitro* transfer of secretory proteins between Golgi cisternae (Sections II.F.3 and V.F) also reflects the sequence of modification events (39,40,268,334,966) (Table V.1)

F. Intra-Golgi movement of secretory proteins

Having established that secretory proteins are progressively processed as they migrate through the Golgi stacks, we now turn our attention to how the proteins themselves migrate between cisternae, and how Golgi proteins can be specifically retained within specific cisternae.

1. Intra-Golgi shuttle vesicles

Of all of the models explaining intra-Golgi movement of secretory proteins (300), only that invoking shuttle vesicles fits the experimental data showing that Golgi cisternae are distinct entities. Numerous protein-coated vesicles are produced by Golgi stacks *in vitro* under conditions which favor the intra-Golgi movement of secretory proteins (see below) (40,820). These uniformly sized vesicles were shown to contain at least one secretory protein, VSV G protein, indicating that they could be bona fide intra-Golgi shuttle vesicles (820). Similar vesicles were also found around the Golgi apparatus *in situ* (996,1088). Furthermore, different secretory proteins destined to different locations and exported or secreted at different rates were present in the same vesicles (1088), which agrees with the idea that secretory proteins are not segregated from each other during intermediate stages in the secretory pathway. The protein coat on these vesicles is distinct from the clathrin-containing coats present on endocytic and lysosomal vesicles and on budding secretory granules (see Section V.G.6). The only regions of the Golgi apparatus which have protein coats are those from which the Golgi shuttle vesicles bud.

2. Physiological conditions for intra-Golgi movement

A number of different conditions prevent intra-Golgi movement of secretory proteins *in vivo* and *in vitro*. The ionophore monesin (Section V.E.4) probably prevents movement though the trans cisternae by disrupting a proton gradient maintained by an ATP-dependent protein pump present in Golgi membranes (48), thus causing the pH of the normally acidic trans Golgi cisternae to rise. Saraste and Hedman (994) demonstrated that migration between different Golgi cisternae was blocked at different critical temperatures and suggested that this might be caused either by ATP depletion or by changes in membrane fluidity. ATP may be required to maintain the acidic pH of trans cisternae, and vesicle fission and fusion, which must occur at the cisternal membranes, can probably only occur when membrane lipids are in a "fluid" state above the phase transition temperature.

ATP and soluble cytosolic factors including a 74-kDa, *N*-ethylmaleimide-sensitive protein (93a) are required in both early and late stages of intra-Golgi movement *in vitro* (39,40,685a,966) and in permeabilized cells (62). Intriguingly, soluble yeast cell extracts can replace endogenous cytoplasmic components in a mammalian cell-derived assay system for intra-Golgi movement of proteins (268), raising the possibility of using extracts from yeast *sec* mutants blocked at different stages in the secretory pathway to define the role of the corresponding wild-type gene products in intra-Golgi transport. Indirect evidence based on the effects of GTP analogs suggests that GTP is also needed for intra-Golgi movement of secretory proteins (710), and studies with antibodies against a yeast GTP binding protein indicate that a similar protein is apparently located in the Golgi apparatus in multicellular eukaryotes (1030). The significance of this observation is not understood, but GTP and GTP binding proteins (G proteins) may be involved in maintaining vectorial movement of shuttle vesicles or in vesicle recycling (see below). Surface components on Golgi membranes are also required for intra-Golgi transport (3,39,40). A receptor may recognize the protein coat on the Golgi transport vesicles.

3. Current model for intra-Golgi transport

Balch *et al.* (40) considered that the migration of secretory proteins from one cisterna to another depends on three separate events:

(i) Secretory proteins are primed to make them competent or available for transport. Priming probably involves the migration of secretory proteins to regions of the cisternae where budding occurs, principally at the outer rims. This step might depend on interactions between the secretory protein and receptors migrating to the budding areas or may rely on free diffusion within the cisternal membrane or lumen. Segregation may depend on signals generated by processing enzymes in individual Golgi cisternae, but most experiments in which processing inhibitors have been used do not support this idea and show instead that processing is nonessential for secretory protein targeting. Some mutationally altered secretory proteins, however, transit normally through the early stages of the secretory pathway and yet are not transported through the Golgi (344,1040). Perhaps secretory proteins must fold into a particular conformation to be competent for movement between Golgi cisternae. The accumulation of such abnormal proteins in the Golgi may cause it to become distended, but this does not drastically affect intra-Golgi movement and secretion of other secretory proteins (344).

(ii) Vesicles move from one cisterna to another. Vesicle migration between Golgi cisternae may be either vectorial or random. Vectorial movement is more compatible with traditional views on the strict sequence of events in secretory protein processing and is supported by the observation that the likelihood of forward transfer is at least five times greater than that of lateral movement in Golgi of fused cells (969,970). This implies that there are receptors on the surfaces of Golgi cisternae which recognize vesicles budding from the preceding cisterna in the chain. Nonetheless, the topology in the Golgi complex is well maintained, possibly because the cytoskeleton prevents cisternae from coming into direct contact. Another puzzling feature of vesicular intra-Golgi transport is that small transport vesicles would be expected to diffuse away from the Golgi complex. The cytoskeleton may restrict the movement of Golgi vesicles and may even play a more positive role in directing vesicles between cisternae. However, although movement along microtubules has been well documented for large organelles (1136) and may be important in the sorting of some proteins leaving the Golgi complex (Section V.G.3), there is no evidence that it could play more than a minor role in the movement of vesicles over the very short distances which separate Golgi cisternae (139). Furthermore, cell fusion studies by Rothman *et al.* (969) show that inter-Golgi movement of VSV G protein can occur, indicating either fusion of the two Golgi complexes or, more likely, that movement between Golgi cisternae is dissociative; i.e., the vesicles do indeed diffuse into the cytoplasm.

(iii) Vesicle fusion with the membrane of the acceptor cisterna and release of vesicle contents into the lumen of the cisterna occurs.

This model for intra-Golgi transport fits many experimental observations, but further studies are required to define clearly the steps involved. For example, ATP may be required for vesicle fission and fusion, as well as for reducing the pH in trans Golgi compartments, but this has not been proven, and the nature and role of some of the cytosolic components required in *in vitro* assays have yet to be determined. Further work is needed to define how proteins find their way to budding regions of the Golgi membrane and what triggers vesicle formation and vesicle fusion with acceptor membranes.

4. Retention of resident Golgi proteins

How are proteins specifically retained in individual Golgi cisternae? We saw earlier that receptor-dependent recycling of ER proteins from the cis Golgi or an intermediate compartment could explain how these proteins remain almost entirely in the ER (Section V.B.4). Golgi residents probably also have signals which are recognized by some kind of receptor. Indeed, coronavirus E1 membrane glycoprotein appears to have a Golgi retention signal in one of its transmembrane domains (670). However, recycling of escaped proteins is a far less attractive model for explaining how Golgi proteins are retained than it is for the case of ER proteins. One possibility, proposed by Pfeffer and Rothman (872), is that the membranes of Golgi cisternae have two domains: a fluid domain, close to the budding rims, and an immobile phase, in which endogenous membrane proteins are anchored. A protein would require a signal to associate with a receptor in the immobile phase or to become anchored to it, whereas all other proteins would enter the mobile membrane phase or the bulk phase of the cisternal lumen. Pfeffer and Rothman point out that this model reduces the number of proteins which need to have signals for routing through or retention in the Golgi, because fewer proteins are retained in the Golgi than transit through it. Cytoskeletal structures, including possibly the cytoplasmic matrix, which "glues" the cisternae together, were proposed to limit movement in the immobile phase. An alternative idea, also considered by Pfeffer and Rothman, is that endogenous Golgi proteins interact to form patches which, by virtue of their size, are too small to fit into transport vesicles.

G. Post-Golgi sorting of secretory proteins

Secretory proteins transit through the Golgi apparatus and arrive in the trans cisternae together. The trans Golgi compartment is therefore the point at which the different branches of the secretory pathway diverge.

The following sections deal with the site at which sorting occurs, the ways in which proteins are sorted, and what happens when vesicles carrying secretory proteins arrive at their destinations.

1. Protein sorting in the trans-Golgi network

As its name suggests, the trans-Golgi network (TGN, also called Golgi endoplasmic reticular lysosomes or GERL) is the most distal compartment of the Golgi apparatus (relative to the RER). It differs from the Golgi cisternae in that it has a distended, reticular appearance rather than that of a flattened dish. Early morphological studies suggested that the TGN was a reticular adjunct of the Golgi specifically involved in lysosome biosynthesis [lysosomal enzymes were originally thought to bypass the Golgi (398)]. The TGN is now known to be distinct from endosomes or lysosomes. Endocytosed horseradish peroxidase does not accumulate in the TGN (401), although some endocytosed proteins may be recycled to the cell surface via the trans Golgi and especially via the TGN (319) (see Chapter VIII). The TGN probably contains tyrosinyl sulfotransferase (33), acid phosphatase, sialotransferase, and galactosyltransferase (365,957) (Table V.1, Fig. V.3), although TGN-derived vesicles are difficult to distinguish from those derived from the trans cisternae. The TGN also marks the site at which assembled clathrin, one of the proteins which coat some secretory and endocytic vesicles, appears along the secretory pathway (see Section V.G.6).

The TGN is the most acidic Golgi compartment, although the pH is almost certainly not as low as in secretory granules (823) or in lysosomal sorting vesicles, in which low pH causes the dissociation of lysosomal proteins from the mannose-6-phosphate receptor (577) (see Section V.G.5). Furthermore, influenza virus hemagglutinin, which is activated at low pH, reaches the cell surface as a nonactivated form, indicating that it does not spend an appreciable period (more than 2 min) in a compartment with a pH of less than 6 (107).

Different secretory proteins accumulate together in the TGN (1088,1123), which can become further distended when the load of secretory or lysosomal proteins increases (398). Transport of secretory proteins from the TGN is blocked at 20°C, and numerous protein-coated and naked vesicles accumulate as buds on the surface of the TGN (401). Different types of vesicles seem to be involved in sorting secretory proteins into different branches of the secretory pathway. Thus, different classes of soluble secretory proteins may be segregated into different domains of the TGN according to their interaction with specific receptors (Fig. V.4). Although this model does not explain how receptor proteins

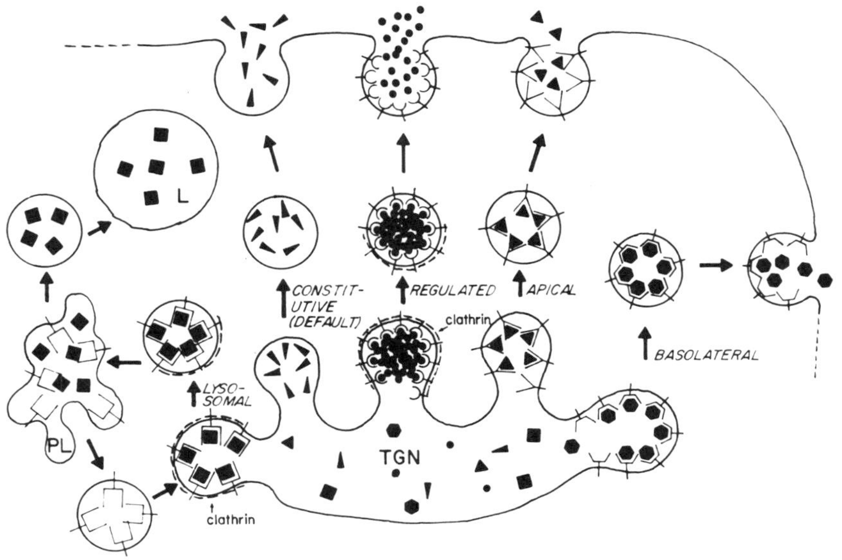

Fig. V.4. Receptor-mediated, post-Golgi sorting of soluble secretory proteins into the lysosomal, regulated, apical, and basolateral terminal branches of the secretory pathway. Note that constitutive default sorting is not receptor-mediated and that sorting into the regulated secretory branch of the pathway may result from protein aggregation and granule formation rather than receptor interactions. TGN, trans-Golgi network; L, lysosome; PL, prelysosome.

segregate into these domains, it does serve to illustrate the possible mechanisms involved in receptor-dependent segregation and packaging secretory proteins discussed in the following sections.

2. Default sorting and constitutive secretion

Default sorting is the final stage in the secretion of proteins which lack specific sorting (lysosomal, vacuolar, or polarity) signals and which are not accumulated within specific secretory storage granules of the regulated branch of the secretory pathway, i.e., those which are constitutively secreted (Fig. V.4). (Note, however, that some constitutively secreted proteins carry sorting signals.) Immunocytochemical studies show that proteins secreted by the default pathway accumulate in secretory vesicles which do not have protein coats (401,821). Plasma membrane proteins are transported to the cell surface in the same vesicles as secreted proteins (114,462,1088).

Default sorting and secretion in an *in vitro* assay system developed by Woodman and Edwardson (1204) (see Section II.F.3) required ATP and was sensitive to trypsin, implying that at least one cytoplasmic or vesicle–cytoplasmic membrane surface protein is involved. ATP might be needed for fusion between the vesicle and cytoplasmic membrane.

Saccharomyces cerevisiae strains carrying a temperature-sensitive mutation in the *SEC4* gene accumulate large numbers of secretory vesicles. These were purified by Walworth and Novick (1165), who found them to contain three dominant proteins (110 kDa, 40–45 kDa, and 18 kDa) together with invertase, the secretory marker protein. These three proteins were made during the period in which the vesicles accumulated, i.e., after secretion had been shut down at the nonpermissive temperature. The 110-kDa protein was the most abundant protein in the lumen of the secretory vesicles, whereas the other two proteins were membrane-associated and had cytoplasmic domains which could interact with the cytoskeleton, the cytoplasmic membrane, or cytoplasmic components (1165). It should be noted, however, that exocytosis in *S. cerevisiae* might be considered as polarized rather than default sorting because secretory vesicles are directed towards the growing bud rather than being randomly distributed over the entire cell surface (Section V.G.2 and Fig. II.5).

The *SEC4* gene was cloned and sequenced by Salminen and Novick (987), who found it to be homologous to GTP-binding RAS regulatory proteins of higher eukaryotes. Whether this is significant for the role of *SEC4* in secretion is unclear, especially since it is not known whether GTP is required for exocytosis in yeasts. Recent studies show that SEC4 protein binds GTP (391). Overexpression of *SEC4* suppresses the effects of mutations in three other *SEC* genes, but strains carrying *sec4*ts mutations rapidly become secretion-defective at the nonpermissive temperature, suggesting a direct role in secretion. The cytoplasmic and secretory vesicle membrane-associated *SEC4* product does not itself appear to be a secretory protein because its predicted primary sequence does not include a potential secretory routing signal. One possibility raised by Salminen and Novick is that the C-terminal cysteine residue of SEC4 is acylated, as are the GTP-binding RAS proteins and that the acyl groups anchor the polypeptide in the membrane, but this could not be confirmed directly. Another GTP binding protein, YPT1 (which is 50% homologous to SEC4) was detected close to the bud as well as in ill-defined structures (possibly the ER and Golgi) of *S. cerevisiae* cells (1030). Mutations in the *YPT1* gene affect several stages in the secretory pathway, but YPT1 protein probably plays an indirect role in protein transport as a result of its role in Ca^{2+} regulation (1014).

3. Polarized export and secretion

The plasma membrane of epithelial cells is divided into two distinct domains. The apical surface, which may have microvilli, is oriented toward the outside (e.g., lumen of the intestine), whereas the basolateral surface is on the inside, facing the basolateral surface of other cells or resting on an extracellular matrix of basal lamina (Fig. V.5). The two membrane domains have distinctly different lipid contents in their outer leaflets [the apical membrane has a higher glycolipid and cholesterol content, and a lower phosphatidylcholine content than the basolateral membrane (1050)], although the lipid contents of their inner leaflets may be identical (inset to Fig. V.5). This implies that only the outer leaflets of the two membranes are separated by tight junctions, morphologically distinct structures rich in nonbilayer lipids and containing unique proteins which form the junction between apical and basolateral surfaces and probably prevent the movement of outer leaflet lipids and proteins between them, as well as acting as ion gates between adjacent cells (Fig. V.5) (177,260,409,706–708,1050). There may also be differences in basal and lateral membrane composition. Cell–cell interactions are required for efficient formation of the basolateral surface of polarized cells, but not for sorting to the apical zone (1143). Adjacent cells may be held together by *desmosomes*.

The major breakthrough in studies on protein sorting in polarized cells

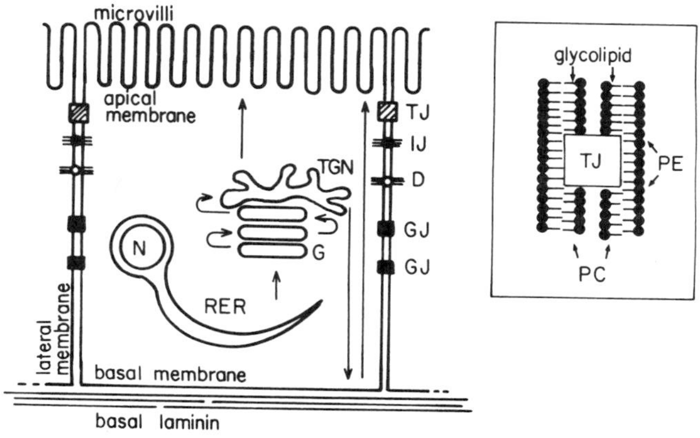

Fig. V.5. Schematic representation of a polarized epithelial cell showing principal secretion-related organelles and the directions of protein movement. N, nucleus; RER, rough endoplasmic reticulum; G, Golgi apparatus; TGN, trans-Golgi network; TJ, tight junction; IJ, intermediate junction; D, desmosome; GJ, gap junction. Inset shows detail of lipid organization on either side of the tight junction. PE, phosphatidylethanolamine; PC, phosphatidylcholine. Redrawn from Ref. 1050.

came when Rodriguez-Boulan (942,943) showed that MDCK cells target different viral glycoproteins to different cell surfaces. The lipid contents of viruses budding from basolateral and apical surfaces are also similar to those of the membrane from which they bud (1050). Further studies indicated that endogenous proteins were also asymmetrically sorted in polarized cells. Examples of endogenous basolateral proteins include laminin and collagen receptors, which probably anchor the basolateral surface to the underlying extracellular matrix (1050), gap junction proteins (600) (see Fig. V.5), transferrin receptor (see Chapter VII), and an ATPase (141). Examples of endogenous apical proteins include a secreted 81-kDa glycoprotein (390), an 80-kDa sulfated glycoprotein (1135), and degradative enzymes, including aminopeptidase N and alkaline phosphatase, secreted by intestinal epithelial cells (878). In general, the steady-state distribution of polarized membrane proteins indicates that sorting of basolateral protein is highly efficient (>97% fidelity) (873), whereas apical proteins may be found in significant amounts in the basolateral membrane. However, the surface area of the apical membrane is usually much smaller than that of the basolateral membrane, which means that the fidelity of apical targeting is actually quite high (about 88%) (873).

Basolateral and apical sorting seem to represent two distinct pathways, but most polarized cells also secrete some proteins from both surfaces, implying that neither basolateral nor apical sorting are default pathways (390,574). Furthermore, lysosomal proteins are secreted by the default pathway from both apical and basolateral surfaces when lysosomal sorting is blocked (140). This may not be the case in Caco-2 cells in which high-level basolateral secretion of normally nonpolarized lipoproteins implies that the basolateral sorting predominates over all other sorting pathways, including the default pathway (136,930a).

a. *Direct sorting of polarized proteins*

Do polarized cells sort and direct proteins directly to their target membranes, or are they all first randomly targeted to both domains, or specifically to one or other domain, and then transcytosed? Although some studies designed to answer this question have been conducted with endogenous proteins, the most detailed results come from studies on the basolaterally targeted VSV G protein and the apically targeted influenza virus hemagglutinin (HA) in MDCK cells. Furthermore, it is through derivatives of these proteins that most recent studies have attempted to identify apical and basolateral targeting signals (see next section).

The results of pulse–chase experiments combined with a trypsin sensitivity assay and tests with anti-HA antibodies applied to the basal surface of MDCK cells led Matlin and Simons (695) to conclude that HA was targeted directly to the apical surface. Conversely, Pfeffer *et al.* (873) showed that pulse-labeled G protein went directly to the basolateral membrane, where it could be detected with specific monoclonal antibodies. Another approach to studying the sorting of HA and G proteins is to "freeze" them in the Golgi apparatus by cooling the cells to 20°C (which specifically blocks protein exit from the TGN) or to use cells infected with viruses carrying temperature-sensitive mutations in the HA or G genes at the restrictive temperature. The bulk movement of these proteins can thus be followed by immunocytochemistry when the cells are restored to the permissive temperature. Rindler *et al.* (929) found that Gts protein was transported directly from the Golgi to the adjacent lateral cell surface, whereas wild-type HA accumulated at 20°C was sorted directly to the region of the apical surface closest to the TGN. Similarly, Pfeiffer *et al.* (873) found that G protein accumulated in the TGN went to the basolateral membrane 67 times faster than to the apical surface when the cells were warmed to 37°C. These studies are compatible with the idea that apical and basolateral proteins go directly to their respective target membranes. Different results were reported by Bartles *et al.* (50), who found that endogenous apical proteins in hepatocytes appeared first in the basolateral membrane and were subsequently redistributed to the apical surface by transcytosis (Section VIII.A.5.b). Whether this result indicates the existence of totally different mechanisms for sorting apical proteins in kidney and liver cells could be tested by expressing hepatocyte genes for apical proteins in MDCK cells.

b. Apical and basolateral sorting signals and their receptors

Sorting of polarized secreted and membrane proteins is assumed to depend on sorting signals (probably signal patches) [see Section I.D.2 and (88)] present in the sorted proteins and on their interaction with receptors. Sorting presumably occurs in the TGN, where at least one set of cognate receptors are probably located. To date, the primary approach used to locate sorting signals has been to introduce major sequence alterations including the deletion of cytoplasmic or transmembrane domains from the HA, G, or other polarized viral glycoproteins, or to create hybrids between them. Even these relatively unsophisticated studies give conflicting results. For example, McQueen *et al.* (702) and Roth *et al.* (959) found that either deleting the transmembranous and cytoplasmic tail of HA (to make a soluble, secreted protein) or replacing them with the corresponding regions of G did not affect apical targeting and concluded that the apical sorting signal was located in the N-terminal, extracytoplasmic domain of HA. Gonzalez *et al.* (387) reported, however, that truncated HA was secreted by the default pathway and concluded that the sorting signal was in the C-terminal transmembranous or 10-amino-acid cytoplasmic domains. There is also disagreement concerning the location of the G protein basolateral sorting signal, which, according to the deletion and HA gene fusion studies of Paddington *et al.* (835) and Gonzalez *et al.* (387) is located in the 29-amino-acid, C-terminal cytoplasmic domain. Stephens and Compans (1080) reached a similar conclusion in their studies on the basolaterally sorted GP70 protein of Friend mink cell focus virus. Recent studies by McQueen *et al.* (703), however, indicate that the basolateral sorting signal of G protein is in the N-terminal extracytoplasmic domain. The palmitoylated cysteine residue in the C-terminal cytoplasmic tail of G protein is not required for basolateral targeting (580).

McQueen *et al.* (703) have suggested that the origin of these conflicting results may be the fact that recombinant genes are not stably expressed in MDCK cells. Their studies (702,703) were carried out shortly after infection by recombinant viruses and are, they claim, more likely to reflect the true sorting pathway of the recombinant gene product. Other groups may in fact be studying the sorting of hybrid or truncated proteins which have been further modified by genetic selection for more stable cell lines. Another argument in favor of the idea that sorting signals are in the extracytoplasmic domain is that this part of the signal is initially localized in the lumen of the TGN, where sorting is presumably initiated, and could therefore bind to hypothetical sorting receptors which segregate sorted proteins from those destined for default export or secretion. These receptors may themselves have cytoplasmic domains recognized by receptors on the basolateral or apical surfaces. Furthermore, such a system could handle both secreted and exported (membrane) proteins. However, there is some evidence, based on the effect of NH_4Cl (or pH) on basolateral sorting of laminin and heparin sulfate proteoglycan, that soluble and membrane proteins may be sorted by different mechanisms (142).

Clearly, more work remains to be done before we can identify polarity sorting signals with any precision. One of the difficulties may be that these signals are probably not linear sequences of amino acids and are therefore less amenable to gene fusion studies, which have been so useful in identifying routing signals. Even very subtle conformational changes may alter sorting signals and render them nonfunctional; this could explain why prevention of glycosylation with tunicamycin may cause normally polarity-sorted proteins to enter the default pathway (1135).

Another aspect of the work on polarity sorting signals which needs to be pursued is the identity of receptors and the stage at which sorted proteins dissociate from them. Sorting receptors are presumably recycled. Little is known about how the vesicles are targeted to different regions of the plasma membrane. One possibility is that sorting of lipids

(Fig. V.5) is involved (709,1050a). Alternatively, sorting vesicles could interact with components of the cytoskeleton, along which they are driven by ATP-dependent motors attached to microfilaments; further receptor–ligand interactions might complete the sorting process when vesicles arrive at the plasma membrane (1136). Although there is no evidence that microfilaments are required for default sorting of secretory proteins (1136), Rindler *et al.* (930) have reported that colchicine and other microtubule-disorganizing agents abolished specific apical sorting of HA and caused influenza virus to bud from both cell surfaces in polarized cells. Salas *et al.* (986) obtained the opposite result, however, and neither group observed any effect on basolateral targeting of VSV G protein. Rogalski (944), however, found that agents which caused microtubule disassembly caused random sorting of G protein. Thus, the role of the cytoskeleton in the targeting of polarity-sorted secretory proteins in complex eukaryotes' proteins remains unclear. A temperature-sensitive mutation in the *S. cerevisiae* actin gene results in abnormal exocytosis and bud formation at the nonpermissive temperature, suggesting that actin filaments may direct secretory vesicles to the growing bud in yeast cells (793).

There is evidence to show that the cytoskeleton may be important in maintaining polarized distribution in complex eukaryotic cells (410). A polarized ATPase has recently been shown to bind directly to ankyrin, a protein which is known to link an integral erythrocyte membrane protein to spectrin and actin of the erythrocyte cytoskeleton (770). Thus, ankyrin might link the ATPase to microfilaments and thereby maintain its polarized distribution.

4. Regulated secretion

Two features distinguish regulated exocytosis from other branches of the secretory pathway:

(i) Protein release only occurs when the cells are stimulated by secretagogues (e.g., cyclic AMP or Ca^{2+}).
(ii) Proteins accumulate within the cell before their release.

Proteins accumulate in a special class of secretory vesicles called *secretory granules,* wherein protein concentrations reach such high levels that they become electron-dense and can be clearly identified by electron microscopy. The secretory pathway is restricted to certain cell types (e.g., endocrine and exocrine cells, mast cells), and only certain types of proteins (examples include hormones, albumin, and some degradative enzymes including proteases and lipases) enter into the pathway. Another feature of the regulated secretory pathway is that proteins may be proteolytically processed before release or as they are released from the cell, although this feature is also found in some constitutively secreted proteins.

a. Sorting into regulated and constitutive pathways

Regulated and constitutive branches of the secretory pathway can coexist in the same cell (395,567,1109), implying that proteins destined for storage in secretory granules must be sorted from other secretory proteins. Electron microscopic studies show that both classes of secreted proteins transit through the Golgi cisternae together and are segregated in the TGN, where proteins destined for the regulated branch of the pathway condense into specific areas coated with the protein complex clathrin (see Figs. V.4 and 6; see also Section V.G.6) (824,1122). The clathrin coat is subsequently removed as the condensing granules bud from the TGN and mature (818) (see Fig. V.6). These areas probably represent the sites at which secretory granules mature and are released from the TGN. It should be noted, however, that some proteins which are normally secreted by the regulated pathway may "escape" and be secreted "constitutively" (31). It remains to be determined whether this is due to incorrect sorting or to low-level, secretagogue-independent secretion from storage granules, as suggested by studies by von Zastrow and Castle (1231). Burgess and Kelly (136) propose that this "spillover" secretion may be due to inefficient sorting in the specific cell lines tested, since Rhodes and Halban (921) observed much more efficient sorting into the regulated pathway.

Efficient segregation of proteins into regulated and constitutive branches of the secretory pathway implies that the former have sorting signals (see Section V.G.4.b for an alternative explanation). The existence of these signals was suggested by Moore and Kelly (731), who transfected a pituitary tumor cell line with a hybrid gene comprising the 5′ end of the VSV G protein and the 3′ end of the gene for human growth hormone. The hybrid protein was diverted into the regulated secretory pathway, indicating that the constitutive pathway had been bypassed due to sorting signals present in the growth hormone part of the hybrid. There are no obvious sequence similarities between proteins secreted by the regulated pathway, implying that the sorting signal is probably a patch signal rather than any identifiable linear stretch of amino acids. Several proteins secreted via the regulated pathway are proteolytically processed prior to their release from the cell (see below). This raises the possibility that the sorting signal could reside in that part of the secretory polypeptide which is eventually cleaved off. This possibility was tested by Burgess *et al.* (135), who found that deleting DNA coding for the propeptide part of trypsinogen had only a minor effect of the targeting of the enzyme into secretory granules. Thus, they concluded, there must be at least one sorting signal in the mature part of the trypsinogen.

One surprising feature of regulated pathway sorting is that the putative sorting signal seems to be universal. Moore *et al.* (732), for example, found that human proinsulin was packaged into secretory granules in mouse adrenocorticotropic hormone (ACTH)-secreting cells, that it was correctly processed (see below), and that its release from these cells was stimulated by the same secretagogues that stimulated ACTH release. Similar results were obtained in studies on the expression of human kidney renin DNA in the same cell line (337). Fibroblast L cells, however, which do not have the regulated pathway, secreted (unprocessed) proinsulin via the constitutive pathway (732).

Receptors localized in specific domains of the TGN membrane may segregate proteins into the regulated branch of the secretory pathway, but this has not been proven, and other factors may be important. For example, treatment with the weak base chloroquinone causes ACTH to be secreted by the constitutive pathway, indicating that low pH is required for sorting into the regulated pathway (733). Low pH in condensing secretory granules may dissociate proteins from their receptors, which can then be reused (but see Section V.G.4.b). Another feature of secretory granules which appears to have been largely overlooked is that proteases involved in post-TGN processing of secretory proteins must also carry sorting signals. It remains to be seen whether the membrane content of secretory granules differs significantly from that of vesicles of the constitutive branch of the secretory pathway.

b. Condensation and granule formation

In certain exocrine cells, the concentration gradient of a secretory protein between the RER and secretory granules can be as high as 200 (988). The dense core of aggregated protein is sometimes seen to be separated from the membrane of the secretory granule (136) and may remain intact when the membrane is removed (1230) or upon exocytosis (19,1122). Although receptor-mediated sorting into specific regions of the TGN may assist the condensation process, proteins secreted by the regulated pathway may aggregate spontaneously and be packaged into secretory granules when they reach a critical size. Thus, the packaging of G protein–growth hormone hybrids into secretory granules discussed above (731) may be due to the presence of "aggregation" sequences in the growth hormone segment of the polypeptide, rather than to the presence of a specific sorting signal. The low pH of the condensing granule (19,824) may be important for protein aggregation, since aggregates dissociate at high pH (534). Pfeffer and Rothman (872) suggest that this could explain the failure of chloroquinone-treated cells to package secretory proteins into secretory granules (see above). Secretory granules of exocrine (hormone-secreting) cells remain acidic during maturation, but those of endocrine (enzyme-secreting) cells return to neutral pH as they mature (534). ATP is also needed for secretory granule formation.

There are conflicting views as to whether different secretory proteins cosegregate and coaggregate into the same secretory granule. Detailed studies by Fumagalli and Zanini (341) revealed that bovine growth hormone and prolactin could be present either in different aggregates in the same secretory granule or in mixed aggregates in the same granule or in pure aggregates in different granules. The ratios of the three type of granules varied from animal to animal. Similar results were reported by Mroz and Lechene (744), who showed that the enzyme content of individual secretory granules derived from single cells from the same gland can vary enormously. The simplest interpretation of these data seems to be that segregation is a random process and that the formation of mixed or pure aggregates depends on the local concentration of the respective proteins and on their preference for forming homo- rather than heteroaggregates. Packaging of proteins into different secretory granules, however, might permit their release to be stimulated by different secretagogues, but there is only limited evidence for such a phenomenon at the level of individual cells (11,136,317a).

c. *Proteolytic processing*

Many of the proteins secreted by the regulated branch of the secretory pathway are proteolytically processed and activated, usually in secretory granules. Processing often involves the proteolytic removal of an N-terminal propeptide and can be mimicked by exogenous proteases such as trypsin (336). Alternatively, short spacer peptides may be removed from polyprotein precursors (187). In the latter cases, cells in different tissues can process the precursors to give different "mature" forms. This is the case for prosomatostatin, which is processed to a 28-amino-acid form by cells in the gut and to a 14-amino-acid form by brain and pancreatic cells (790). Islet tissue from angler fish pancreas contains at least two proteases which process prosomatostatin to give products of different lengths. One of these proteases can also process proinsulin (674). Other examples of processing of heterologous secretory proteins (188,444,732) indicate the existence of only a limited number of processing proteases, which may also be found in some cell types which do not have a regulated secretory pathway (444,1168). When secretory proteins which would normally use the regulated pathway are produced in cells which do not have the regulated pathway, however, they are constitutively secreted as unprocessed (pro-) forms (889).

The site of proinsulin processing was determined cytochemically by Orci *et al.* (816,825), using monoclonal antibodies specific for mature insulin. Fully processed insulin was first detected in clathrin-coated vesicles budding from the TGN, and subsequently in naked granules. Processing was coincident with condensation and acidification and was inhibited at higher pH, indicating that the two proinsulin-processing proteases have low pH optima (821). Therefore, this and other proteolytic processing steps probably occur in a late "compartment" of the TGN.

Although propeptides may not play a role in protein sorting, they may prevent enzymes such as proteases from folding into active conformations prior to their release, thereby protecting secretory granules from endoproteolytic attack. Proinsulin and other prohormones may be loosely membrane-associated (789,819). In these cases, bridge sequences may contribute to the patch signal which shunts proteins into budding secretory granules.

d. Release of proteins from storage granules

Relatively little appears to be known about the events which accompany protein release from secretory granules. The general view seems to be that secretagogues directly, or more likely indirectly, stimulate fusion between the granule membrane and the cytoplasmic membrane, resulting in the release of granule contents to the outside of the cell. Secretagogues bind to specific cell-surface receptors and promote Ca^{2+} influx, which seems to be intimately related to the fusion event (756). GTP binding proteins also seem to be involved in generating the signal which stimulates secretion (138), and microtubules may play a minor role in directing storage granules to the cell surface (112). Inhibitors of metalloprotease and dipeptide protease substrates inhibit exocytosis, suggesting that proteolytic cleavage of a membrane protein may be essential for exocytosis. Mundy and Strittmatter (756), who found that metalloprotease activity was highest in the plasma membrane, propose that proteolysis may unmask the active site on a fusogenic membrane protein.

Breckenbridge and Almers (121) have recently studied exocytosis-associated changes in membrane capacitance in a mouse mast cell mutant with enlarged secretory granules. Small fluctuations in capacitance preceded larger increases, which themselves preceded granule swelling and the release of a fluorescent tracer dye from the lumen of the granule. The large increase in capacitance probably results from productive fusion between the plasma and granule membranes (312), whereas capacitance "flutters" may represent nonproductive membrane association. The most plausible interpretation of these data is that release of secretory granule contents is preceded by the formation of a narrow channel, the *fusion pore,* (121) between the two membranes, leading eventually to the opening of the granule membrane to the outside of the cell and the subsequent swelling and dissociation of the granule contents and their release as soluble proteins.

5. Lysosomal and vacuolar proteins

Lysosomes (in animal cells) and vacuoles (in plant and fungal cells) contain most of the cells' degradative enzymes, which function not only in general "housekeeping" but also in the degradation of endocytosed material (Chapter VIII). The following sections review the evidence for lysosomal and vacuolar sorting signals, special features of the sorting pathways, and differences between the lysosomal and vacuolar routes.

a. *Mannose-6-phosphate-dependent sorting of lysosomal enzymes*

Specific sorting of soluble lysosomal enzymes is determined by mannose-6-phosphate (M6P) residues on N-linked core oligosaccharides (Section V.D.1.a). Two receptors have been identified. The major M6P receptor (275 kDa, formerly called the 215-kDa receptor) was detected predominantly in the cis Golgi compartment (131), leading to the proposal that the lysosomal pathway diverged from the main secretory pathway at the cis end of the Golgi stack rather than in the TGN. This idea is incompatible with the observation that some lysosomal proteins are terminally processed by enzymes located in medial and trans Golgi compartments. Other cytological studies indicate that M6P receptors are located in the TGN, as well as throughout the Golgi stack (208), in coated vesicles and the plasma membrane (365,366) (see below for explanation), and in a Golgi-proximal vesicular structure (402), but not in lysosomes. Thus, lysosomal enzymes probably do transit through the TGN, possibly already complexed with their receptor, although Farquhar (299) argues strongly in favor of multiple lysosomal sorting pathways and in particular for sorting from the cis Golgi in certain cell lines. Thus, the site of accumulation of M6P receptors along the secretory pathway may have little relevance for lysosomal enzyme sorting.

Two M6P receptors (275 kDa and 46 kDa) have been identified and characterized. The receptor activity of the 275-kDa protein, but not that of the 46-kDa protein is cation-independent, and the 46-kDa receptor recognizes only phosphomonoesters whereas the 275-kDa protein also binds methylphosphomannosyl residues (461). The 275-kDa protein appears to be present in most mammalian cell lines tested so far (982), but the distribution of the 46-kDa receptor has not been determined. Some mutant cell lines lack the 275-kDa M6P receptor yet still target lysosomal enzymes normally, which suggests that both receptors are involved in lysosomal sorting (461). Studies with such cell lines revealed another difference between the two receptors, however. Although part of the cellular pool of both receptors is located on the cell surface, only cells with the 275-kDa protein can endocytose secreted lysosomal proteins (1077). This "defect" in the 46-kDa protein appears to be due to a failure to bind the ligands, since antibodies against the 46-kDa protein are endocytosed normally (see Chapter VIII for more details on endocytosis). Thus, the 46-kDa protein seems to be specifically involved in the sorting of endogenous lysosomal enzymes.

The genes for both M6P receptors have been cloned and sequenced. Although the predicted sequence of the two gene products are generally different, the two proteins have a region of moderate sequence similarity in their lumenal domains (645,826) which Dahms *et al.* (208) propose could be the M6P binding domain.

The sorting of the lysosomal enzymes cathepsin C and cathepsin D was studied directly by Lemansky *et al.* (626), who found that lysosomal enzyme precursors occurred only in coated vesicles. Proteolytically matured forms were found in lysosomes. Schulze-Lohoff *et al.* (1022) also observed the transient accumulation of one of these enzymes in coated vesicles, which, they proposed, are specifically involved in the sorting of lysosomal proteins from the secretory pathway. Lemansky *et al.* (626) devised procedures which allowed vesicles derived from the secretory pathway to be separated from those derived from the endocytotic pathway. They found that both classes of vesicles contained precursor forms of cathepsin. These studies have two profound implications:

(i) The fact that vesicles involved in direct sorting of cathepsins to the lysosome contain clathrin implies that they had passed through the TGN, the first site along the secretory pathway at which clathrin is detected (Section V.G.1).

(ii) The fact that some lysosomal enzymes are "fished" out of the surrounding medium and retargeted to the lysosome implies that some lysosomal enzymes are incorrectly sorted, probably into the constitutive secretory pathway (see also Section V.D.1.a). Higher levels of incorrect sorting occurs in NH_4-Cl-treated cells, probably because low pH is required to dissociate M6P from its receptor in the prelysosome (134,953, Section VIII.A.4.b). Whether or not this has any physiological significance remains to be determined, but at least it explains the presence of 275-kDa M6P receptors on the cell surface (see above), which Braulke *et al.* (119) have shown are in equilibrium with intracellular receptors.

Another interesting observation concerning the M6P receptor is that it specifically binds to one of the components (the 100-kDa accessory protein) of the clathrin cage which coats the sorting vesicles (854) (see below). This may have particular relevance for the sorting of lysosomal enzymes because different classes of clathrin-coated vesicles appear to have different types of accessory proteins (855).

Coated vesicles almost certainly do not transport proteins directly into lysosomes. Instead, the vesicles are targeted to endosome-like reticular organelles (prelysosomes or secondary endosomes), where the receptor probably dissociates and recycles back to the Golgi cisternae. This organelle is also the site to which endocytosed lysosomal proteins are targeted (134,402) (Fig. V.6 and Section VIII.A.4.b). A different class of vesicles may complete the transport of lysosomal enzymes once they have dissociated from their receptor, but von Figura and Hasilik (313) consider it more likely that there is a gradual transition from tubular prelysosomes to lysosomes proper.

b. M6P-independent lysosomal protein sorting

Although M6P is undoubtedly the major, and in some cells the only, sorting signal on lysosomal enzymes, some lysosomal proteins do not have M6P residues. Owada and Neufeld (829) found that a human liver cell line devoid of *N*-acetylglucosamine-1-phosphotransferase [and therefore unable to phosphorylate mannose residues (see Section V.D.1.a)], still targeted some lysosomal enzymes correctly with, at most, only slightly reduced efficiency. These cells may have a completely M6P-independent system for sorting lysosomal enzymes. All lysosomal membrane proteins are also devoid of M6P residues. Therefore, some lysosomal enzymes may have membrane-associated intermediates which are sorted to the lysosome together with authentic lysosomal membrane proteins. Barriocanal *et al.* (49) used immunocytochemistry to follow the fate of three lysosomal membrane proteins which they detected in lysosomes, the Golgi apparatus, and coated and uncoated vesicles in the region of the TGN. They found that oligosaccharide modifications were not required for lysosomal targeting (although they may be required to protect against proteolysis). Thus the sorting pathway for these proteins remains to be determined; they could be sorted completely independently of lysosomal enzymes or could be colocalized to the same coated vesicles by an M6P-independent receptor and then segregated from recycling vesicle membrane components in the prelysosome. Green *et al.* (397) have recently found that newly synthesized lysosomal membrane proteins appear in lysosomes with the same kinetics as newly synthesized plasma membrane proteins appear at the cell surface, making it unlikely that the former pass via the plasma membrane en route to the lysosome.

c. *Vacuolar proteins of plants and fungi*

Although vacuoles are the functional equivalents of lysosomes in animal cells, the sorting of vacuolar enzymes is completely independent of M6P receptors. This was demonstrated most simply by the fact that tunicamycin treatment did not affect vacuolar protein targeting (572,1025). However, at least one vacuolar protein does have phosphorylated mannose residues.

Most of the work on the sorting of vacuolar proteins has concentrated on the identification of sorting signals in yeast vacuolar proteins. Early studies showed that vacuolar proteases were proteolytically processed in two distinct stages, the first of which corresponded to the removal of a signal peptide (711). The second processing step is catalyzed by a vacuolar protease, proteinase A, which removes an additional N-terminal polypeptide segment, the propeptide, from other vacuolar enzymes. Proteinase A is also autoactivated by the same mechanism (15,1206). The second processing step is blocked by certain *sec* mutations, which cause secretory proteins to accumulate in the RER or Golgi apparatus, whereas mutations which affect the final stage of the secretory pathway from the Golgi to the cell surface do not affect vacuolar protein sorting or processing (1082). This implies that the Golgi is the site of sorting of vacuolar and secreted or plasma membrane proteins in yeast cells.

Gene fusion studies conclusively demonstrated that propeptides are vacuolar sorting signals. Bankaitis *et al.* (44) and Johnson *et al.* (541) found that at most 50 N-terminal residues of preprocarboxypeptidase Y (CPY), including the 20-residue propeptide, could target the normally secreted enzyme invertase into the vacuole. Similar results were obtained with proteinase A–invertase hybrids (572). Valls *et al.* (1137) subsequently found that mutations affecting the sequence of the proCPY propeptide caused the enzyme to be secreted in an inactive form. There appears to be no sequence similarity between the propeptides of different yeast vacuolar enzymes, even though genetic studies described below suggest that they are sorted into the vacuole via a common pathway. Part of the propeptide may be required to maintain vacuolar enzymes in an inactive form until they reach the vacuole and may also maintain the precursors in a competent conformation for transport through the secretory pathway.

The overproduction of vacuolar proteinase A (972) and of CPY–invertase hybrids (44) causes them to be secreted into the medium, suggesting that some component of the vacuolar sorting pathway (e.g., a receptor) had been saturated. These observations led to the development of techniques for selecting mutants which secreted CPY or CPY–invertase without overproduction (Section II.F.1.b). Mutations in over 50 different genes (called *VPL* or *VPT*) have been identified (44,939a,971). The extent of the sorting defect varied in different mutants: Some of them, for example, did not affect the targeting of proteinase A, and none of them affected the sorting of the vacuolar membrane protein α-mannosidase, which is presumably sorted to the vacuole by an alternate pathway (971). It is unlikely that any of the mutations affected protein retention in the vacuole because vacuolar enzymes were not terminally processed, and the kinetics of CPY secretion were comparable to those of a normally secreted protein. The characterization of the *VPT* or *VPL* gene products and their localization in the cell could provide revealing insights into the mechanisms of vacuolar protein targeting, but at present we can only speculate on their roles. Obvious candidates are the propeptide receptor, vacuolar or Golgi ATPases which might produce a low pH environment necessary for sorting or receptor dissociation, or proteins involved in vesicle fission and fusion.

Much less work has been done on plant vacuolar proteins. Tague and Chrispeels (1100) found that the plant vacuolar storage protein phytohemagglutinin was targeted mainly to the vacuole when its structural gene was expressed in yeast cells. This protein does not have a cleavable propeptide (it does have a signal peptide, which was at least partially processed in yeast cells), which means that the vacuolar plant sorting signal which is recognized by the yeast vacuolar sorting pathway is located in the mature part of the phytohemagglutinin polypeptide. mRNA coding for a second plant storage protein, globulin P, has been microinjected into frog oocytes, which secreted the protein into the medium (51). This confirms that plant vacuolar proteins are bona fide secretory proteins and that plant cells have a special branch of the secretory pathway which shunts storage proteins into vacuoles.

6. The role of clathrin in the secretory pathway

As we have seen, protein-coated vesicles are involved in transporting secretory proteins at various stages of the secretory pathway. Some vesicles (e.g., lysosomal sorting vesicles and immature secretory granules), are coated with a protein complex called *clathrin,* which also coats endocytotic vesicles (Chapter VIII). Other secretory vesicles (e.g., those mediating protein transport through the Golgi) have a different type of protein coat (797).

Clathrin, which is composed of equimolar amounts of heavy and light chains, forms a three-layered cage which envelops vesicles in a shell with fibrous interconnections, which give it mechanical strength and stability. The vesicle membrane is thought to have receptors which anchor the clathrin cage to the surface via a number of ancillary assembly proteins which probably act as bridges (1148,1149). Different assembly proteins are found in different classes of clathrin-coated, TGN-derived, and endocytic vesicles, suggesting that they might contribute to their respective specificity for particular membrane targets (7).

The clathrin coat probably prevents intimate contact between fusing membranes and must thus be removed to allow fusion to occur. This may explain, for example, the disappearance of the clathrin coat from maturing secretory granules (Section V.G.4.b) and the presence of lysosomal protein precursors in naked as well as coated vesicles. Vesicles coated with other proteins are also presumably uncoated to allow fusion to occur. The uncoating of clathrin-coated vesicles is mediated by the ATP-dependent cytosolic "uncoating" protein, which remains attached to the released clathrin (1010). Soluble clathrin retains its typical triskelion conformation. Uncoating protein is a member of a highly conserved group of stress proteins and may be related to HSP70 heat shock proteins involved in other stages in secretory protein transport and in mitochondrial protein import (Sections III.C.3 and VI.B.5). The significance of this observation remains to be determined (855,967), but one possibility is that the ATPase (HSP70) is required to activate an uncoating enzyme which is already present in the clathrin complex.

The action of uncoating protein must be triggered in some way to prevent it from destroying protein coats on immature vesicles or on coated buds, but the nature of the signal remains to be determined. The requirement for ATP for uncoating activity could in part explain the observed requirement for the same nucleotide during the transport of proteins between different stages of the secretory pathway if a similar activity is required to remove other, nonclathrin coats.

Attempts to determine the role of clathrin in protein targeting in yeast cells have given ambiguous results. Yeasts are known to have clathrin, and coated vesicles have been observed, but it is not known whether clathrin coats vesicles involved in secretory protein targeting (1165). The gene for the clathrin heavy chain was independently cloned by two groups who used it to inactivate the chromosomal gene to study the effect of the absence of clathrin on cell growth and protein secretion. Payne and Schekman (852) and Payne *et al.* (853,853a) reported that their mutants grew somewhat more slowly than wild-type cells, secreted invertase at a slightly reduced rate, and were partially defective in prepro-α-factor processing. The mutants accumulated unusual vacuoles, vesicles, and Golgi-derived structures. These results suggested that the absence of clathrin did not completely impair plasma membrane growth and protein secretion but that there was nevertheless a reduced rate of transit of secretory proteins through the later stages of the secretory pathway. Different results were obtained using exactly the same approach by Lemmon and Jones (627). They found that cells lacking the clathrin heavy chain were not viable unless they also carried a suppressor mutation. Even with this mutation, the cells grew slowly, were larger and rounder, had an unusual granular appearance, tended to aggregate in liquid culture, and were polyploid. These results suggest that the absence of clathrin is highly detrimental to yeast cells, making it difficult to determine whether clathrin plays a specific role in protein secretion in this organism.

H. Secretory pathway-independent export and secretion

The secretory pathway is almost certainly the route by which the vast majority of secreted and plasma membrane proteins are exported by eukaryotic cells. However, there is increasing evidence that some plasma membrane and secretory organelle proteins may reach their final destinations directly rather than via the secretory pathway. Examples of this class of proteins include RAS-like GTP binding proteins such as SEC4 (987) and RAS2 (232) of the yeast *S. cerevisiae* and similar proteins from complex eukaryotes (1194), mating pheromones in yeast and some fungi (726,886,983), capsid proteins of picornaviruses (851), and src proteins of Rous sarcoma and other transducing viruses (1029). Most if not all of these proteins are fatty acylated. Two different amino acids seem to be modified: N-terminal glycines, which are substrates for myristoyl CoA protein *N*-myristoyltransferase (1124), and C-terminal cysteines in RAS-like proteins (232). These Cys residues are reported to be palmitoylated, but a farnecyl residue has been found in basidiomycete pheromones (983). The absence of the fatty-acylated amino acid disrupts membrane association of RAS and src proteins (232,555,1029,1194), indicating that fatty acids probably anchor these proteins in their respective membranes. It remains to be determined how fatty-acylated proteins actually cross the plasma membrane (as in the case of the fungal pheromones) or what determines their specificity for certain membranes. A further example of a secreted protein which does not have a secretory routing signal is interleukin 1, but very little appears to be known about how this protein crosses the plasma membrane.

I. Concluding remarks

There is general agreement concerning the events which lead to the sorting of secretory proteins into different terminal branches of the secretory pathway, as illustrated in Fig. V.6, but we are clearly a long way from understanding exactly what directs proteins to their specific targets. Patch signals are undoubtedly necessary for sorting soluble proteins (other than lysosomal enzymes) into the various branches of the secretory pathway, but these will be difficult to identify by gene fusion techniques. At present, we have no clear idea of the extent to which the sorting vesicles have different membrane contents, but it seems probable that specific

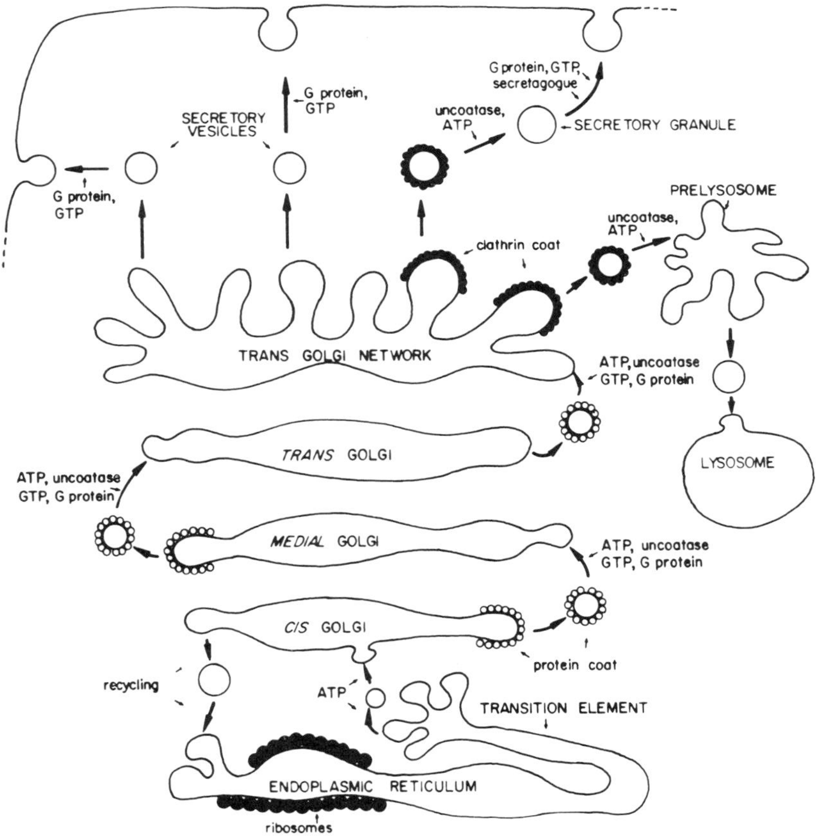

Fig. V.6. Model showing post-RER sorting of secretory proteins through the Golgi and on to the cell surface or lysosomes.

groups of membrane proteins (lysosomal, secretory granule, apical, and basolateral) accumulate at different sites in the membrane of the TGN from which sorting vesicles bud. This specialization is presumably also determined by protein–protein interactions, but the possibility that other interactions (e.g., protein–lipid) might be involved should not be overlooked.

Another interesting observation is that ATP seems to be required at almost every stage in the secretory pathway. ATP has been proposed to act in a variety of ways, including acidification of secretory organelles and vesicles, phosphorylation of receptors or ligands, protein folding and "proofreading," activation of cytoskeletal motors, and vesicle uncoating activity. There is increasing evidence that GTP and GTP binding proteins (G proteins) are also involved at several stages in the secretory pathway. GTP binding proteins are also known to be involved in the generation of other intracellular signals, such as the activation or inactivation of adenylate cyclase, the activation of cyclic GMP phosphodiesterase, and the control of phospholipase C action (769). By analogy, GTP binding proteins may act as signals or to activate receptors or ligands to ensure vectorial transport through the secretory pathway (109).

Further reading

Burgess, T. L. and Kelly, R. B. (1987). Constitutive and regulated secretion of proteins. *Annu. Rev. Cell Biol.* **3,** 243–293.

Farquhar, M. G. (1985). Progress in unravelling pathways of Golgi traffic. *Annu. Rev. Cell Biol.* **1,** 447–488.

Kornfeld, R., and Kornfeld, S. (1985). Assembly of asparagine-linked oligosaccharides. *Annu. Rev. Biochem.* **54,** 631–634.

Matlin, K. S. (1986). The sorting of proteins to the plasma membrane in epithelial cells. *J. Biol. Chem.* **103,** 2565–2568.

Pfeffer, S. R., and Ruthman, J. E. (1987). Biosynthetic protein transport and sorting by the endoplasmic reticulum and Golgi. *Annu. Rev. Biochem.* **56,** 829–852.

Schekman, R. (1985). Protein localization and membrane traffic in yeast. *Annu. Rev. Cell Biol.* **1,** 115–143.

von Figura, K., and Haslik, A. (1986). Lysosomal enzymes and their receptors. *Annu. Rev. Biochem.* **55,** 167–193.

CHAPTER VI

The targeting of mitochondrial, chloroplast, and peroxisomal proteins

The targeting of nuclear-encoded mitochondrial, chloroplast, and peroxisomal (microbody) proteins is independent of the secretory pathway. Studies on the targeting of mitochondrial and chloroplast proteins have reached an advanced stage; structural features of routing signals have been determined, their receptors identified, and stages in protein import and processing defined and characterized. Early stages in protein import by chloroplasts and mitochondria are very similar, and they will therefore be considered together. Later sections deal with the targeting of endogenous mitochondrial and chloroplast proteins and with more complex stages in chloroplast thylakoid biogenesis. Much less is known about the biogenesis of peroxisomes and related organelles, but the available evidence points to several unique features which distinguish peroxisomal protein import from the import of mitochondrial or chloroplast proteins. Peroxisomes are therefore considered separately at the end of this chapter.

A. Mitochondrial and chloroplast organization

Mitochondria have four independent compartments, the outer and inner membranes, the intermembrane space, and the matrix, (Fig. I.1), each of which contains proteins. Different mitochondrial fractions can be separated by physical methods similar to those used to separate the components of Gram-negative bacteria (217). Plant and animal mitochondria are regularly shaped, individual organelles, whereas those of the yeast *S. cerevisiae* are larger and less uniform, although they break up into smaller, typical mitochondria upon cell fractionation. The vast majority of mitochondrial proteins are encoded by nuclear genes, but a limited number are encoded by the mitochondrial chromosome, which is located in the matrix. Some of these proteins are "exported" into or across the inner membrane.

Chloroplasts are similarly organized except that the membrane-limited *thylakoids* within the stroma (the equivalent of the mitochondrial matrix) represent two additional protein-containing compartments (Fig. I.1). Thylakoids have a distinct organization which includes granular (stacked) lamellae containing photosystem II, and unstacked, interconnected lamellae, which contain most of photosystem I. Chloroplasts also have an endogenous genome, located in the stroma; it is generally larger than the mitochondrial genome and encodes more proteins, many of which are transported into or across either the inner or the thylakoid membranes.

In both mitochondria and chloroplasts, nuclear and organelle-encoded proteins interact to form active enzyme complexes (256). Both types of organelle may have originated from prokaryotic organisms, possibly from parasites living in the cytoplasm of primitive eukaryotic cells. Space limitations do not allow specific details to be discussed, but there are arguments both in favor of and against this attractive hypothesis. The fact that genes coding for organelle proteins are present either in the nucleus or in the organelle, depending on the organism (110), may reflect the progressive movement of genes from the organelle to the nucleus, probably by gene duplication, transposition, and eventual deletion of the gene from the organelle genome. This process still seems to be in progress because some mitochondrial proteins can be encoded by either mitochondrial or nuclear genomes, depending on the species. Of course, changing the location of the gene has profound effects on the mechanisms of protein targeting, which has to be changed from the "export" or "neutral" (nontargeted) mode to the "import" mode (440).

B. Protein import into mitochondria and chloroplasts

This section deals with all of the early stages in protein import into mitochondria and chloroplasts up to and including their arrival in the matrix or stroma.

1. Is protein import co- or posttranslational?

One of the reasons for our extensive knowledge of the mechanisms of protein import into mitochondria and chloroplasts is that import can be studied relatively easily *in vitro* because it is independent of protein synthesis; i.e., it occurs posttranslationally (numerous examples are given in the following sections). Posttranslational protein uptake can also be observed *in vivo* by allowing radiolabeled, nuclear-encoded mitochondrial proteins to accumulate in cells treated with energy uncouplers and then re-energizing the organelles (915,916,1239). However, cytoplasmic pools of mitochondrial protein precursors are rapidly degraded and cannot be detected under normal circumstances (163,639,915).

The N-termini of most mitochondrial or chloroplast proteins includes a signal which routes them to their respective organelles. This signal is presumably able to bind its cognate receptor (Section VI.B.3) as soon as it emerges from the ribosome, which would route the entire nascent chain–ribosome complex to the organelle surface. The surfaces of mitochondria from growing cells were reported by Fellems *et al.* (307) to be covered with polysomes which could be released by EDTA. Mitochondria from resting cells did not have bound ribosomes, although they could bind ribosomes extracted from active cells. Suissa and Schatz (1090) also found that polysomes synthesizing mitochondrial proteins were specifically bound to the organelle surface, although they proposed that such an association was not essential for protein import, and Verner and Schatz (1144) have reported that nascent mitochondrial precursor proteins with attached tRNA have a lower energy requirement for mitochondrial import than do completed polypeptides. Altogether, these results suggest that mitochondrial proteins may be imported cotranslationally *in vivo,* although posttranslational import may also occur. Polysomes were not observed to be associated with the chloroplast envelope (246), suggesting that chloroplasts import proteins entirely posttranslationally.

2. Mitochondrial and chloroplast routing signals

The following sections review the evidence for the existence of organelle routing signals at the N-termini of mitochondrial and chloroplast proteins. Most of these routing signals (mitochondrial prepeptides or chloroplast transit peptides) are 20–40 residues long and are processed during translocation (see Section VI.B.2.f. for exceptions). Processed proteins are therefore smaller than the precursors (246,668), and the site of processing can be identified by comparing the determined N-terminal sequence of the mature protein with that of the precursor (1146).

Some mitochondrial proteins also have identical or almost identical nonmitochondrial counterparts (70,469,652,815). For example, yeast alcohol dehydrogenase isoenzymes are encoded by three different genes, the major difference between the mitochondrial (*ADH3*) and the two cytoplasmic isomers being the presence on the former of an N-terminal extension, the prepeptide (1227). Deletion of the sequence coding for this extension from the *ADH3* gene causes its product to remain in the cytoplasm (652). In contrast, mitochondrial and nonmitochondrial α-isopropylmalate synthetases are encoded by the same nuclear gene, which has two potential translation initiation sites, one for the precursor (mitochondrial) form and one for the cytoplasmic form (70). Alternate mRNA splicing, the third mechanism for producing different proteins from the same transcript, appears to be used in the cases of a yeast tRNA methylase (469) and possibly the chloroplast protein pyruvate orthophosphate dikinase (1012). These and other studies discussed below underscore the importance of the N-terminal extensions in mitochondrial routing.

a. Organelle routing of nonorganelle proteins

A large number of studies have shown that normally cytoplasmic reporter proteins can be routed into organelles if they are synthesized with N-terminal extensions corresponding to authentic prepeptides or transit peptides (see Table II.2). Imported hybrid proteins cofractionated with authentic matrix proteins, were protected against exogenous proteases, and in some cases were processed by matrix prepeptidase and interfered with mitochondrial metabolism (255,282,481,482,493,660).

In many cases, only the extreme N-terminus of the precursor is required to target a cytoplasmic protein into the mitochondrial matrix or the chloroplast stroma (129,282,660). Routing signals which lack the processing site from their C-termini are also active in hybrid proteins, although they are not processed (see below). As few as 12 amino acids of a normally 25-residue-long prepeptide can target dihydrofolate reductase (DHFR) into the mitochondrial matrix (495), and the extreme N-terminal 9 residues of yeast 8-aminolevulase prepeptide are sufficient to target β-galactosidase into the yeast matrix (569). In both of these cases, N-terminal residues of the reporter proteins may contribute to prepeptide activity. Import efficiency may vary with different reporter proteins, probably because some of them adopt tight secondary or tertiary structures which cannot be unfolded to allow the polypeptide to cross the organelle envelope (601,1075,1170). In general, however, these studies show that the routing signal is all that is required to target a protein into an organelle. The use of reporter proteins to assess the effects of mutations affecting routing signal structure is discussed in Section VI.B.2.c.

b. Prepeptide and transit peptide structure and membrane activity

One of the most remarkable features of prepeptides and transit peptides is the almost complete lack of similarity between signals from different precursor proteins. It will be recalled that signal peptides also show a very low level of sequence similarity (Section III.A.1). However, prepeptides and transit peptides, of which some examples are shown in Fig. VI.1,

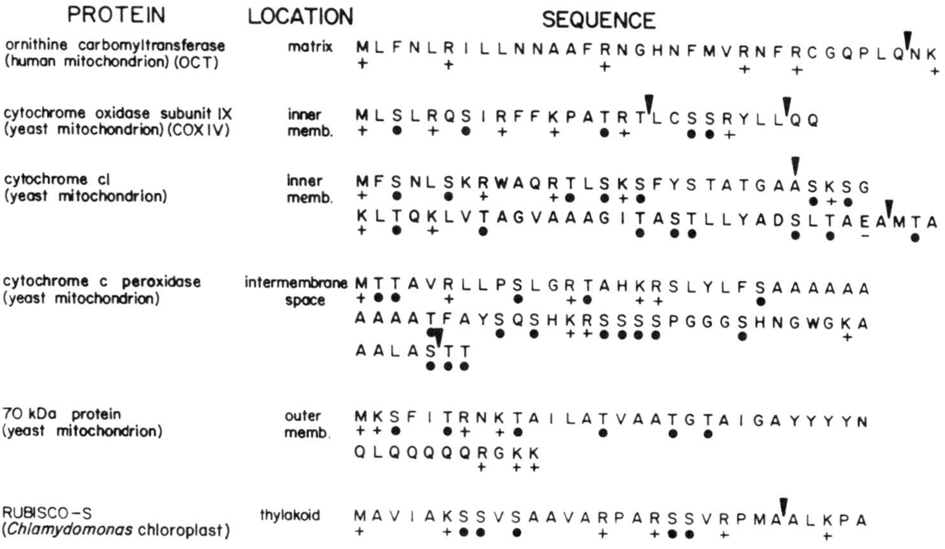

Fig. VI.1. Sequences of typical routing signals located at the extreme N-termini of nuclear-encoded mitochondrial and chloroplast proteins. +, Basic charges; ●, hydroxylated amino acids; ▼, processing cleavage sites.

differ significantly from signal peptides in having basic residues periodically arranged along their length and also in having few hydrophobic residues. Indeed, the presence of basic residues (usually Arg), the absence of acidic residues, and a preponderance of Ser and Thr seem to be the major characteristics of transit peptides and prepeptides (Fig. VI.1). The only known exception to this rule seems to be the imported, human mitochondrial inner membrane, hinge protein subunit of ubiquinol–cytochrome-*c* reductase, which has four acidic residues and one basic residue in its 13-residue prepeptide (804). Studies suggest, however, that even relatively minor changes in the sequence of a signal peptide can endow it with prepeptide activity and that the same sequence can route a protein both into the secretory pathway and into mitochondria (13). Thus, although other studies discussed below suggest that there are only minimal constraints on prepeptide sequences, these sequences must have been optimized for specific mitochondrial targeting.

A detailed analysis of 23 mitochondrial prepeptides revealed the interesting feature that they could all form amphiphilic α helices with the periodically arranged basic amino acids on one face of the helix (439). One prediction made on the basis of this and similar analyses is that prepeptides should have a high affinity for the surfaces of negatively charged phospholipids. This was tested experimentally by Roise *et al.* (948), who found that the prepeptide of yeast cytochrome oxidase subunit 4 (COX4) was water-soluble, bound to and de-energized mitochondria, spontaneously inserted into lipid monolayers, and disrupted unilamellar liposomes. Spectroscopic measurements indicated that the prepeptide had an essentially random structure in water and was α-helical in detergent micelles. A C-terminally truncated form of this prepeptide behaved similarly except that it did not disrupt liposomes. Tamm (1106) also observed high-affinity interaction between COX4 prepeptide and anionic lipids, and Epand *et al.* (289) and Skerjanc *et al.* (1056) obtained similar results with the prepeptide of ornithine carbamyltransferase (OCT).

These results suggest that the amphiphilic, α-helical structure of prepeptides may be important in the initial stages of precursor uptake across the mitochondrial envelope. Genetic studies discussed in the next section are fully compatible with this notion. Prepeptide amphiphilicity may be masked by the mature part of the polypeptide, necessitating partial unfolding or a highly acidic lipid environment to unmask its activity (283). Amphiphilic helices are also present at the N-termini of some yeast cytoplasmic proteins, but they are generally acidic rather than basic (1142).

The sequences of an insufficient number of transit peptides are known to allow such detailed analyses as is possible with prepeptides. Nonetheless, it appears that they too have a similar periodicity of positive charges and that they could form α helices. According to Karlin-Neumann and Tobin (556), transit peptides seem to have well-conserved N-terminal, central, and C-terminal domains separated by two "interblocks" which vary in length and sequence except for the fact that the C-terminal interblock has a conserved Pro residue. However, these conserved domains are present neither in the subsequently identified transit peptide of a chloroplast heat shock protein (1147) nor in the plastocyanin transit peptide (1062).

c. *Genetic studies on routing signal structure and function*

Mutations affecting prepeptide and transit peptide sequences have been used to demonstrate that they are essential for organelle routing and to test the importance of the conserved structural elements discussed in the preceding section. Mitochondrial (483,495,639,876) or chloroplast (601,724,918,1170) proteins remain in the cytoplasm *in vivo* or are not imported *in vitro* when synthesized as truncated forms completely or partially devoid of prepeptides or transit peptides. However, several precursors with only partially truncated routing signals are imported, indicating that their integrity is not necessarily required for activity. One possibility raised by Bedwell *et al.* (66) is that different prepeptide segments act cooperatively to achieve maximum efficiency.

Particular attention has been focused on the significance of the periodically arranged Arg residues in prepeptides. The number of Arg residues can be reduced to a certain extent without affecting prepeptide function (481–483,639), whereas the addition of further Arg residues may destroy prepeptide activity (174,484). Some Arg residues may be more important than others, presumably because they occupy key positions in the amphiphilic structure (483,484). Robinson and Ellis (939) also noted the importance of Arg residues in their studies on the effects of amino acid analogs on chloroplast protein import. Conversely, the introduction of acidic residues into the neutral regions of prepeptides reduces their activity (174,483), indicating the importance of an overall charge balance.

Allison and Schatz (12) used a somewhat different genetic approach to study the essential features of prepeptides. They fused synthetic DNA coding for artificial prepeptides to the 5′ end of a *COX4* gene devoid of its natural 5′ (prepeptide) end and assayed the gene fusions for their ability to complement a *COX4* mutation *in vivo*. They found that a prepeptide composed entirely of periodically arranged Ser, Thr, and Arg residues was active but that prepeptides in which the normal equilibrium of charge, hydrophobicity, and polarity were not respected were not active. Horwich *et al.* (484) used a similar approach with the human *OCT* gene. Their results underscored the importance of Arg residues but suggested that human and yeast mitochondrial prepeptides might be subject to different sequence constraints. Surprisingly, one of the active prepeptides generated by Allison and Schatz was not predicted to form an α helix, although it was membrane-active and therefore probably amphiphilic. This and two other similar prepeptides were subsequently shown to form amphiphilic β sheets, indicating that amphiphilicity, rather than α-helical structure, is the essential prepeptide characteristic (949). Horwich *et al.* (484) also found that the introduction of helix-breaking residues into the OCT prepeptide did not destroy its activity *in vivo*.

The structural constraints on prepeptides are so low that random DNA fragments might have a relatively high probability of coding for pseudoprepeptides. This was tested experimentally by Baker and Schatz (35), who used the *COX4* complementation test described above to screen for DNA coding for pseudoprepeptides in a random *E. coli* DNA bank. A significant proportion (>2.7%) of inserted fragments encoded functional (but noncleavable) prepeptides. The sequences of the peptides encoded by these DNA fragments were similar to those of prepeptides, most notably in that they all carried an overall positive charge contributed entirely by Arg residues except for the initiator Met residue. Rather surprisingly, some DNA fragments from the mouse DHFR gene also encoded pseudoprepeptides when tested in this way. One of these peptides was extensively characterized by Hurt and Schatz (492), who found that it de-energized mitochondria and could direct full-length DHFR as well as COX4 into the mitochondrial matrix. Similarly, Vassarotti *et al.* (1142) found that the removal by mutations of acidic residues from the extreme N-terminus of the mature part of yeast mitochondrial F1 ATPase β subunit allowed this protein to be routed into mitochondria in the complete absence of its natural prepeptide. The proteins encoded by genes with these "cis-complementing mutations" all had positively charged, amphiphilic N-termini.

In conclusion, genetic studies show that prepeptides and transit peptides are essential for routing of mitochondrial and chloroplast proteins and that the constraints on their structure are limited to amphiphilicity and net positive charge. The possibility that specific sequences are required for processing is discussed in Section VI.B.2.e.

d. *Organelle specificity is determined by the routing signal*

Chloroplasts and mitochondria coexist in plant cells. Specific targeting must therefore depend on some feature which distinguishes between nuclear-encoded mitochondrial and chloroplast proteins. Hurt *et al.* (496,497) reported that a transit peptide could direct DHFR or prepeptide-deleted COX4 protein into yeast mitochondria *in vitro* and that a prepeptide could route the nuclear-encoded small subunit of chloroplast ribulose-1,5-biphosphate carboxylase/oxygenase (RUBISCO-S) into mitochondria. These surprising results suggested that there is nothing intrinsic in chloroplast proteins which directs them specifically into chloroplasts. However, the RUBISCO transit peptide was much less efficient than the authentic COX4 prepeptide, and plant mitochondrial protein prepeptides may have additional features which ensure organelle specificity. This suggestion was strengthened by studies by Boutry *et al.* (111), who found that a plant mitochondrial protein prepeptide and the RUBISCO-S transit peptide targeted the reporter protein chloramphenicol acetyltransferase only into the appropriate organelle in transgenic tobacco plants.

e. *Primary processing of prepeptides and transit peptides*

Primary processing of mitochondrial and chloroplast routing signals occurs in the matrix or stroma and is distinct from any secondary processing of prepeptides or transit peptides which subsequently removes the C-terminal sorting signal (Section VI.C.1). Almost all imported proteins are subjected to primary processing, irrespective of their final location (some exceptions are described in the next section). The nuclear-encoded yeast prepeptidase was partially purified by Böhni *et al.* (95), who demonstrated the requirement for Zn^{2+} or Cu^{2+} as cofactor. Copper or zinc chelators prevent precursors' processing (189,218,699) but not their uptake into mitochondria. The site of precursor accumulation was not determined, however, although adding Cu^{2+} to chelator-treated cells at 4°C allowed processing of preaccumulated precursors, suggesting that they had reached the matrix (1244). However, some of the results obtained in a study of a temperature-sensitive mutation in a gene coding for a 51-kDa subunit of yeast prepeptidase (*MAS1*) (1195) suggested that precursors accumulating outside the mitochondrial inner membrane blocked subsequent processing (1211).

The idea that prepeptide processing is required for import is incompatible with numerous observations that genetically engineered precursors lacking the processing site or other sequences necessary for processing are imported into the mitochondrial matrix (see below). Prepeptidase is essential for yeast viability, however (1195), and its absence [in a *MAS1* mutant (1210)] blocks protein import, possibly because precursor accumulation in the mitochondrion prevents further uptake (1195). The *Neurospora* and *Saccharomyces* prepeptidases are composed of two structurally related subunits (430,538a,881a,1218a). The larger subunit has catalytic activity (881a) while the smaller, apparently less abundant and partially membrane-associated subunit (430), may be involved in precursor transport or in presenting prepeptides to the catalytic subunit (881a).

Accumulated prepeptides and transit peptides might be toxic (947). It therefore seems likely that there are one or more peptidases which degrade these routing peptides once they have been cleaved from the mature part of the imported polypeptide.

There is little or no sequence similarity around prepeptide processing sites of mitochondrial protein precursors, even among those from the same species, in which all precursors are likely to be processed by the same enzyme. Furthermore, denatured mitochondrial protein precursors are not substrates for prepeptidase (498), suggesting that a specific secondary structure is required for processing. Almost the entire length of some prepeptides can be deleted without affecting processing (66), whereas other studies show that some precursors are not processed *in vitro* if the Arg residues are replaced (481,482). There is similar confusion regarding the role of the cleavage site itself. For example, Horwich *et al.* (483) found that the integrity of the preOCT cleavage site was essential for processing, whereas others found that sequences within the prepeptide, and not the processing site, were essential (780). The artificial prepeptides constructed by Allison and Schatz (12) were not processed. These results could be explained if prepeptidase recognizes and binds to a site within the prepeptide which is then processed on the C-terminal side. Mutations causing conformational changes, including those affecting the area around the processing site, could prevent the peptidase from recognizing the prepeptide, and some changes to the processing site might prevent cleavage.

Two-step processing of "complex" prepeptides of mitochondrial inner membrane and intramembranous space proteins is now known to be an important part of the sorting process (Section VI.C). Two-step processing of some matrix proteins, notably preOCT, has also been reported, but the significance of the intermediate form is not fully understood (554a). *In vitro* tests with preOCT have recently shown that the intermediate form of rat OCT results from processing by an enzyme which is distinct from prepeptidase. The site of primary processing is eight residues to the N-terminal side of the second processing site. The same two processing steps were observed when preOCT was injected into oocytes. Some mutations affecting the preOCT prepeptide sequence displaced the processing site. Sztul *et al.* (1097) suggest that the intermediate form can only be detected in their system because the rat OCT prepeptide is processed more slowly than other prepeptides. Two distinct prepeptidases have been identified in the matrix of rat liver mitochondria (554a).

The situation with regard to the processing of imported chloroplast proteins is somewhat less clear. Robinson and Ellis (937) partially purified a 180-kDa stromal peptidase which processed transit peptides of both stromal and thylakoid proteins of pea chloroplasts. The enzyme, like prepeptidase, was inhibited by metal chelators. They also observed two-step processing of RUBISCO-S and proposed that both steps were catalyzed by the same transit peptidase (938). Multi-site processing of the wheat thylakoid light-harvesting chlorophyll binding protein was also observed by Lamppa and Abad (609), but it is not clear whether their enzyme is the same as that studied by Robinson and Ellis. A nuclear-encoded peptidase isolated from the alga *Chlamydamonas* processed algal but not pea RUBISCO-S (1011), suggesting that chloroplasts from different plant species may have different transit peptidases with different sequence requirements for processing. Whether the differences in sequence around the transit peptidase cleavage sites of different chloroplast precursors indicate that they are processed by different peptidases (1062) remains to be determined. Genetic analyses of sequence requirements for transit peptide processing are currently limited to studies of the effects of deletions. Reiss *et al.* (918) found that deletion of the C-terminal half of the RUBISCO-S transit peptide abolished processing, whereas the N-terminus and the central region were not required. They also found that changing the Met residue at the Cys-Met processing site did not affect processing by transit peptidase.

f. Imported mitochondrial proteins without cleavable prepeptides

The N-terminus of a number of imported mitochondrial proteins is not proteolytically processed. These proteins fall into two classes.

The first of these includes outer membrane proteins such as the integral (protease-resistant) mitochondrial outer membrane porins of the fungus *Neurospora crassa* (327) and the yeast *S. cerevisiae* (352) and the 70-kDa peripheral outer membrane protein of *S. cerevisiae* (426). Deletion analyses have shown that the N-termini of at least some of these proteins are essential for mitochondrial targeting (426), although as discussed in Section VI.B.3, the initial stage in their insertion into the outer membrane depends on their binding to a different receptor than proteins with prepeptides. This segment is referred to here as the mitochondrial outer membrane (OM) routing signal. OM routing signals may not be processed for two reasons: Either they never reach the mitochondrial matrix or they lack the recognition or processing site(s) for prepeptide peptidase. The OM routing signal of the 70-kDa protein is similar to that of classical prepeptides except that many of the basic residues are lysines rather than arginines. Its extreme 21 N-terminal residues can target the reporter protein β-galactosidase into the matrix, whereas hybrids containing more than 60 residues of the 70-kDa protein are localized to the outer membrane (427). Furthermore, short deletions after residue 11 also cause the 70-kDa protein to be matrix-targeted (426), which confirms the similarity between this routing signal and prepeptides. The N-terminus of the *N. crassa* mitochondrial porin also resembles prepeptides in that it has an amphiphilic-type helical structure, but it also has acidic residues, which are not normally found in prepeptides (571). It seems probable that outer membrane proteins of mitochondria, and presumably of chloroplasts, use some of the early stages of the prepeptide-dependent import route but have additional signals which prevent them from crossing the outer membrane.

The second class of unprocessed, imported mitochondrial proteins includes adenylate kinase from the intramembranous space of chicken liver mitochondria (1171), matrix 3-ketoacyl CoA thiolase (KAT) from rat liver (739), cytochrome *c* oxidase subunit VIIa (COX7a) (1207), inner membrane ADP/ATP carrier (AAC) (868), and apocytochrome *c* (925). COX7a is usual in that the mature forms lack one N-terminal residue and four C-terminal residues. This and other enzymes in this group may have N-terminal or internal signals which allow them to use the same import machinery as proteins with prepeptides (see next section), but others, such as apocytochrome *c*, may use an entirely different pathway. Recent studies with AAC–DHFR hybrids suggest that the routing signal may be located between residues 72 and 111 of AAC in a region containing both a hydrophobic α-helix and an amphiphilic helix. However, sequences located in the C-terminal part of AAC are required for complete import into mitochondria because hybrid proteins lacking this part of AAC remain extractable at high pH (1059).

3. Receptors for imported proteins

How are transit peptides and prepeptides able to distinguish between the surfaces of chloroplasts or mitochondria and all other membrane surfaces facing the cytoplasm? The first evidence for receptors was the observation that normally imported proteins bound to de-energized or chilled (4°C) mitoplasts but remained accessible to proteases (180,874,1238,1245), that they could also bind to isolated outer membranes (874), and that large amounts of one species of precursor, or even isolated prepeptides or transit peptides, could prevent the uptake of other precursors (372). Early studies suggesting that there might be several different types of receptor (927,1240) have been confirmed by the identification of at least four receptors, one each for proteins with prepeptides, AAC, porin, and apocytochrome *c* (371,477,866a). Competition between different classes of imported proteins (739,865,866,868) probably occurs at a postreceptor stage which is common to several import pathways (866a).

Competition experiments such as those of Gillespie *et al.* (372), as well as numerous studies with reporter proteins (Section VI.B.2.a), indicate that the receptors must recognize and bind sequences within the routing signal. This is rather surprising considering their low level of sequence similarity, but it will be recalled that the signal peptide receptor also recognizes peptides which have widely different sequences (Section III.C.2.a). As in the case of signal peptides, it will be difficult to dissect prepeptides and transit peptides to determine which segment, if any, is specifically required for receptor recognition. Hurt *et al.* (498) showed that removal of the first four residues of yeast COX4 precursor blocked uptake and *in vitro* processing by prepeptide peptidase but not receptor recognition. This suggests that in this case the receptor recognition region of the peptide does not include the extreme N-terminus. However, the fact that proteins with totally artificial prepeptides are imported (and are therefore recognized by the receptor) (13) throws serious doubts on the possibility that any particular sequence is specifically required for receptor recognition. Like the signal peptide receptor of the RER, prepeptide and transit peptide receptors may recognize a specific secondary structure rather than a sequence of amino acids.

The mitochondrial prepeptide receptor is firmly anchored in the outer membrane, from which it cannot be released by high salt concentrations (24). It is sensitive to trypsin (353) and can be extracted in active form and reconstituted into liposomes (927). The chloroplast transit peptide receptor is also a trypsin-sensitive integral membrane protein (180). Antibodies raised against pooled yeast mitochondrial outer membrane proteins also reduce protein uptake to a certain extent, particularly if the mitochondria are pretreated with low doses of trypsin, which may unmask the receptor and allow antibodies to bind (801). The major 45-kDa protein recognized by these antibodies could be the prepeptide receptor or a protein involved in the second import step (866a). A different approach was used by Gillespie (371) to identify the rat mitochondrial receptor for the OCT prepeptide. She found that an abundant 30-kDa protein could be specifically cross-linked to this prepeptide and that the binding activity had all of the characteristics described above for authentic prepeptide receptor. The prepeptide receptor is not the outer membrane porin, which has the same molecular mass and is equally abundant but is highly trypsin-resistant. The transit peptide receptor of pea chloroplasts has also been identified as a major 30-kDa protein by its reactivity with anti-idiotype antibodies raised against antibodies to a chemically synthesized RUBISCO-S prepeptide and by cross-linking to a synthetic transit peptide (547,838). The abundance of this protein in chloroplasts agrees well with an earlier estimate of the number of receptors (~3000 per chloroplast) (874).

Evidence discussed in Section VI.B.6 shows that import of mitochondrial proteins occurs at sites where the inner and outer membranes are juxtaposed or even fused. Transit peptide receptors were shown by Pain *et al.* (838) to be located at these sites in chloroplasts. *Mitoplasts* (mitochondria from which the outer membrane has been stripped off) are still able to import proteins (807). Although this may indicate that mitoplasts retain outer membrane over the membrane fusion sites (1024), Ohba and Schatz (801,802) also found that protein uptake into mitoplasts was not affected by the anti-outer membrane protein antibodies which prevented uptake in whole mitochondria. The implications of these results are discussed in Section VI.B.6.

This section has dealt only with results pertaining to the high-affinity prepeptide or transit peptide receptor. Initial contact between these routing signals and the organelles may involve low-affinity binding to lipids (1056) or other surface components. These interactions are unlikely to be organelle-specific, since the same components probably exist in other membranes.

4. Energy requirements for protein import

One of the earliest observations made on protein import by mitochondria was that uncouplers of the transmembrane energy potential blocked import *in vivo* (1239) and *in vitro* (353,1009). The transmembrane energy potential across the mitochondrial envelope is composed of a chemical (electrical) gradient ($\Delta\psi$) and a pH gradient (ΔpH). Pfanner and Neupert (867) reported that the chemical gradient (ΔK^+) rather than the ΔpH is utilized for protein import. This fits well with theoretical arguments showing that the electrical gradient should make the transfer of basic charges in the prepeptide energetically favorable and provide a net energy gain for every basic residue imported, provided the initial activation energy was not too high (439).

The chemical gradient was originally assumed to be the only source of energy required for mitochondrial protein import. Studies from three different groups show that this is not the case, however. Eilers *et al.* (274) developed a simple *in vitro* test for protein import using purified prepeptide–reporter protein hybrid polypeptides and isolated yeast mitochondria. Import of the hybrid protein in this system requires both $\Delta\psi$ and ATP or GTP. The ATP could not be replaced by a nonhydrolyzable analog, indicating that the phosphate group is transferred to another ligand or is consumed to release energy essential for protein uptake, for the release of cytosolic "competence factors," or for protein unfolding (see below). Phosphodiester energy was also shown by Chen and Douglas (165) to be required for uptake of the yeast F_1 ATPase β subunit. Even more surprising is the report by Pfanner *et al.* (869) that nucleoside triphosphate (NTP) is required at two distinct stages in the uptake process. The significance of this observation and the stages at which the various energy sources are consumed are discussed below. It is also significant that the insertion of porin into the outer membrane requires NTP, but is completely independent of $\Delta\psi$ (571).

The initial demonstration that protein import into mitochondria was apparently ATP-independent was at variance with observations that protein import into chloroplasts utilizes hydrolyzable ATP as the sole source of energy (180,321). According to Flügge and Hinz (321), NTP is apparently used at an early stage in chloroplast protein import, possibly at the face of the outer membrane, since import can be prevented with exogenous phosphatase inhibitors. Pain and Blobel (837) obtained rather different results by using an exogenous ATP-consuming system (glucose and hexokinase-apyrase), which should have destroyed all ATP at the organelle surface, but which did not affect RUBISCO-S uptake *in vitro*. These studies do not rule out the possible involvement of ATP in the intermembranous space or the transfer of ATP through the membrane at contact sites where proteins are imported (Sections VI.B.6 and 7). ATP consumption during protein import into chloroplasts may be explained in part by phosphorylation. A 25-kDa autophosphorylated outer membrane protein has been shown to phosphorylate at least two other imported proteins when they are bound to the outer membrane (1070).

5. Protein unfolding and cytosolic factors

Studies with gene fusions suggested that the mature part of imported mitochondrial and chloroplast proteins did not play a direct role in their uptake. Other studies showed, however, that their import efficiency and energy requirements *in vitro* or *in vivo* varied significantly depending on the nature of the reporter being transported and on the amount of natural imported protein present in the hybrid (601,1075,1170). Other examples of "conformational effects" on protein import include the observation by Ness and Weiss (771) that derivatives of *Neurospora* matrix precarbonyl phosphate synthase subunit A which are truncated at their C-termini compete inefficiently with full-length precursors for import into mitochondria *in vivo,* and the fact that water-soluble porin extracted from the outer membrane can reinsert spontaneously, whereas porin produced *in vitro* requires NTP for insertion (865).

A detailed investigation of the role of the mature sequence in protein import indicated that F_0-ATPase subunit 9 (Su9)–DHFR hybrids bound to the surface of *Neurospora* mitochondria in the absence of $\Delta\psi$ only when the hybrid contained the mature part of Su9 (870). This could be either because the shorter hybrids adopt a conformation in which the Su9 prepeptide is not exposed on the surface or because sequences in the mature part of Su9 are required for binding in the absence of $\Delta\psi$. The latter explanation is favored by Pfanner *et al.* (870), who point out that some other imported proteins exhibit only very low binding in the absence of $\Delta\psi$ and that all of these proteins, like the shorter Su9–DHFR hybrids, lack hydrophobic segments of significant length, whereas proteins which show significant $\Delta\psi$-independent binding have such a segment. These segments are proposed to assist specific, high-affinity binding to the prepeptide receptor or to another membrane component in the absence of $\Delta\psi$.

These results could explain one source of the "conformational effect" on protein import as well as at least part of the requirement for $\Delta\psi$ in protein uptake. Another form of "conformational effect" on protein import was discovered by Eilers and Schatz (273), who showed that a prepeptide–DHFR hybrid protein was not imported into yeast mitochondria *in vitro* if it was complexed with methotrexate, an analog of the natural

DHFR substrate folate. Similar results were obtained with a prepeptide–copper metallathionin hybrid, which was not imported in the presence of Cu^{2+} (164), and with 5-enopyruvylshikimate-3-phosphate synthase, which was inefficiently imported into chloroplasts *in vitro* when complexed with shikimate-3-phosphate and the herbicide glyphosphate (170). These results indicate that precursor proteins must be in an unfolded or partially folded state to be import-competent and that proteins can be "locked" into import-incompetent states corresponding to irreversible conformations such as those induced by methotrexate, Cu^{2+}, or disulfide bridges (1145a).

These observations raise questions concerning the extent of secondary structure formation which can be tolerated without destroying import competence (or whether loosely folded proteins can be imported more efficiently) and whether import competence can be maintained or restored prior to uptake. The experiments on outer membrane porin described above could be explained if extracted and partially denatured porin was in an import (insertion)-competent state whereas that synthesized *in vitro* needed NTP to attain import competence or was bound to another protein (competence factor) whose dissociation required ATP (865,871). The experiments of Verner and Schatz (1144), which indicated that nascent chains of a prepeptide–DHFR hybrid with attached tRNA were more sensitive to protease than completed polypeptide chains and were imported in the absence of ATP, also indicated that NTP might be required to unfold polypeptide chains to make them import-competent. Natural precursors may thus adopt secondary structures prior to import, as was suggested by Chen and Douglas (163), who found that the precursor of F_1-ATPase β subunit of yeast mitochondria formed tetramers which were incompetent for import *in vitro,* but which could be converted into an import-competent (monomeric) state after the addition of ATP and a component of reticulocyte cell lysates. Mutant subunits which could not tetramerize remained import-competent in the absence of ATP (166). Pfanner *et al.* (869) also noted that the level of NTP required for the import of natural mitochondrial proteins into isolated *Neurospora* mitochondria varied according to the protein under study, which suggests that NTP consumption may vary according to the degree to which the precursors must be unfolded. They also found evidence that NTP was required for a subsequent stage in protein import after the initial NTP-dependent interaction (see Section VI.B.7).

Further evidence for the importance of protein unfolding comes from work on the import of preCOX4–DHFR hybrids. A urea-denatured hybrid was imported in the total absence of ATP and at a lower temperature than was normally possible; also, methotrexate no longer inhibited import once the hybrid had undergone $\Delta\psi$-dependent insertion into the mitochondrial envelope (275). Furthermore, mutations which destabilized COX4–DHFR by reducing secondary structure formation in the DHFR part of the hybrid also reduced the temperature at which import occurred (1145). Thus, efficiency of uptake seems to be inversely correlated to the degree of secondary structure. ATP-dependent unfolding of the pre-COX4–DHFR hybrid occurred when it was mixed with acidic lipid vesicles, suggesting that unfolding occurs spontaneously (i.e., without additional cytosolic or membrane proteins) upon insertion of the prepeptide into a lipid environment (283).

The requirement for cytosolic components noted above in the study by Chen and Douglas (163) is not unique; several earlier articles also reported that *in vitro* uptake of mitochondrial precursors did not occur in the absence of soluble cytoplasmic components (24,725). Ohta and Schatz (803) found that the cytosolic components required for *in vitro* uptake into yeast mitochondria could be supplied either by yeast or by reticulocyte lysates and proposed that they were necessary to confer import competence upon the precursor. The same laboratory also noted that uptake of purified COX4 prepeptide–DHFR hybrid did not require cytosolic factors, perhaps because competence factors do not bind to DHFR, although NTP was required (274).

One possibility raised by Chen and Douglas (163) is that the cytosolic factors bind to precursors and change their conformation to permit productive interaction with the mitochondrial outer membrane. This conformational change might, for example, ensure that the prepeptide is exposed on the surface of the polypeptide. The prepeptide may be predisposed to surface exposition in some precursors, like the COX4–DHFR hybrid, which would therefore not require cytosolic components for uptake. It has also been proposed that cytosolic factors might prevent the prepeptide from crossing the outer membrane through the porins (372).

Yeast cytosolic factors required for mitochondrial protein import have molecular masses of about 40 kDa and 200 kDa, as measured on size fractionation columns, and are protease-sensitive (759a,803,814). Ono and Tuboi (814) have recently shown that the 200-kDa competence factor is probably a complex of several proteins which binds to prepeptides. Other studies have shown that one of the heat shock family of proteins, HSP70, is involved in protein import into yeast mitochondria. Depletion of HSP70 levels in *S. cerevisiae* leads to the accumulation of at least one mitochondrial protein precursor (234). As in the case of secretory proteins, HSP70-type proteins may prevent tight folding of import precursors in the cytosol. By analogy with other systems (857), the release of HSP70 from the precursor complex may require ATP hydrolysis. Thus, as suggested by Pfanner *et al.* (869), ATP may be required at two stages in the import process: first for the release of HSP70 from the precursor (probably after it had bound to its receptor), and second for some additional stage in import. It is not clear whether HSP70 is one of the components of the 200-kDa competence factor complex, but if it is, then it presumably binds to the prepeptide. One possibility would be for it to bind to the hydrophobic face of the amphiphilic prepeptide, leaving the hydrophilic face free to interact with the prepeptide receptor.

6. Import occurs at sites of inner–outer membrane contact

Points of contact between the mitochondrial or chloroplast inner and outer membranes are clearly visible in plasmolyzed organelles examined by electron microscopy. The exact nature of these contact points remains obscure. Van Venetië and Vertleij (1141) proposed two models involving lipids in nonbilayer configuration which could explain their formation. Both models predict at least some degree of mixing of lipids between the two membranes, but the polar lipid contents of chloroplast and mitochondrial inner and outer membranes are significantly different (253), implying that mixing does not occur. Protein import into mitochondria has often been proposed to occur through inner–outer membrane contact sites (307,1245), but it is not clear whether import sites are the same as those observed previously or whether they are induced during the translocation of a prepeptide or transit peptide through the mitoplast envelope.

The most direct evidence for the involvement of contact sites in protein import is the observation that the pea chloroplast receptor for preRUBISCO-S is located over the contact sites (838). This supports several observations indicating that mitochondria import proteins through membrane contact sites. These studies stem from the observation that it is possible to "freeze" protein import by chilling *in vitro* uptake assays to 11°C. The import intermediates were cleaved by matrix prepeptidase and yet remained accessible to exogenous proteases. Uptake continued when the mitochondria were warmed to 25°C and, significantly, no additional source of energy was required to complete uptake (1008). Mobility through the membrane may have been negligible because the lipids were not in a "fluid" state at 11°C, but this does not necessarily mean that the precursor was in direct contact with envelope lipids. The envelope-spanning segment may have included the hydrophobic segment immediately C-terminal to the prepeptide in the F_1-ATPase β subunit and cytochrome c_1 precursors tested (see preceding section).

Ohba and Schatz (802) and Schwaiger *et al.* (1024) both found that mitoplasts (mitochondria stripped of their outer membrane) could transport mitochondrial precursors, but they reported differing requirements for residual, protease-sensitive outer membrane components (receptors?) still present at the membrane contact sites. Schwaiger *et al.* (1024) also found that a detergent-extractable component was required for protein import into mitoplasts and noted that import intermediates obtained at 11°C remained associated with the mitoplasts, probably at the contact sites, when the outer membrane was stripped off. Preliminary results suggest that it might be possible to separate the two membranes and the contact sites containing the frozen intermediates. This could be a useful approach to identify any unique envelope components such as proteins or nonbilayer-forming lipids in contact sites. The noncatalytic subunit of prepeptidase could be associated with contact zones (see Section VI.B.2.e).

7. Summary and model for the early stages in organelle protein import

Most of the observations discussed above provide a coherent view of the early stages in protein uptake by mitochondria and chloroplasts (Fig. VI.2); some exceptions are discussed later in this section. Proteins destined for import into the matrix or for insertion into the outer membrane bind to the same receptor, which is specifically located at contact sites or which migrates there once the protein has bound. The receptor recognizes the prepeptide or transit peptide or the functional equivalent sequence in unprocessed proteins (stage 1). Low-affinity interactions with polar membrane lipid head groups may precede this stage. ATP hydrolysis may be required at this stage to unfold proteins and make them competent for import, to expose the prepeptide or transit peptide on the protein surface, or, more likely, to release cytosolic competence factor(s) for the import precursor (stages 1 and 2).

The next stage in the incorporation of outer membrane proteins is independent of further energy imput and probably involves the spontaneous insertion of transmembrane segments into the membrane (not shown in Fig. VI.2, see also Ref. 780a) because uptake is independent of $\Delta\psi$ (see below) and because porin inserts spontaneously into isolated yeast outer membrane vesicles (352). Ono and Tuboi (813) could not obtain spontaneous porin insertion into rat liver outer membrane vesicles, however, and proposed that porin inserted into the inner membrane or contact zones prior to its final assembly in the outer membrane. Porin may thus share most of its import pathway with proteins destined to the matrix and inner membranes (866a). In mitochondria, $\Delta\psi$ is utilized to catalyze the vectorial insertion and transport of the prepeptide across the envelope and into the matrix (stage 2). The prepeptide is proposed to penetrate headfirst into

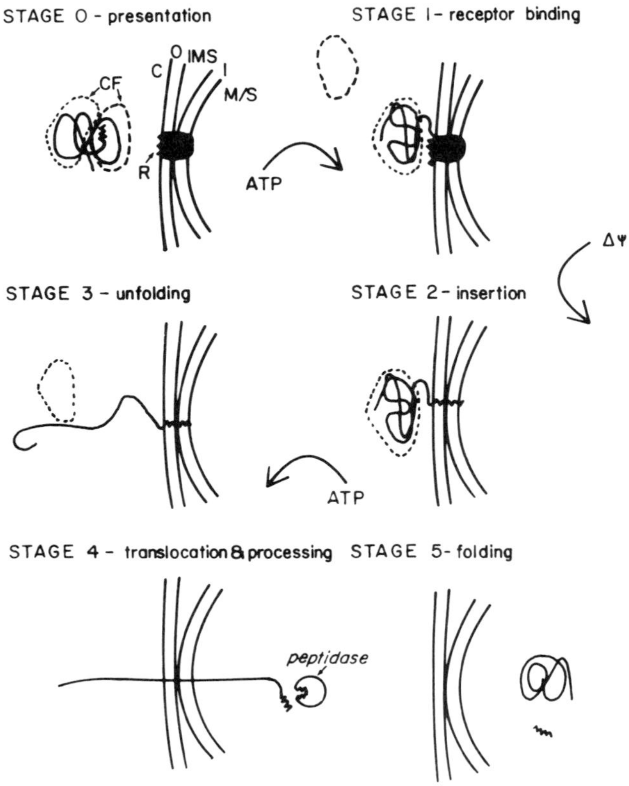

Fig. VI.2. Speculative model for the import of a matrix or stromal protein across mitochondrial/chloroplast envelopes. (Stage 0) Presentation of the routing signal (〰) to the receptor on the surface of the organelle. Competence factors (CF) are bound to the mature part of the import precursor and to the routing signal. (Stage 1) Binding of routing signal to receptor and release of one of the competence factors; (Stage 2) insertion of routing signal across contact sites in the organelle envelope; (Stage 3) unfolding of the mature part of the polypeptide with concomitant release of second competence factor; (Stage 4) translocation and proteolytic processing of routing signal by prepeptidase or transit peptidase; (Stage 5) folding. C, Cytoplasm; O, outer membrane; I, inner membrane; IMS, intermembrane space; R, receptor; M/S, matrix or stroma.

the envelope. This is in contrast to signal peptides, which are thought to insert as loops (Section III.C.1). Insertion of the prepeptide into the membrane may stabilize initially weak prepeptide–receptor interactions (868). Isolated prepeptide of rat OCT is not translocated across lipid bilayers, even when a transmembrane voltage potential is applied, despite the fact that it binds to and probably inserts into these membranes (1056). This may indicate that prepeptides do not move through contact sites by electrophoresis. Other membrane components besides lipids may be involved, and import pathways may converge at this stage (866a).

Additional energy may be required to release the prepeptide from its receptor; this may be supplied either by ATP or by the $\Delta\psi$. Competence factors may remain bound even at this intermediate stage to prevent the precursor from folding at the organelle surface; their release (stage 3) might require ATP and probably occurs simultaneously with translocation (stage 4). In chloroplasts, release of the transit peptide from its receptor and its penetration into the matrix is entirely catalyzed by ATP-dependent reactions. Phosphorylation of the prepeptide or transit peptide or the mature sequence may be involved in either of these two stages (1070). In the next stages (4 and 5), the prepeptide or transit peptide is processed by the specific matrix/stromal peptidase, and import is completed as the entire polypeptide chain is translocated. This stage is independent of $\Delta\psi$ (1024) but requires ATP, probably to complete the unfolding reaction and possibly also to release the bound cytosolic competence factors (869). The folding of imported proteins in the matrix or stroma may provide part of the energy required to pull the polypeptide through the organelle envelope (i.e., stages 4 and 5 occur simultaneously).

As discussed in the next section, some mitochondrial inner membrane proteins are imported into the matrix and then inserted into the inner membrane from the inner face. All of these proteins seem to be made as precursors with prepeptides. This is not the case for the ADP/ATP carrier (AAC) of the *Neurospora* inner membrane, which probably has an internal sequence which binds to a unique receptor (866a). It is subsequently inserted into the outer membrane, where it is fully protected from external proteases, and thence, in a $\Delta\psi$-dependent step, spans the contact site as a loop; the N- and C-terminal domains remain in the outer membrane. At low temperatures, AAC spans the contact sites as a loop with the N- and C-terminal domains exposed on the outside and accessible to protease. The final stage of complete insertion into the inner membrane and dimerization occurs in the absence of $\Delta\psi$ (868).

It is not known whether additional proteins besides competence factor(s), the receptor, and the chloroplast protein kinase are involved in protein import into organelles. It should be possible to reconstitute at least some aspects of the import pathway using purified components, and further molecular genetic studies should allow the role of proteins in the organelle envelope to be defined more clearly. One of the current difficulties is the acquisition of large amounts of natural import precursor for *in vitro* reconstitution studies. Two proteins, porin and apocytochrome *c*, can be obtained in large amounts in water-soluble form by extracting them from mitochondria. Porin is not ideal for two reasons, however: it is partially denatured during the extraction procedure, so its import requirements are not the same as those of nascent polypeptides, and it is not imported into the matrix. Apocytochrome *c* can be obtained from cytochrome *c* haloenzyme by removing the heme group. Its import pathway is apparently different from that of almost all other mitochondrial proteins (782); it binds to a different receptor and its uptake is $\Delta\psi$-independent. Apocytochrome *c* can spontaneously penetrate and cross membranes containing high levels of negatively charged phospholipids (which are common in mitochondria). Whether this enzyme is unique in this property remains to be determined, but some facets of the import model proposed by Rietveld *et al.* (925), particularly the role of lipids, may be applicable to other imported proteins.

One other important aspect of organelle biogenesis which seems to have been largely overlooked is how lipids are synthesized and how they are transported to the organelle. The majority of them are presumably made in the endoplasmic reticulum and are transported to organelles by lipid-carrier proteins. Nothing appears to be known about how these lipids reach internal organelle membranes.

C. Sorting of imported and endogenous mitochondrial and chloroplast proteins

Mitochondria and chloroplasts contain many inner membrane and intermembranous space proteins which are processed by prepeptidase or transit peptidase. This implies that they penetrate at least partially into the matrix or stromal compartments in which these peptidases are located. Furthermore, many imported chloroplast proteins continue their journey into and through the thylakoid membranes. This section will consider the possibility that both of these aspects of protein import involve sorting mechanisms which operate once the protein has crossed the organelle envelope. As we shall see, these studies indicate that sorting may occur by the same pathway as that used to target proteins encoded by the endogeneous genome.

1. Sorting or re-export of imported mitochondrial proteins

There is no doubt that a number of mitochondrial inner membrane and intermembranous space proteins are processed by matrix prepeptidase (354,424,653,917), but there is some dispute as to whether they penetrate fully into the matrix. This difference of opinion stems largely from the observation that imported intermembranous space proteins in this class have unusually long, "complex" prepeptides which are processed in two stages, the first by matrix prepeptidase and the second by an enzyme located in the intermembranous space (see below). The second part of the prepeptide, which is exposed after cleavage by prepeptidase, is usually a long, uninterrupted sequence of hydrophobic and neutral amino acids (Fig. VI.1). This has been proposed to function as a stop transfer signal which anchors the protein in the inner membrane, with only the N-terminus of the prepeptide penetrating into the matrix (408,653,800). This mechanism would be fundamentally similar to that by which stop transfer signals in secretory proteins anchor them in the membrane of the rough endoplasmic reticulum or in the bacterial cytoplasmic membrane (Section III.F.1). The observation that DHFR is targeted to the intermembranous space when fused to the complex prepeptide of yeast cytochrome c_1, an inner membrane protein which faces the intermembranous space (635,653,654), could be explained by this model, but other explanations are possible (see below and Ref. 780a). The major drawback to the model is that proteins are now thought to be imported across the envelope at membrane contact sites, so the C-terminus of proteins with complex prepeptides would never be exposed in the intermembranous space unless the entire polypeptide diffused away from contact sites before or while it was being translocated through the outer membrane. Another feature which remains to be tested is whether stop transfer-like hydrophobic segments would ever come into direct contact with lipids during translocation through the mitochondrial envelope.

The alternative proposal, that the entire protein enters the matrix and is then re-exported, is attractive because the second stage might be carried out by the ancestral protein export pathway which existed in the mitochondrion when it had a far more complex genome and synthesized many more of its own proteins (424). The Fe–S protein subunit of *Neurospora* ubiquinol–cytochrome-*c* reductase studied by Hartl *et al.* (424) may be unusual in that it has an unusually short complex prepeptide, the second segment of which contains only eight amino acids, and appears to be processed entirely in the matrix (425). It is also only weakly anchored in the inner membrane (424). Nevertheless, Hartl *et al.* (424,425) were able to detect the precursor forms of Fe–S protein and of cytochrome b_2 and cytochrome c_1 (both of which have longer, complex prepeptides) in the matrix. However, van Loon and Schatz (651) were unable to detect cytochrome c_1 prepeptide–DHFR intermediates in the matrix but found instead that the intermediates remained membrane-associated. The reason for this discrepancy remains unclear, but the evidence now seems to be strongly in favor of successive translocation into and out of the matrix for at least some imported intermembranous space and inner membrane proteins with complex prepeptides.

The peptidase responsible for secondary processing of most complex prepeptides is located in the intermembranous space (425). It has not been characterized in any detail. A temperature-sensitive mutation blocking secondary processing of imported yeast cytochrome b_2 (*pet2858*) has been proposed to affect the peptidase itself or an activator. The same mutation affects the processing of cytochrome oxidase subunit II (COX2), which is encoded by a mitochondrial gene (887). Sequence comparisons predict that processing of these proteins occurs between Asn and an acidic residue (Asp or Glu). This again suggests that sorting of imported proteins into or across the inner membrane occurs by an ancestral protein export pathway. However, processing of another sorted protein, cytochrome-*c* peroxidase, is not affected by the *pet2858* mutation.

Sorted proteins may be modified as well as processed in the intermembranous space. The most common modification involves the addition of a heme group to cytochromes of the inner membrane (354,800,916). Apocytochrome *c*, which is imported by a different pathway, is also heme-modified in the intermembranous space. The enzyme which attaches this heme group, cytochrome-*c*–heme lyase, is located in the intermembranous space, and requires a heat-stable, protease-insensitive low-molecular-weight component (782).

2. Targeting of endogenous mitochondrial proteins

Little appears to be known about the targeting of endogenous mitochondrial proteins, other than that at least some of them are synthesized as precursors with an N-terminal "matrix export signal" (887). One of the major difficulties in studying the targeting of this class of proteins is that the mitochondrial genetic code is slightly different from the universal genetic code. Gearing and Nagley (357) changed the "unusual" codons in the gene for ATPase subunit 8 and fused it to DNA coding for the prepeptide of ATPase subunit 9. The normally endogenous polypeptide could now be imported from outside the cell, but it could not rescue a mutation in the ATPase 8 structural gene *in vivo*, indicating that it was not correctly inserted into the inner membrane. The prepeptide of COX4 was unable to catalyze import of ATPase 8 into isolated mitochondria, possibly because it is shorter and therefore less "powerful" than the ATPase 9 prepeptide (357).

3. Targeting and sorting of chloroplast proteins

Relatively little seems to be known about the sorting of imported proteins to the chloroplast inner membrane and intermembranous space. However, studies have centered on another interesting aspect of protein sorting in these organelles, namely, their insertion into and across the thylakoid membrane.

Many thylakoid membrane proteins, including cytochrome *f* of pea chloroplasts, are synthesized as precursors with a long N-terminal segment which is cleaved off during targeting (1191). Some nuclear-encoded thylakoid proteins undergo two-step processing during import and sorting, but studies by Chia and Arntlein (168) on a chloroplast genome mutation affecting the second processing step suggest that it is not required for correct localization. Do these sequences represent a true thylakoid targeting signal, or are thylakoid proteins routed first to the inner membrane and then selectively sorted into the thylakoid? The latter idea came from the observation that the inner chloroplast membrane has numerous invaginations which seem to pinch off into the stroma as vesicles. These vesicles could migrate to and fuse with the thylakoid membrane and thereby deliver their load of soluble (thylakoid lumen) and membrane proteins to the thylakoid (144,253,1012). Such a model supposes that proteins destined to the thylakoid are segregated into specific domains of the inner membrane prior to transport to the thylakoid and that this might involve some form of signal. This signal need not be present on all thylakoid proteins, since both exogenous and endogenous proteins could interact to form complexes in the inner membrane prior to being transported to the thylakoid. The lipid contents of the thylakoid and inner membranes are markedly different; patching of thylakoid lipids in the inner membrane might therefore provide a selective focus for the assembly of thylakoid proteins prior to their delivery to the thylakoid.

Gene fusion studies indicate that the information responsible for the targeting of nuclear-encoded thylakoid proteins resides in the transit peptide. Smeekens *et al.* (1063) used molecular genetic techniques to exchange the transit peptides of the stromal protein ferredoxin and the soluble thylakoid lumen protein plastocyanin and found that the ferredoxin transit peptide directed plastocyanin away from its normal destination. The complex plastocyanin transit peptide has two domains and is processed in two stages, the first by stromal transit peptidase and the second by a thylakoid membrane peptidase (570). The C-terminal part of its transit peptide may therefore contain the thylakoid targeting signal. However, the plastocyanin transit peptide–ferredoxin hybrid was not directed into the thylakoid. Instead, a minor amount of this hybrid was found bound to the outside of the thylakoid membrane and a large proportion remained in the stroma. No secondary processing of this hybrid was detected. Failure of the plastocyanin transit peptide to direct ferredoxin into the thylakoid lumen may have been due to the inability of ferredoxin to cross the thylakoid membrane (which assumes direct sorting and transport across this membrane), to its failure to interact with other thylakoid proteins as they assembled in the inner membrane, or because it interacted with other proteins in the stroma. Thus, interesting as these studies are, they fail to give clues as to the sorting pathway for nuclear-encoded thylakoid proteins.

An alternative pathway for the sorting of thylakoid proteins involves their temporary accumulation in the stroma in a complex with a specific binding protein. Endogeneous (L) and exogeneous (S) subunits of RUBISCO have been shown to associate with the same 57-kDa stromal protein (BP, or chaperonin) prior to their assembly in thylakoids (45,356,445,762). This situation is very reminiscent of the action of the endoplasmic reticulum protein BiP, which binds to newly imported secretory proteins until they associate into oligomers and are transported to the Golgi (Section V.B.3). Further studies on this chloroplast competence factor should reveal whether it is specific for RUBISCO subunits. Antibodies against chaperonin are reported to react with a mitochondrial protein (701), and chaperonin is related to the bacterial heat shock protein GRO E (Section III.C.3), raising the intriguing possibility that it is one of a family of proteins which facilitate protein folding and oligomerization.

The possibility that imported proteins can be targeted directly into the thylakoid was tested in an *in vitro* assay system by Cline (181). He found that the precursor of light-harvesting chlorophyll a/b protein (LHCP) was preferentially inserted into the thylakoid membrane, not into the envelope membrane, when whole or disrupted chloroplasts were used. The protein was inserted into the thylakoid membrane in its correct configuration, and the process was ATP-dependent. Insertion into purified thylakoids was also demonstrated, although in this case, import required additional stromal factors. Unlike the case with plastocyananin, no processing intermediate was detected; in fact, preLHCP was the predominant species integrating into the thylakoid membranes when chloroplast homogenates were used. Low-level accumulation of an LHCP processing intermediate has been observed *in vitro* (609), and multiple forms of LHCP have been detected *in vivo* (1013), but their significance remains to be determined. These results argue strongly in favor of direct targeting to the thylakoid and against secondary sorting via an inner membrane intermediate.

The 32-kDa herbicide-binding protein of thylakoid photosystem II is subject to proteolytic processing in the thylakoid lamellae (448). This protein has an N-terminal hydrophobic segment, which may be important for targeting to the thylakoid membrane. It is also palmitoylated, as are three other thylakoid proteins. The protein is made in the stroma on polysomes which appear to be attached to the surface of unstacked thylakoid lamellae. Processing and modification appear to occur in these lamellae, and the protein then migrates to stacked thylakoids, its final destination. The role of palmitoylation in this process is unclear, but the fatty acids could act as a membrane anchor or as the signal which targets the protein to the stacked lamellae (696). Phosphorylation has also been shown to have a strong influence on thylakoid protein location (73). Chlorophyll-containing protein diffuses laterally when phosphorylated by a protein kinase. The negative charge which is thus introduced may disrupt the critical membrane adhesion properties of the protein necessary to form the granulated, stacked lamellae, consequently causing migration and formation of unstacked lamellae and regulating the amount of light delivered to the two photocenters (1072).

D. Import of proteins into peroxisomes

Peroxisomes (also called glyoxysomes or microbodies) are almost ubiquitous in eukaryotic cells, yet their biogenesis has received much less attention than that of mitochondria or chloroplasts. They perform a variety of metabolic functions including respiration, alcohol oxidation, and glyoxalate cycle reactions. Peroxisome proliferation, which occurs by division, is inducible; for example, peroxisomes are larger and more abundant in methylotrophic yeasts such as *Candida* sp. during growth on methanol. Almost all microbodies are bordered by a single membrane; the only exception seems to be the hydrogenosomes of trichomonads, which have two membranes and are probably functionally quite different from other microbodies. Peroxisomes do not have endogenous DNA; their biogenesis therefore depends entirely on the import of proteins from the cytosol [reviewed in (614)]. The peroxisomal membrane is permeable to small solutes, suggesting the presence of a pore, but it is unlikely that this is involved in any way in protein import (1140).

Although it was originally proposed that peroxisome biogenesis occurred by budding of vesicles from the RER (614,1239), the overwhelming evidence is now in favor of posttranslational import of mature-sized polypeptides. The most plausible working model, based on studies of mitochondria and chloroplasts, is that peroxisomal proteins contain sequences which route them specifically to the peroxisome surface, where they recognize and bind to a surface receptor (protein) and are subsequently imported. The import process seems to be driven by the energy potential because uncouplers of oxidative phosphorylation block import *in vivo* and can even cause recently imported and octamerized alcohol oxidase to monomerize, suggesting that both import and tertiary structure formation require energy (69). Small *et al.* (1060) claimed, however, that import of acyl-CoA oxidase into rat liver peroxisomes occurred in the absence of an additional energy source *in vitro*. However, the assay mixture used in these experiments contained ATP included for efficient translation of the acyl-CoA oxidase mRNA, and Imanaka *et al.* (514) subsequently demonstrated that ATP hydrolysis, and not a transmembrane energy potential, was required for import of this protein. The uncouplers used by Bellion and Goodman (69) may have reduced cellular ATP levels to such an extent that peroxisomal protein import was arrested, rather than having a direct effect on membrane energy potential-driven uptake. The presence of a permanently open pore in the peroxisome membrane (1140) would be incompatible with the generation of a transmembrane potential.

It is not clear whether ATP is used to drive protein import or to unfold import-incompetent proteins. At least one imported peroxisomal protein, fructose biphosphate aldolase of *Trypanosoma brucei,* is known to fold in one configuration in the peroxisome and another in the cytosol (178), and furthermore, tetramerization and octamerization of some peroxisomal proteins probably only occur in the peroxisome. Thus, import is accompanied by conformational changes and refolding, which may be important for the retention of proteins in the lumen of the peroxisome. Some peroxisomal proteins are also modified (e.g., by heme) after import. Protein import *in vitro* can be extremely inefficient, leading to the suggestion that tightly folded polypeptides might not be import-competent (1061).

Peroxisomal proteins have been shown to ind specifically to peroxisomes and not to mitochondria *in vitro* (514,1060), supporting the idea that these organelles have specific receptors on their surfaces. The absence of precursor forms of imported peroxisomal proteins has been repeatedly noted in several plant, fungal, and animal systems (178,338,906,932, 945,1239). The imported proteins are synthesized on free rather than membrane-associated polysomes (69); a transitory pool of peroxisomal proteins has been detected in the cytoplasm by pulse–chase experiments (388,934); and they are not taken up by RER-derived microsomes (384). There are very few reported examples of peroxisomal proteins which are synthesized as larger precursors (451,725,1156) and even in these cases, processing is thought to be unrelated to protein import (451,614).

The nature and location of the peroxisomal routing signal is now beginning to attract attention. Evidence suggests that the signal may be universal. For example, alcohol oxidase from the methylotrophic yeast *Hansenula polymorpha* is routed to the *S. cerevisiae* peroxisome when its structural gene is expressed in this organism, although it does not adopt its usual octameric structure, probably because of differences in the environment of the two types of peroxisomes (244). Even more surprising is the observation that firefly peroxisomal luciferase is also located in the peroxisome when its structural gene is expressed in mammalian cells (566). There are, however, no obvious identical sequence segments in the relatively few examples of peroxisomal proteins for which the primary structure has been determined.

Luciferase may be a suitable reporter protein for future studies on peroxisome biogenesis. Gould *et al.* (393) have recently identified two regions of the luciferase polypeptide which are necessary for import into peroxisomes. One of these, the C-terminal region, can target DHFR into peroxisomes, indicating that it carries the peroxisomal routing signal (in this case, the luciferase-derived sequences were fused to the C-terminus of the DHFR polypeptide). As few as 13 residues from the extreme C-terminus of luciferase constituted the peroxisomal routing signal in this assay system. The sequence (LIAKAKKGGKSKL) is unusually rich in lysines and neutral or hydrophobic residues and probably has a sharp turn around the two glycine residues, but it is not present in some other peroxisomal proteins. However, four other mammalian peroxisomal enzymes were subsequently found by gene fusion techniques to have C-terminal routing signals similar to that found in luciferase. In fact, the last three residues of the putative luciferase routing signal (Ser-Lys-Leu) are also present at or close to the C-terminus of these other peroxisomal proteins (His replaces Lys in one protein). One of these routing signals became nonfunctional when Lys was replaced by Asn, confirming that the integrity of the tripeptide is essential for its activity (393a). Efforts are also being made to locate routing signals in yeast peroxisomal proteins, which are also amenable to study by molecular genetic techniques. Small *et al.* (1060) recently used DHFR fusions to show that one such protein, acyl-CoA oxidase of *Candida tropicalis,* has two peroxisomal routing signals, one close to the N-terminus and one near the center of the polypeptide. The C-terminal half of the polypeptide is itself routed into peroxisomes (1060), as is the central region (1061).

Another approach to searching for routing signals is to compare the sequences of peroxisomal and nonperoxisomal isoenzymes. Glyoxisomal phosphoglycerate kinases of *Trypanosoma brucei* and *Crithidia fasciculata* differ from their cytosolic counterparts (which are encoded by a different gene) in having a longer C-terminal tail and a higher net positive charge (827,1094). Similarly, peroxisomal 3-oxo-acyl CoA thiolase has an additional N-terminal sequence which is absent from the mitochondrial isoenzyme (the mitochondrial enzyme is unusual in that it does not have a cleavable prepeptide, and the peroxisomal enzyme is unusual in that it is proteolytically processed, although processing is apparently not involved in peroxisomal routing) (23,451). Further comparisons may indicate whether such sequence differences are important for peroxisomal routing. There is little doubt that the mechanisms of targeting of this important and interesting class of proteins, and the mechanisms of peroxisomal membrane biogenesis, will receive increasing attention in the next few years.

Little appears to be known about the targeting of peroxisomal membrane proteins, although they are known to be made on free polysomes (614), or of the origin of peroxisomal membrane lipids. It would be interesting to see whether peroxisomal membrane proteins have transmembrane hydrophobic segments, and if they do, what prevents them from acting as secretory routing signals. Patients with Zeeweger syndrome have recently been shown to have peroxisomal ghosts with intact membranes but lacking soluble peroxisomal proteins. This indicates that the pathways for membrane and soluble protein import into peroxisomes may be different (992). Further analysis of the defect in cells from these patients may provide useful insights into the requirements for protein import. It should also be possible to select similar mutants peroxisomal protein import-defective in eukaryotes such as yeasts.

Further reading

Cashmore, A., Szabo, L., Timko, M., Kausch, A., Van den Broeck, G., Schreier, P., Bohnert, H., Herrera-Estrella, L., Van Montagu, M., and Schell, J. (1985). Import of polypeptides into chloroplasts. *Bio/Technology* **3,** 803–808.

Hurt, E. C., and van Loon, A. P. G. M. (1986). How proteins find mitochondria and intramitochondrial compartments. *Trends Biochem. Sci.* **11,** 204–207.

Lazarow, P. B., and Fujiki, Y. (1985). Biogenesis of peroxisomes. *Annu. Rev. Cell Biol.* **1,** 489–530.

Nicholson, D. W., and Neupert, W. (1988). Synthesis and assembly of mitochondrial proteins. *In* "Protein Transfer and Organelle Biogenesis" (R. C. Das and P. W. Robbins, eds.), pp. 677–746. Academic Press, San Diego, California.

Schatz, G. (1987). Signals guiding proteins to their correct locations in mitochondria. *Eur. J. Biochem.* **165,** 1–6.

Schmidt, G. W., and Mishkind, M. L. (1986). The transport of proteins into chloroplasts. *Annu. Rev. Biochem.* **55,** 879–912.

CHAPTER VII

The targeting of nuclear proteins

A. The structure of the nucleus

The nucleus is surrounded by a double bilayer membrane which is contiguous with the endoplasmic reticular membrane, and they consequently share many lipids, proteins, and enzymes. The nuclear membrane may even be the site of protein synthesis, judging from the fact that polysomes have been observed on the surface of the nuclear envelope. These are probably involved in secretory protein synthesis (see Section III.C.2), whereas targeting of nuclear proteins is completely independent of the secretory pathway. Ultrastructure studies have indicated that the nuclear inner and outer membranes are contiguous (see below and Fig. VII.1), but it is not known whether they have identical protein and lipid compositions. The inner membrane is adjacent to the underlying lamina or *karyoskeleton,* meshwork of intermediate-type microfilaments composed of proteins called *laminins,* which line the inside of the nuclear envelope.

The nuclear envelope is perforated by pores which play a central role in protein targeting and the movement of other macromolecules into and out of the nucleus. Ultrastructure analyses reveal that the nuclear pore has a well-defined structure composed essentially of a ring of eight spokes, which protrude into the lumen of the pore, and a central, possibly transiently associated, electron-dense plug of unknown nature which seems almost to fill the lumen of the pore. The pore is flanked on its upper and lower surfaces by eight globular structures forming a ring with an outer diameter of 120 nm and an inner diameter of 80 nm, the internal diameter of the pore (523,779,1134) (Fig. VII.1). The spokes of the pore may act as a diaphragm. The pores, which are not uniformly distributed over the surface of the nucleus (779), may be anchored in place via interactions with the underlying laminin. Pore complexes are also found in the ER membrane, but the significance of this is not understood. The inner and outer membranes appear to be fused at the periphery of the pore complex, as shown in Fig. VII.1. Chromosomes seem to be attached to the inner face of the nuclear membrane, and specific associations between chromosomal domains and pore complexes may allow mRNA transcripts to exit rapidly from the nucleus (89) (see below).

Protein components of the pore and nuclear envelope are now beginning to be identified. An ~190-kDa, relatively abundant glycoprotein which binds to the lectin concanavalin A has been identified as an integral membrane protein which remains associated with laminin when nuclei are dissociated in detergent (316,362). A group of less abundant proteins ranging in size from 35 to 110 kDa were identified by monoclonal antibodies raised against nuclear pore proteins. These antibodies were directed against an antigenic determinant which included O-linked *N*-acetylglucosamine, which also causes the proteins to bind to the lectin wheat germ agglutinin (225,463,845).

In most cells, the nucleus breaks down during mitosis and meiosis and reforms afterwards (exceptions include yeasts and *Drosophila*). Fragments of membrane with some, but not all, species of laminin remaining intact can be detected in the cytoplasm of cells during meiosis and mitosis. These fragments presumably represent core elements which, together with chromatin–DNA complexes, seed the reforming nuclear membrane around the DNA. During mitosis or meiosis, pore complexes can be de-

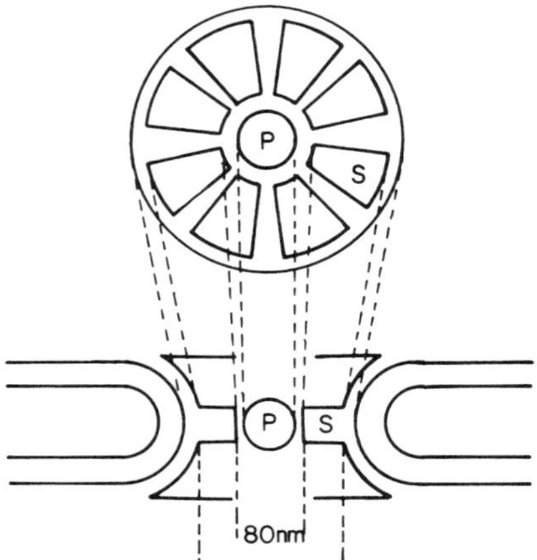

Fig. VII.1. Diagrammatic representation of the ultrastructure of a nuclear pore. The lower part of the figure shows a side view of a section through the pore (note that the two membranes of the nuclear envelope are contiguous), and the upper part shows an enlarged top view. P, plug; S, spoke. Redrawn from Refs. 323 and 1134.

tected as minute foci in the cytoplasm by using monoclonal antibodies directed against nuclear pore protein (1069). *In vitro* reconstitution experiments show that the formation of the nuclear envelope requires ATP and that the reassociation of laminins in the nucleus is accompanied by a dramatic change in their level of phosphorylation (318,777). Microinjected heterologous DNA is also enveloped by laminin and nuclear membranes with pore complexes, indicating that formation of the nucleus does not depend on specific DNA sequences (322). Normally, nonnuclear macromolecules are excluded during nuclear reassembly, possibly because condensed chromatin–DNA is very compact and does not expand until it is surrounded by the nuclear envelope in postmitotic cells (1093).

B. Targeting and retention of nuclear proteins

The nucleus contains a wide variety of proteins, most of them interacting directly with the DNA as regulatory proteins controlling gene expression, as enzymes involved in transcription or replication and in DNA repair, and as nonspecific DNA binding proteins (histones, etc). The nucleoplasm also contains soluble proteins and the laminins. Protein concentrations in the nucleus may reach levels approaching those in protein crystals. The remainder of this chapter is mainly concerned with how these proteins enter the nucleus and how they become selectively enriched. Components of the nuclear membrane, including those found in the pore, must also be targeted to the nucleus, but very little is known about the biogenesis of the nuclear envelope.

1. Size limitations on protein diffusion into the nucleus

Early studies showed that small, radioiodinated tracer proteins slowly reach an equilibrium between the nucleus and the cytoplasm when microinjected into oocytes, whereas larger proteins (>66 kDa) are virtually totally excluded from the nucleus. Large histones and proteins extracted from the nucleus, however, entered the nucleus at least as freely as small proteins when microinjected into oocytes (98,99,411). These experiments led to the concept of selective enrichment of nuclear proteins (230), which accounted for the observation that the nucleus and cytoplasm contained two almost entirely different sets of proteins (99). From the data on the exclusion of proteins and dextrans (839) it was calculated that free diffusion was limited to macromolecules of less than 9–10 nm in diameter, about one-tenth of the diameter of the pore (241). Furthermore, the rate of import decreases markedly as the size of the macromolecule increases.

Two models have been proposed to explain the selective uptake and retention of nuclear proteins. One possibility is that nuclear proteins have karyophilic signals which route them to the nucleus. The second possibility, which *a priori* seems less likely, is that nuclear proteins have an elongated rather than globular structure and that there are therefore fewer constraints on their free diffusion into the nucleus. Both models predict that nuclear proteins might have additional "nuclear retention" signals to keep them in the nucleus.

2. Nuclear proteins have karyophilic signals

This section reviews some of the extensive literature describing different approaches used to define and study karyophilic signals.

a. *Karyophilic signals demonstrated by limited proteolysis*

Microinjection techniques provided the first concrete evidence for karyophilic signals in nuclear proteins. Nucleoplasmin, a pentamer composed of five identical 21-kDa subunits, rapidly accumulates in the nucleus when injected into frog oocytes (Fig. VII.2F) (241a). Core pentamers devoid of C-terminal tails (obtained by limited proteolysis of assembled pentamers) remain in the cytosol when microinjected (Fig. VII.2B), whereas isolated tails, or pentamers retaining only one tail, are directed to the nucleus (Fig. VII.2D and E). This shows that the C-terminal tail of the nucleoplasmin monomer contains a karyophilic signal, a conclusion which was subsequently confirmed by different approaches (see below). When core pentamers without karyophilic signals were microinjected directly into the nucleus, they remained there and did not leak out into the

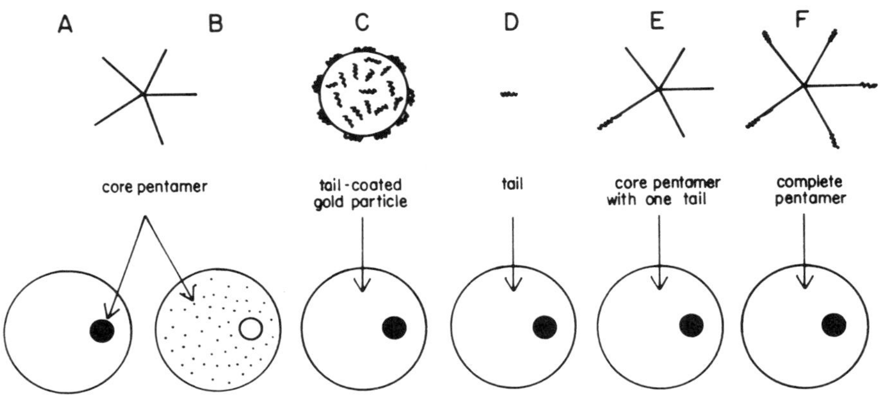

Fig. VII.2. The fates of microinjected nucleoplasmin pentamers and derivatives. Arrows indicate the site of injection. The nucleus is the only organelle shown.

cytosol (Fig. VII.2A). This led to the suggestion that nucleoplasmin has two signals, one near the C-terminus for nuclear uptake and another in the N-terminal part for nuclear retention.

b. Karyophilic signals defined by gene fusion studies

The first reports of the application of gene fusion techniques to define karyophilic signals appeared in 1984, 2 years after the studies on proteolyzed nucleoplasmin. The advantage of this technique is that the karyophilic signal can be taken out of its normal context and significantly reduced to define its maximum size. The pioneering work was carried out by Hall *et al.* (416) on a yeast regulatory protein, MATα2, and by Kalderon *et al.* (552) on the large T antigen of SV40 virus. Both groups used β-galactosidase as a reporter protein (Kalderon *et al.* also used pyruvate kinase). Hall *et al.* found that the karyophilic signal of MATα2 was located within the extreme N-terminal 13 residues, whereas Kalderon *et al.* found that the karyophilic signal of SV40 T antigen was located close to the center of the polypeptide. Furthermore, the two sequences were apparently quite different (Table VII.1). Subsequent studies with other proteins confirmed that karyophilic signals may be located almost anywhere along the length of a nuclear polypeptide, including the extreme C-terminus, as in the case of the adenovrius E1a protein (Table VII.1).

Although gene fusions are potentially very useful for defining karyophilic signals, the context in which they are placed may affect their ability to route nonnuclear proteins into the nucleus. For example, certain yeast ribosomal protein L3–β-galactosidase hybrids were transported into the nucleus only when the two parts of the hybrid were separated by a turn-inducing glycine or proline bridge (735), and the karyophilic signal of the SV40 virus large T antigen efficiently routed pyruvate kinase into the nucleus only when fused into the sequence of the normally cytoplasmic protein at certain sites (933). Given the known structure of the pyruvate kinase monomer, hybrid proteins in which the karyophilic signal was nonfunctional were probably folded in such a way that the signal was buried within the protein, rather than exposed on the surface. Correct positioning and spacing of the minimal karyophilic signal of nucleoplasmin in pyruvate kinase hybrids is also essential for nuclear targeting (242a).

Another, similar approach to the characterization of karyophilic signals is to cross-link them onto other proteins or even inert particles (Fig. VII.2C), and to study their localization following microinjection (382,611). This technique has been extensively used to study the activity of chemically synthesized karyophilic signals (see next section) and to study the role of the pore in nuclear protein uptake (see Section VII.B.3).

Table VII.1. Karyophilic signals in nuclear proteins

Protein (organism)	Reporter protein	Location	Minimal sequence	Reference
MATα2 (yeast)	β-Galactosidase	N-terminal	K-I-P-I-K	416
Large T antigen (SV40 virus)	Pyruvate kinase	Central	P-K-K-K-R-K-V	553
Large T antigen (polyoma virus)	Pyruvate kinase	Central	V-S-R-K-R-P-R-P	923
		C-terminal	P-P-K-K-A-R-E-D	
VP1 (SV40 virus)	VP1 of influenza V	N-terminal	A-P-T-K-R-K-G-S	1209
Ela (adenovirus)	Galactokinase	C-terminal	K-R-P-R-P	667
GAL4 (yeast)	β-Galactosidase	N-terminal	Not defined	1048
L3 (yeast)	β-Galactosidase	N-terminal	P-R-K-R	735
Nucleoprotein (influenza virus)	α-Globin	C-terminal	A-A-F-E-D-L-R-V-L-S	219
Histone 28 (yeast)	β-Galactosidase	N-terminal	K-K-R-S-K-A	736
Glucocorticoid receptor (rat)	β-Galactosidase	Central	K-T-K-K-K-I-K-G	875
Nucleoplasmin (frog oocytes)[a]	β-Galactosidase and pyruvate kinase	C-terminal	R-P-A-A-T-K-K-A-G-Q-K-K-K-K-L-D-K	137,242

[a] Detected by deletion analysis; flanking sequences are required for nuclear targeting of nucleoplasmin–pyruvate kinase hybrids (242).

c. *Common features of karyophilic signals*

Karyophilic signals from different nuclear proteins do not exhibit significant sequence similarity, but many share several characteristics, including an overall basic charge due to the presence of numerous Lys and to a lesser extent Arg residues, and a turn-inducing Pro or Gly residue. The only exception seems to be the karyophilic signal of influenza virus nucleoprotein (see below). Goldfarb *et al.* (382) found that antibodies raised against synthetic SV40 T antigen karyophilic signal reacted with some, but not all, nuclear proteins but not with cytoplasmic proteins, suggesting that some endogenous nuclear proteins may have a similar sequence.

d. *Mutagenesis defines key residues in karyophilic signals*

Two approaches have been used to determine which residues in the SV40 T large T antigen karyophilic signal play a crucial role in its activity. The first, pioneered by Kalderon *et al.* (552,553) and by Lanford and Butel (610), was to use site-directed mutagenesis to change specific residues in the karyophilic signal selectively, either in the context of the large T antigen itself or in hybrid proteins. Replacement of Lys-128, the second of the four Lys residues shown in Table VII.1, by Thr, Asn, Met, Gln, Leu, or His caused the complete loss of karyophilic activity, whereas its replacement by the similarly charged Arg caused only a partial loss of activity (184). The loss of activity caused by the replacement of Lys-128 could not be compensated by the introduction of an additional Lys residue elsewhere within the signal. Lys-129 could, however, be replaced by Arg without affecting karyophilic activity, and the change from Lys-131 to Met had only slight effect on karyophilic activity (148). Sequence changes in the T antigen region flanking the karyophilic signal (as defined by gene fusion studies) did not affect karyophilic activity (552).

These studies not only confirmed that the integrity of the karyophilic signal is important for its activity, they also demonstrated the critical importance of Lys-128. This was confirmed by a second technique, namely cross-linking a chemically synthesized signal and its Asn- or Thr-128 variants to ferritin, bovine serum albumin, or IgG and showing that only the wild-type sequence could route this reporter protein into the nucleus after microinjection (382,611). Further studies are required to define the role of other residues in this and other karyophilic signals.

e. *Multiple karyophilic signals in nuclear proteins*

Do nuclear proteins have several karyophilic signals which act cooperatively to promote efficient protein targeting? Cross-linking studies show that the efficiency of nuclear targeting rises as the number of peptides coupled to the reporter protein is increased (269a,611). There are also examples of nuclear proteins which have been shown by gene fusion or deletion experiments to have at least two karyophilic signals. For example, polyoma virus large T antigen has two karyophilic signals, the sequence of one of which is quite similar to that in the corresponding protein of SV40 virus. The deletion of either of the two signals from polyoma large T antigen sequence reduces the efficiency of nuclear localization, whereas the deletion of both of them causes T antigen to remain cytoplasmic (923). Mutations outside the karyophilic signal of SV40 T antigen also cause at least a partial loss of nuclear targeting (1178). The region defined by these mutations (around residue 186) may be a second, dormant karyophilic signal or may be required to ensure the correct exposure of the karyophilic signal on the surface of the folded polypeptide. Two potential karyophilic signals have been identified in the glucocorticoid receptor (875), and nucleoplasmin contains four sequences which resemble karyophilic signals (137,242). Silver and Hall (1047) have also recently reported that yeast MATα2 protein may contain a second karyophilic signal.

As discussed in Section VII.B., karyophilic signals seem to function at least in part by recognizing a receptor present in the nuclear pore. It has also been proposed that nuclear retention signals might ensure that nuclear proteins do not leak back into the cytoplasm (229,230,241,757). The unusual karyophilic signal of influenza virus nucleoprotein (Table VII.1) may fall into the latter class (219). It has been noted that β-galactosidase hybrids carrying the SV40 T antigen karyophilic signal remain close to the periphery of the nucleoplasm (100), suggesting that although the karyophilic signal may be sufficient for nuclear routing, distribution and retention in the nucleus may depend either on specific protein–protein interactions within the nucleoplasm or on protein–DNA interactions (1178).

If nuclear proteins do indeed have retention signals, then their retention in the nucleus should not depend on an intact nuclear membrane. This possibility was addressed by comparing the protein contents of intact nuclei and of those which had been punctured with a glass needle (305). Only 10–15 of the ~300 nuclear proteins identified by SDS-PAGE were present in reduced amounts in the punctured nuclei, which is strong evidence in favor of nuclear retention signals.

f. Regulated and "piggyback" uptake of nuclear proteins

The 80-kDa polyA binding protein of yeasts is located in both the cytoplasm and the nucleus. Both forms of the protein are encoded by a single, unspliced gene, and the nuclear protein is apparently identical to its cytoplasmic counterpart except that it is proteolytically nicked to give 53-kDa and 17-kDa polypeptides, which remain associated (981). Numerous other proteins are known to remain in the cytoplasm for long periods before they are transported into the nucleus, usually under hormonal or developmental stimuli (229). The transport of these proteins into the nucleus may result from some sort of activation such as proteolysis, which might induce conformational changes, thereby exposing or activating a normally cryptic karyophilic signal. Another possibility is that the uptake of these proteins depends on the synthesis and transport of another protein which itself carries a karyophilic signal.

Free ribonuclear proteins are stockpiled in the cytoplasm but can be induced to accumulate in the nucleus by microinjecting ribonuclear RNA (1232). Mutant transcripts which cannot bind to a sequence-specific ribonuclear protein are not transported into the nucleus, suggesting that the interaction between the two might be required to expose a karyophilic signal in the sequence of the ribonuclear protein (698). Glucocorticoids have also been proposed to unmask the karyophilic signal(s) in the glucocorticoid receptor (875). Other examples of "piggyback" transport include the nuclear localization of adenovirus DNA polymerase (1234a) and microinjected antibodies raised against nuclear proteins (72,678,1127), the nuclear targeting of a ribosomal protein–β-galactosidase hybrid which depends on its association with the wild-type ribosomal protein in the cytoplasm (403), and the similar nuclear targeting of a yeast histone H2A–H2B–β-galactosidase hybrid complex (736). Thus piggyback transport of proteins into the nucleus could provide a simple method for the regulated movement of proteins into the nucleus. Ribosomal proteins also exit from the nucleus as ribonuclear particles. These particles may have specific "cytophilic" signals which direct them out of the nucleus; they are presumably inactive until ribosomal proteins have associated into ribosomes.

g. Processing and modification of nuclear proteins

It is now widely accepted that karyophilic signals are permanent features of nuclear proteins and that the absence of proteolytic processing allows signals to be reused following nuclear membrane dissociation during mitosis or meiosis. Some nuclear proteins are apparently subject to post-translational processing and modification, however. These modifications include the proteolytic processing of yeast polyA binding protein (see previous section) and of some of the laminins (623), glycosylation of nuclear pore components (225), and myristoylation of the VP2 proteins of polyoma and SV40 viruses (1087). Some of these modifications may be required to unmask karyophilic signals or for the correct assembly of proteins in the nucleus.

3. The role of the nuclear pore in protein import

The key role played by the nuclear pore in the accumulation of nuclear proteins was elegantly demonstrated by Feldherr *et al.* (306), who found that gold particles coated with nucleoplasmin accumulated in the nucleus when microinjected into oocytes. Gold particles coated with trypsin-treated nucleoplasmin (lacking the C-terminal tail containing the karyophilic signal) did not accumulate in the nucleus. Imported gold particles were seen in the pore when oocytes were examined by electron microscopy, confirming the pore as the site of nuclear import. Coated gold particles as large as 26 nm in diameter appeared to pass quite freely through the pores (269a). Although these particles are still considerably smaller than the pore diameter as measured by electron microscopy (80 nm), they are larger than the smallest "excluded" cytoplasmic proteins (~9 nm in diameter). More recent studies indicate that gold particles coated with tRNA, 5S RNA, or polyA reach the cytoplasm via the nuclear pores if microinjected into the nuclei (269). Thus, the pore is involved in the bidirectional movement of macromolecules through the nuclear envelope.

The second line of evidence for the direct involvement of the nuclear pore in protein import comes from studies on the effects of a lectin (wheat germ agglutinin) which reacts with the *N*-acetyl glucosamine residues on nuclear pore proteins. Finlay *et al.* (317) found that this lectin reversibly inhibited nucleoplasmin import, whereas other lectins were without effect. Dextran uptake, which also occurs through the pores (1006), was unaffected by the lectin. Wheat germ agglutinin–ferritin complexes were shown by electron microscopy to adhere to the cytoplasmic face of the nuclear pore.

Feldherr and Ogburn (305) demonstrated that an intact nuclear membrane was not required for the retention of most nuclear proteins within the nucleus (see previous section). They also found that only one of the five cytoplasmic proteins they tested (actin) entered punctured nuclei more freely than intact nuclei, suggesting that diffusion through pores is not a limiting factor for the uptake of nuclear proteins. A somewhat different result was obtained, however, by Newmeyer *et al.* (778), who found that cytoplasmic proteins entered damaged nuclei *in vitro* and by Zimmer *et al.* (1263a) who found that nuclear proteins leaked out of injured nuclei. Thus, nuclear pores seem to provide the only means of access to the nucleoplasm for proteins, and cytoplasmic proteins are probably excluded both because they are too large to diffuse through the pores and because they lack karyophilic signals which would allow them to interact with the pore and to be actively accumulated.

4. ATP- and receptor-dependence of protein import

The *in vitro* studies of Newmeyer *et al.* (778) were the first to show the ATP-dependence of nuclear protein import. They also showed that no import occurred at 4°C, suggesting that protein import is an active process. Similar results were reported in studies of chilled or azide plus deoxyglucose-treated tissue culture cells (924). Newmeyer and Forbes (776) showed that human serum albumin (HSA) cross-linked to the karyophilic signal of SV40 T antigen bound to the nuclear pore in the absence of ATP, whereas HSA carrying the nonfunctional Thr-128 variant of the karyophilic signal did not. High numbers of coupled karyophilic signals were required for efficient binding and subsequent ATP-dependent uptake, suggesting that multivalent interactions between the signals and receptors might increase the efficiency of protein import (see Section VII.B.2.e).

These *in vitro* studies confirm other observations suggesting that nuclear proteins interact with a receptor in the pore during uptake. Nucleoplasmin-coated gold particles were shown by Feldherr *et al.* (306) to be clustered around the cytoplasmic face of the pore, suggesting that they might be bound to receptors and queueing up to cross the pore. Similar results were obtained by Richardson *et al.* (924), who noted that nucleoplasmin–gold complexes associated with fibrils around the aperture of the oocyte nuclear pore. Another line of evidence suggesting the existence of receptors for nuclear protein import is the fact that uptake of karyophilic signal–protein conjugates is saturable and inhibited by free karyophilic signal (382). According to one report, wheat germ agglutinin does not prevent nuclear protein–receptor interactions (776), specific inhibition of nuclear protein import by wheat germ agglutinin must therefore occur at a stage after receptor binding. However, a second report describes inhibition of receptor activity by wheat germ agglutinin (513a), although it is not clear whether there are two distinct receptors (one each for SV40 T-antigen and nucleoplasmin karyophilic signals), only one of which is blocked by the lectin.

Although the first stage in protein uptake is now reasonably well defined, much work remains to be done before we understand the entire process in detail. For example, it has been proposed that the diaphragm of the pore may enlarge to allow the passage of large nuclear proteins (or nuclear protein–gold conjugates) (269a). One might envisage the pore operating as a kind of trapdoor whose opening is triggered by protein bound to the receptor in the neck of the pore. *N*-acetylglucosamine-modified proteins might be involved in this process. The role of ATP in protein import also needs to be more thoroughly investigated. By analogy with other systems, ATP might be required to phosphorylate proteins as they penetrate the pore, to unfold proteins to make them competent for protein import, or to provide energy for the import process. Interestingly, ATP is also required for dextran uptake (1006), which is otherwise independent of pore components (receptor and *N*-acetylglucosamine-modified proteins) required for protein import. A further possibility is that pores are only activated for nuclear protein uptake when one or more of their components binds ATP but that ATP hydrolysis does not occur. Finally, it remains to be determined whether cytoplasmic components are required for protein import.

Further reading

Dingwall, C., and Laskey, R. A. (1986). Protein import into the cell nucleus. *Annu. Rev. Cell Biol.* **2,** 386–390.

Silver, P. A., and Hall, M. N. (1988). Transport of proteins into the nucleus. *In* "Protein Traffic and Organelle Biogenesis" (R. A. Das and P. W. Robbins, eds.), pp. 747–770. Academic Press, San Diego, California.

CHAPTER VIII

Endocytosis

Eukaryotic cells of many different types internalize a variety of proteins and other macromolecules from the surrounding environment and from the cell surface. In most but not all cases, internalization occurs via endocytosis, a process involving the invagination of specific regions of the plasma membrane to form vesicles which then migrate into the cytoplasm and away from the plasma membrane. Endocytosis is known or has been proposed to play an essential role in several aspects of cell physiology.

(i) Nutrient uptake: many micronutrients are accumulated by endocytosis. Among the best characterized examples is iron, which is internalized as a complex with its specific binding protein, transferrin, from which iron is subsequently released in an internal compartment (see Section VIII.A.5.a).
(ii) The clearance of unwanted molecules from the blood stream by macrophages and hepatocytes: The endocytosed molecules are targeted to the lysosome, where they are degraded.
(iii) The transcytosis of secretory class (IgA) and maternal (IgG) immunoglobulins by epithelial cells
(iv) The capture of "escaped" lysosomal proteins
(v) Antigen processing and presentation
(vi) The control of cell dimensions involving the recycling of plasma membrane material into late stages of the secretory pathway
(vii) Pseudopodal locomotion
(viii) The control of cell growth and development by growth factors and hormones
(ix) DNA uptake by transformation

In addition, many viruses, bacteria, and toxins fortuitously gain entry into eukaryotic cells by processes which are similar if not identical to endocytosis.

Indeed, two different but probably convergent endocytic pathways can be distinguished by the requirement for a cell surface receptor for endocytosis. In receptor-independent, fluid-phase endocytosis (*pinocytosis*), material in the growth medium is internalized as part of the bulk fluid volume of invaginating vesicles, whereas *receptor-mediated endocytosis* involves ligand recognition by cell surface receptors and usually occurs at specific sites on the cell surface. Specific examples will be used in this chapter to illustrate the mechanisms and cell components involved in endocytosis and the fate of the endocytosed material and associated membrane.

A. Receptor-mediated endocytosis

1. Characterization of endocytic receptors

Endocytic receptors reach the cell surface via the constitutive branch of the secretory pathway and are therefore probably randomly inserted into the plasma membrane. The only well-characterized exceptions are receptors involved in IgA and maternal IgG transcytosis and in transferrin endocytosis by epithelial cells, in which cases the receptors are located in the basolateral surface (Section V.G.3). All of the well-characterized receptors for which the corresponding nucleotide sequences have been published seem to be type I or type II membrane proteins; none of them appear to be polytopic membrane proteins, although some of them are oligomeric (e.g., the homodimeric transferrin receptor and the insulin receptor, which is composed of two α and two β subunits). Furthermore, the sizes of these receptor proteins vary enormously, ranging from 283 amino acids for the asiologlycoprotein receptor (262) to 2455 residues for the bovine mannose-6-phosphate (M6P)–insulin-like growth factor 2 (ILGF2) receptor (826). According to the predicted positions of the single transmembrane domains, the major part of all endocytic receptors is exposed on the cell surface, although some, such as the insulin and epidermal growth factor (EGF) receptors, also have long cytoplasmic segments. Both of these proteins have tyrosine kinase activities in their cytoplasmic domains and are involved in hormonal signaling. None of the endocytic receptors share significant sequence homology (262,270,645,700,738, 742,826,1020,1131,1215).

2. Ligand binding domain

The most extensively characterized endocytic receptor is that for low-density lipoprotein (LDL). The most important domain for ligand binding is the extreme N-terminus, which contains seven repeats of ~40 residues, including six cysteines, all of which form disulfide bridges (386). The C-terminus of each repeat includes a cluster of negatively charged residues which are complementary to and probably interact electrostatically with positively charged residues in apolipoprotein E (386,515). The LDL receptor may thus be able to bind several ligand molecules.

The low-density apolipoproteins, the LDL receptor ligands, bind circulating cholesterol and deliver it via endocytosis to the lysosome. The genetically inherited disease hypercholesterolemia is caused by defects in LDL uptake, leading to the massive accumulation of LDL in the plasma. Several classes of mutations affecting the LDL receptor gene have been characterized. Two of these classes of mutations (2 and 3) reduce LDL–receptor interactions. The class 3 mutants are of particular interest because all other measurable functions including receptor density on the cell surface and reactivity with an LDL receptor-specific monoclonal antibody are normal, although some of the mutant proteins have abnormally low molecular weights, apparently because segments of the receptor polypeptide are absent. Goldstein *et al.* (386) have suggested that internal deletions within the LDL receptor gene might occur by homologous recombination between segments of the gene coding for the repeat units which code for the ligand binding domain. The deletion of several of these segments would be expected to reduce LDL receptor activity. Krieger *et al.* (592,593) have developed two suicide-selection protocols for isolating similar mutants in tissue culture cells. Some of these mutations map in the presumed LDL receptor structural gene. Further analysis of this gene by site-directed mutagenesis should be useful in determining the exact sites of ligand binding; they may also provide information of the role of the receptor polypeptide in endocytosis and lysosome targeting.

Other endocytic receptors also have repeated sequences in their extracytoplasmic domains. One particular receptor is of special interest because it recognizes two different ligands, M6P residues on incorrectly sorted (secreted) lysosomal enzymes and ILGF (1224). According to MacDonald *et al.* (669), M6P increases the affinity of the receptor for ILGF, indicating that the protein has cooperative, high-affinity binding sites for both ligands. The extracytoplasmic portion of this protein has 15 repeated domains which are also present in the cation-dependent M6P receptor (646,826) (see Section III.G.5), suggesting that they may form the M6P receptor domain.

3. Endocytic receptors cluster in coated pits

Receptor-mediated endocytosis occurs at specific regions of the plasma membrane. These regions, which account for up to 2% of the cell surface, were first observed by Roth and Porter in mosquito oocytes (960). They are distinguished from the rest of the plasma membrane by the presence of small invaginations which are coated on the inner, cytoplasmic face by clathrin (Section V.G.6 and Fig. VIII.1). As discussed later, receptors with or without ligands are internalized at these sites as the invaginations pinch off into the cytoplasm as vesicles. Thus, receptors which are randomly inserted into the plasma membrane must migrate to these areas to be efficiently endocytosed. Most endocytic receptors migrate to coated pits without ligand (17,468,594,1157). The EGF receptor is unusual in that it remains evenly distributed over the cell surface until it binds EGF (264). Since only a limited number of plasma membrane proteins are found in coated pits, and since the EGF receptor only accumulates there when it

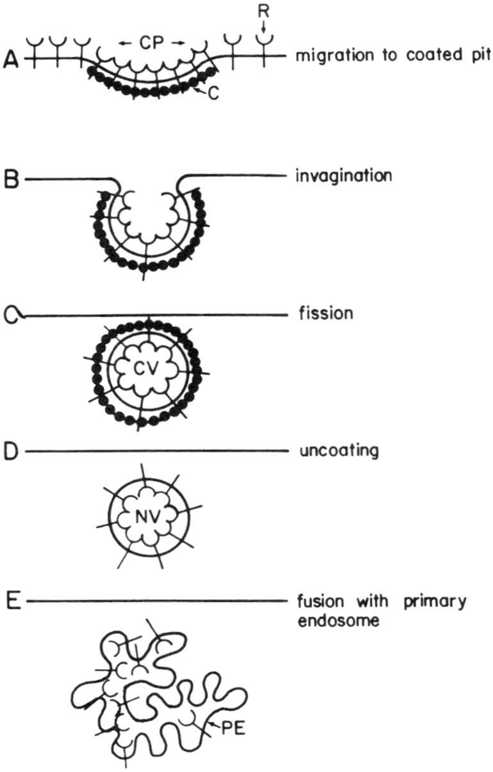

Fig. VIII.1. Early stages in receptor endocytosis. The figure shows the five early stages used by all receptors which are endocytosed from coated pits. The receptors are shown being endocytosed without bound ligand. CP, coated pit; C, clathrin; NV, naked vesicle; PE, primary endosome; R, receptor.

has bound EGF, processes other than random diffusion may be involved (e.g., migration along the cytoskeleton). Alternatively, since membrane lipids are apparently continuously flowing towards coated pits (122), it may be that endocytic receptors remain trapped in them, probably by interactions between their cytoplasmic tails and clathrin or clathrin-associated proteins (see below), whereas all other proteins diffuse away again (123,855a).

The best evidence in favor of specific interactions between the cytoplasmic domains of endocytic receptors and components of the coated pits comes from studies of LDL receptor mutants from patients with hypercholesterolemia. The first example of a mutant LDL receptor (class 4) which did not cluster in coated pits was isolated from patient JD in 1976 by Brown and Goldstein (132). Subsequent genetic and biochemical studies of this and other mutant class 4 LDL receptor mutants are beginning to reveal the nature of the defects which block their accumulation in coated pits. One mutant receptor (FH 274) lacks the C-terminal membrane anchor domain, which causes more than 80% of the receptors to be secreted into the medium. However, the remaining 20% of the receptors are firmly membrane-associated by an unknown mechanism which precludes their accumulation in coated pits (625). These results suggested that the C-terminus of the LDL receptor was essential for interaction with coated pits. This was confirmed by sequence analysis of exons coding for this segment of the LDL receptor in cells from patient JD and from an Arabian patient. The JD LDL receptor has two Cys residues instead of the usual one in the 50-residue, C-terminal, cytoplasmic domain, whereas the LDL receptor produced by cells from the Arabian patient has only 2 of the normal 50 residues in the cytoplasmic domain (223,624). In another patient, a frameshift mutation in the same exon was shown to truncate the C-terminal tail of the LDL receptor, which again abolished accumulation in coated pits (624).

Recent studies have revealed the critical importance of a cytoplasmically located Tyr residue in the LDL receptor. In patient JD, this Tyr residue has been changed to Cys (223). Subsequent studies with *in vitro* mutagenized LDL receptor genes showed that this Tyr-807 residue could be replaced by Phe or Trp without affecting receptor internalization, whereas all other substitutions abolished uptake. Residues around position 807 also seemed to be important for the movement of LDL receptor to coated pits (224). The roles of the cytoplasmic domain of the endocytosed VSV G protein and the nonendocytosed influenza virus HA have also been examined. G–HA hybrids which have the cytoplasmic tail of HE are not endocytosed (958), whereas HA is endocytosed if Cys-543 in the cytoplasmic tail is replaced by Tyr (613a). These results again indicate the critical importance of a Tyr residue in localization to coated pits (but also see Ref. 855a), but in no case is there evidence for tyrosine phosphorylation.

Mutations which reduce the length of the cytoplasmic domain of other endocytic receptors have also been reported to abolish endocytosis (964) and accumulation in coated pits (421a). None of the endocytic receptors for which the primary sequences are known share significant homology in their cytoplasmic domains. The protein which anchors endocytic receptors in coated pits seems to be a clathrin accessory protein (855a).

4. The endocytic pathway

The early stages of the endocytic pathway are shared by all endocytosed proteins, irrespective of their final destination and fate. Endocytic pathways subsequently diverge as ligands and receptors are recycled to the plasma membrane or travel further along the pathway to the lysosome. Studies are beginning to clarify some of the details of the early stages in the endocytic pathway and to provide information on the organelles in which endocytosed proteins accumulate.

a. *Invagination and primary endosomes*

The first stage of the endocytic pathway involves the further invagination of the coated pit and the formation of a clathrin-coated endocytic vesicle (Fig. VIII.1B). Endocytic vesicles lose their protein coats, probably through the action of an ATP-dependent uncoating protein (Section V.G.6), and fuse to form peripheral primary endosomes (Fig. VIII.1D,E) (402). Many of the details of these early stages in the endocytic pathway are still far from being fully understood. The first question which arises concerns the role played by clathrin. As in the case of secretory pathway vesicles (Section V.G.6), clathrin probably masks other proteins in the membrane of the endocytic vesicle. It is therefore unlikely that membrane proteins could play a role in targeting endocytic vesicles to primary endosomes unless vesicles are uncoated prior to fusion with primary endosomes. Early studies showed that microinjected anti-clathrin antibodies were without effect on endocytosis (1177), but this may have been due to the specificity of the antibodies used (855); monoclonal antibodies against the poorly antigenic clathrin heavy chain have been shown to inhibit endocytosis when microinjected at high concentrations (258). Thus clathrin may play an active role in invagination of the coated pit. It is not known whether invagination and fusion are spontaneous events occurring when coated pits become loaded with endocytic receptors and depending solely on clathrin cage–receptor interactions. Low pH also inhibits the invagination of coated pits *in vivo* but does not inhibit ricin uptake or fluid-phase endocytosis (see below) (991).

Another interesting question is whether primary endosomes are formed solely by coalescing endocytic vesicles or whether they contain specific proteins which have independent biosynthetic origins. This question has been addressed by Gruenberg and Howell (404,405), who made use of an *in vitro* assay system to characterize elements involved in the early part of the endocytic pathway (Section II.F.3). Two proteins (224 kDa and 242 kDa) were tentatively identified as endosome-specific because they were not present in the plasma membrane (404). Purified endocytic vesicles were found to retain optimal ability to fuse with primary endosomes for about 5 min after endocytosis (115,405). Polarized endothelial cells, which endocytose different proteins from different cell surfaces, have two classes of primary endosomes which fuse into a common secondary endosome.

Studies by Salzman and Neufeld (989) in which fusion between compartments containing sequentially endocytosed proteins was assayed in Chinese hamster ovary (CHO) cells indicate that newly endocytosed protein mixes rapidly and efficiently with preendocytosed protein only when the period between the two endocytic events is very short. Similar results were obtained in an *in vitro* assay in which fusion required ATP hydrolysis as well as soluble and membrane proteins (237). The fusion event was assumed to occur in a *sorting endosome* compartment, which presumably corresponds to primary endosomes as defined above. Mixing may have been less efficient in secondary endosomes because their tubular structure might restrict fluid-phase diffusion (see below). This would be in line with results from *in vitro* studies showing that endocytic vesicles are able to fuse only with primary endosomes (115,404,405) and *in vivo* studies showing that the pool of intracellular transferrin receptor, which is thought to accumulate in secondary endosomes (see below), does not mix with freshly endocytosed receptors (9).

b. Secondary endosomes and prelysosomes

Several lines of evidence indicate that cells contain two types of endosomes corresponding to two distinct stages in the endocytic pathway which, according to Griffiths *et al.* (402), are probably linked by intermediate organelles (probably large vesicles). Two distinct endosome subpopulations have been resolved by free-flow electrophoresis by Schmidt *et al.* (1018), who showed that secondary endosomes were distinguished from primary endosomes by the kinetics of accumulation of fluid-phase (receptor-independent) and receptor-bound ligands and by the fact that the former did not contain transferrin. Unlike primary endosomes, which contain proteins derived from the plasma membrane as well as unique proteins (see above), secondary endosomes do not contain detectable amounts of plasma membrane-derived proteins. Secondary endosomes also have a lower pH than primary endosomes (712,761). As discussed later, the low pH of the secondary endosome may be important for the dissociation of certain endocytosed ligands from their receptors, thereby permitting endocytosed receptors to be recycled.

Current models for protein sorting in the secretory pathway make no provisions for the targeting of secretory proteins into endocytic compartments other than the lysosome (Chapter V). Thus some of the proteins which are found in endosomes but not in the plasma membrane may be derived from the lysosome. Another possibility is that secondary endosomes are the same compartment as the acidic prelysosome, through which secretory proteins travel en route to the lysosome (402). Both compartments have been reported to have extensive internal membranes, and lysosomal enzymes and lysosomal membrane proteins have been detected in the prelysosomal compartment. If this is the case, then vesicular transport between the primary endosome and the secondary endosome or prelysosome must operate in both directions, since primary and secondary endosomes have common proteins. Endosomes seem to have a pH intermediate between that of endocytic vesicles and that of the lysosome. Low pH is maintained by a membrane ATPase (712).

c. *Lysosomes*

Lysosomes are membrane-bound organelles containing a variety of acid hydrolases together with endocytosed material. Stereological techniques have been used to estimate the cell volume occupied by lysosomes in macrophages, which have ~1000 lysosomes (considerably more than most other cell types) which occupy 2–5% of the cell volume and have a total surface area which approaches 20% of the area of the plasma membrane (1078). Lysosomes have a low pH [estimated at between 4.5 and 5 by fluoresceine fluorescence quenching (807)] (see Section II.F.2) which causes them to accumulate any weak base which gains access to the cytoplasm. This low pH is maintained by a proton ATPase in the lysosome membrane (712).

5. The fates of endocytosed proteins

The primary endosome is the sorting center from which different receptors, receptor-bound ligands, plasma membrane proteins, and probably phospholipids, which accumulate in the same clathrin-coated pits and are endocytosed, are shunted into different pathways leading back to the cell surface or to the Golgi apparatus or the lysosome. The following sections use specific, well-characterized examples to illustrate these pathways and to describe the events which accompany the movement of endocytosed ligands and receptors as they move through the cell (Fig. VIII.2).

a. Recycling of ligand and receptor to cell surface

Certain receptor–ligand complexes typified by transferrin and its receptor recycle to the plasma membrane from an endosomal compartment and never reach the lysosome (799) (Fig. VIII.2A). The transferrin–receptor complex is not dissociated at low pH, although iron is released from transferrin (218a). A network of vesicles and membrane tubules has been proposed to lead the complex directly back to the cell surface (366), where the return to neutral pH causes the apotransferrin to be released from the membrane receptor (Fig. VIII.2A). The cycle is reinitiated when iron binds to the apotransferrin and increases its affinity for the receptor.

There is, however, considerable evidence that transferrin receptor does not always recycle directly to the plasma membrane, but rather passes

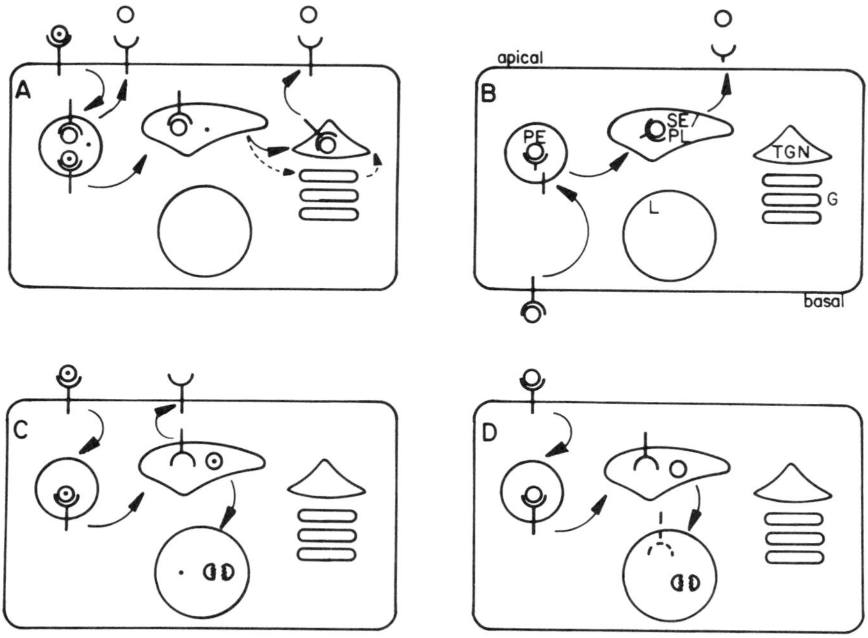

Fig. VII.2. Alternate pathways for protein traffic in the later stages of endocytosis. Receptors are shown being endocytosed with bound proteins. Y, receptor; ○, protein ligand; ·, secondary ligand; PE, primary endosome; SE/PL, secondary endosome/prelysosome; L, lysosome; G, Golgi apparatus; TGN, trans-Golgi network.

through the trans-Golgi network (TGN) where it is resialyated (319) (see Sections V.D and V.G.1) (Fig. VIII.2A). Some cell lines are reported to accumulate as much as 80% of their endocytosed transferrin receptor in secondary endosomes (9). The transferrin–receptor complex is then packaged into secretory vesicles together with secretory proteins including choline esterase (319). Woods *et al.* (1205) detected transferrin receptor in all other Golgi compartments as well as in immunoglobulin-secreting myeloma cells, although at most only very low amounts of this receptor appear in cis or medial Golgi in other cell lines (768a,1068).

These conflicting data on the route taken by recycling transferrin receptor are reconciled by other data showing that there are in fact two independent recycling pathways in some cell lines. Rapid recycling of transferrin–receptor complex occurs from a peripheral cell compartment, probably the primary endosomes. The pH in this compartment is presumably low enough to allow iron to dissociate from the transferrin (467). This particular route is unaffected by treatment with the carboxylic ionophore monesin, which essentially prevents protein movement through trans Golgi compartments. The second transferrin–receptor recycling pathway is sensitive to monesin, however (1076). The reason for the existence of these two pathways is not clear, although it appears that both transferrin–receptor complexes and ligand-free receptor recycle by both pathways. One possibility is that conformational changes resulting from the release of iron are sufficient to target the receptor–ligand complex into the "rapid" recycling route, whereas iron–transferrin–receptor complexes which persist in the primary endosome are eventually transported into the secondary endosome.

Transferrin receptor is a polarized secretory protein in endothelial cells (Section V.G.3). Transferrin–receptor complexes which are endocytosed from the basolateral surfaces of polarized cell monolayers are recycled via endosomes to the basolateral surface, indicating that the basolateral sorting signal in the transferrin receptor is not masked by bound transferrin (340).

b. Transcytosis

Transcytosis is an unusual phenomenon whereby a receptor is first inserted into the plasma membrane of a polarized epithelial cell and then endocytosed and transported to the opposite surface (Fig. VIII.2B). The classic examples of transcytosed receptors are the polymeric IgA receptor and the maternal IgG receptor. Both receptors are initially localized to the basal surface of polarized cells and then transported to the apical surface. IgA and IgG are thus transported from one side of the cell and secreted on the other, although the receptors can be transcytosed without bound ligand.

Mostov *et al.* (743) found that when the IgA receptor was truncated at the C-terminal end, thereby removing the transmembrane anchor domain, the protein was preferentially secreted from the apical surface. Thus, not only had the putative basolateral sorting signal present in the C-terminal tail been removed, but the N-terminal apical targeting signal had been activated. In the normal process of transcytosis, this region of the Ig receptor binds to the immunoglobulin and is subsequently endocytosed in coated vesicles and then cleaved to give a soluble form (*secretory component,* or SC) during transcytosis, probably in primary or secondary endosomes (763) (Fig. VIII.2B). Thus, the truncated IgA receptor constructed by Mostov *et al.* is similar to SC. Transport of SC–Ig complexes proceeds via secondary endosomes (366) and then on to the apical surface in naked vesicles. Further proteolytic processing of SC occurs during this final stage in transcytosis or when the vesicles reach the cell surface (741) (Fig. VIII.2B). There are obvious similarities between transcytosis of Ig receptors and the proposed sorting pathway for hepatocyte apical membrane proteins (50).

c. *Receptor recycling and ligand degradation*

A large number of endocytosed ligand–receptor complexes dissociate when they reach the secondary endosomes or prelysosomes, apparently as a result of low pH (133,443) (Fig. VIII.2C). In many cases, typified by LDL–LDL receptor, the receptor recycles to the plasma membrane and the ligand is transported into the lysosome, where it is degraded. LDL receptors probably leave the secondary endosome in vesicles budding from its rim (366) and appear to go directly to the cell surface rather than passing through the TGN. The asialoglycoprotein receptor, whose ligand also dissociates in the endosome and is transported to the lysosome, also recycles to the plasma membrane in different vesicles from those carrying secretory proteins (367). Thus, it seems that vesicles budding from both the TGN and prelysosomes are targeted to the plasma membrane. LDL receptor has been reported to make up to 150 round trips through the secondary endosome without being inactivated and to recycle every 10–20 min (385,386).

d. *Degradation of endocytosed receptor and ligand*

Epidermal growth factor (EGF) and its receptor are apparently transported into the lysosome, where both proteins are degraded. Since it is known that EGF dissociates from its receptor at low pH (242a), they are presumably transported independently from the prelysosome to the lysosome. It has been proposed that the cytoplasmic domain of the EGF receptor is proteolytically removed and released from the prelysosome or lysosome surface and that it contains a karyophilic signal which routes it into the nucleus (875). This idea is particularly attractive because this domain of the receptor has tyrosine kinase activity (491), which might be required to trigger cell division (386). It will be recalled that EGF receptors do not migrate to coated pits, and are therefore not endocytosed, unless they have bound EGF. This means that the kinase domain cannot be active in the absence of the hormone. Tyrosine kinase is activated as ligand binds and the receptor moves to coated pits, but a mutation which abolishes the tyrosine kinase activity of this protein is reported to be without effect on receptor activity, endocytosis, or EGF-stimulated DNA synthesis (894).

B. Other modes of protein uptake

1. Pinocytosis and fluid-phase endocytosis

It is clearly important for cells to be able to recover excess material reaching the cell surface via the secretory pathway, especially in the case of stimulated exocrine and endocrine cells, in which secretory granules with a total surface area ten times greater than that of the cell surface fuse with the plasma membrane in a single round of secretion (1079). Membrane retrieval is probably specifically designed to recover proteins whose distribution should normally be restricted to secretory granules and vesicles and to allow the reuse of membrane lipids in successive cycles of exocytosis; it therefore need not necessarily involve receptor–ligand interactions and will be referred to here as pinocytosis. However, the membranes of exocytosed chromaffin granules are reported to form patches on the cell surface. These patches then acquire a protein coat and pinch off to give coated vesicles which are subsequently uncoated and spread to various parts of the cell including the TGN but not the Golgi cisternae (849).

Endocytosis of endogenous membrane proteins has also been reported to occur together with bound specific antibodies, suggesting that some plasma membrane proteins are also endocytosed and probably recycled to the plasma membrane in the absence of antibodies. The endocytosed antibodies were also reported to recycle to the plasma membrane (1185), although considerable amounts were also reported to accumulate in lysosomes or endosomes (656).

Tracer proteins such as horseradish peroxidase (HRP) and cationized ferritin as well as a variety of other solutes have also been used to follow the fate of pinocytosed material. Endocytosed HRP has been reported to accumulate in endosomes (401), the lysosome (449), and the medial Golgi cisternae (822), whereas other endocytic tracers are reported to accumulate in pinocytic vesicles and even in secretory granules [reviewed in (1079)]. These differences almost certainly reflect differences in the cell types used in particular studies. Only transport of pinocytosed material through Golgi elements or the TGN is compatible with the recycling of pinocytosed membrane into the secretory pathway, however. Thus, some pinocytosed membrane may inevitably be lost to the lysosome, possibly because recycling depends on the random segregation of membrane lipids and associated proteins into different compartments. However, it is by no means clear that pinocytosed membrane accompanies the fluid phase of the pinocytic vesicle throughout its journey through the cell, and it may therefore never reach the lysosome. It is worth noting that some but not all types of recycling endocytic receptors transit through the TGN (see Section VIII.A.5) and therefore rejoin the secretory pathway, together with accompanying lipids.

2. Toxin uptake

Many toxins are taken up by eukaryotic cells by a process resembling endocytosis of ligand–receptor complexes. The toxins bind first to cell surface receptors, some of which have been identified and partially characterized. For example, chemical cross-linking of diphtheria toxin (DT) to tissue culture cells and immunoprecipitation of toxin–cell surface complexes with anti-toxin antibodies led to the tentative identification of a cluster of 20-kDa proteins as the DT receptor (176). Not all toxin receptors are proteins, however; the cholera enterotoxin and tetanus toxin receptors, for example, are glycolipids (199,730).

DT, which is secreted by toxigenic strains of the bacterium *Corynebacterium diphtheriae,* is among the best characterized of the endocytosed toxins. Analysis of truncated DT peptides and of point mutations in the DT structural gene allowed three different functions to be assigned to different domains of the toxin polypeptide (C-terminal receptor recognition domain, central translocation domain, and N-terminal catalytic domain with ADP ribosylation activity) (774).

Some toxins do not appear to be endocytosed via coated vesicles, probably because they do not use classical endocytic receptors, although receptor–DT clusters have been detected on the cell surface (564). DT uptake is ATP-independent, indicating that clathrin-uncoating activity is not involved in the uptake pathway (488), and ATP-dependent uptake of the plant toxin ricin, like fluid-phase endocytosis (see above) differs from receptor-mediated endocytosis in being unaffected by acidification of the cytosol (991).

It is not clear whether DT reaches secondary endosomes or the TGN or whether it follows an entirely different pathway from other endocytosed proteins, although endocytosed ricin has been shown to accumulate in the same post-Golgi compartment as does secretory G protein in vesicular stomatitis virus-infected cells (261) and as does specific antibody in hybridoma cells (1226). DT activity is unusual, however, in that its action requires a transmembrane energy potential which can be supplied either by a ΔpH or by $\Delta\psi$ (488,990) (the activity of the related plant toxin ricin requires ATP hydrolysis), indicating that the DT uptake pathway may be unique. Hudson *et al.* (488) have proposed that the energy potential is required for the "maturation" of vesicles containing DT and for membrane fusion, which, they suggest, might occur between the plasma membrane and the "matured" vesicle. They propose that the A chain (N-terminal catalytic domain) of DT is released from the rest of DT (B chain) by proteolysis during a lag phase occurring after endocytosis and that hydrophobic segments in the central, translocation domain, which remains attached to A via a disulfide bridge, penetrate into the membrane of the vesicle or endosome. The A chain may then cross the membrane, possibly through a pore formed by the translocase domain (see below), as the vesicle fuses with the plasma membrane. Chain A can cross the cytoplasmic membrane directly (i.e., without being endocytosed) if the cytoplasm is acidified (844a).

DT-resistant cell lines which still bind DT, and are therefore presumably defective in a later stage in DT uptake, have been isolated (383), but the nature of the defects have not been determined. Other groups have proposed that the B chain of DT penetrates into the membrane when the pH is lowered due to conformational changes which bring the hydrophobic segments close to the surface of the polypeptide (186). Exotoxin A of *Pseudomonas aeruginosa* has been shown to undergo similar conformational changes at low pH, leading to the suggestion that this toxin penetrates into the membrane when it arrives in the low-pH environment of the endosome (14,298).

The idea that the B chain of DT could form a pore is strengthened by *in vitro* studies showing that purified B chain forms pores in planar lipid bilayers in multilamellar vesicles and *in vivo* (549,844a). The way in which the A peptide of DT might penetrate through this pore has not been worked out, but photolabeling experiments indicate that both subunits are in contact with the lipids, although the pore formed by combined A + B subunits is smaller than that formed by the B subunit alone (1032).

The uptake of the closely related heat labile enterotoxins of *Vibrio cholerae* (CT) and *E. coli* (HLET) has been partially characterized. Both toxins contain a single A polypeptide which activates adenylate cyclase, together with five copies of the smaller B subunit. Some mutations in the gene for the B subunit have been reported to reduce the binding of these toxins to the cell surface (1125), confirming other studies showing that it has receptor recognition activity. One possibility is that the conformation of the B chain is altered upon contact with its receptor (ganglioside GM_1) to expose a hydrophobic face which penetrates into the membrane to form a pore through which the A chain can enter the cytoplasm.

3. Endocytosis and pinocytosis in yeasts

The yeast *S. cerevisiae* is potentially useful as a model system for studying endocytosis because of the ease with which it can be manipulated. Although the extent to which exocytosed membrane is recycled has not been determined, both fluid-phase and receptor-mediated endocytosis have been reported in this organism. The most widely used marker of fluid-phase pinocytosis is lucifer yellow, which is reported to accumulate in the vacuole (926). Fluid-phase pinocytosis of amylase and fluoresceine-labeled dextran have also been described (683), although a more recent study has shown that the latter phenomenon was probably due to uptake of low-molecular-weight contaminants rather than to dextran pinocytosis (893). Even vesicular stomatitis virus can be endocytosed by yeast cells, provided the cell wall is removed by converting the cells to spheroplasts (684).

The only proteins which are known to enter yeast cells by receptor-mediated endocytosis are the pheromones α and **a**. The α factor is the more extensively characterized of the two. The α-factor receptor is only present on haploid cells of the opposite, a mating type. Its density has been variously calculated at between 8×10^3 and 1×10^6 per cell (175,537,538). Sequence analysis of the α-factor receptor structural gene (*STE2*) indicates that the receptor differs from all other characterized endocytic receptors in being a polytopic membrane protein (7676). Endocytosed receptors seem to be degraded, probably in the vacuole, and are not recycled to the cell surface (536). The pheromone is also degraded shortly after uptake (175). Internalization of unoccupied receptors seems to be stimulated by the addition of α factor to **a** cells, possibly because the receptors form clusters which are endocytosed together, irrespective of whether all or only a few of the receptors have bound α factor. Although yeast cells are known to have clathrin and therefore could have coated pits similar to those found in other eukaryotes, studies on clathrin-deficient mutants have so far failed to reveal any defect in α-factor response. Energy is required for α-factor endocytosis, but the nature of the energy requirement has not been determined (175).

Both fluid-phase and receptor-mediated endocytosis are affected by certain *sec* mutations, which were originally identified by their defects in the secretory pathway (926), suggesting that endocytosis and exocytosis may share certain cellular components (e.g., in membrane fission and fusion). A selection procedure designed to yield mutants which were defective in endocytosis but not in invertase secretion yielded cold-sensitive mutations in two genes, *END1* and *END2*. Both classes of mutants were defective in both fluid-phase and α-factor endocytosis, but no endocytosis occurred at the permissive temperature even though the cells grew normally, indicating that the absence of endocytosis is not the cause of temperature-sensitive growth. The *end1* mutants are also defective in the biosynthesis of some vacuolar proteins and have numerous small vesicles in place of the usual large vacuole, indicating that the effects of the *end1* mutation are not restricted to endocytosis. The *end2* mutants accumulate a large number of membrane-limited organelles which can also be detected in reduced numbers in wild-type cells, although their identity and function have not been determined (926).

The great potential of the yeast system for studying the mechanisms of endocytosis has not been fully realized. Further work may lead to the design of a suicide selection procedure for more endocytosis-defective mutants, and it should be possible to design an *in vitro* assay system with which to study at least the early stages in endocytosis.

4. Bacteriocin uptake in bacteria

Membrane biogenesis and protein secretion in bacterial cells are not known to involve vesicular transport to the cell surface, and bacteria do not import proteins or any other macromolecules by an endocytosis-like process. The only proteins which are known to gain access into bacterial cells are *bacteriocins,* polypeptide toxins which usually only act on strains closely related to the producing strain. The best-studied group of bacteriocins, the colicins, act exclusively on *E. coli* and related strains. Colicins are divided into two major classes depending on their targets. Most of the characterized colicins cross the outer membrane and penetrate into the cytoplasmic membrane, where they either form pores or inhibit peptidoglycan synthesis, whereas the second class of colicins cross both outer and cytoplasmic membranes and degrade nucleic acids in the cytoplasm (895).

All of the characterized colicins have three distinct domains concerned, respectively, with translocation into susceptible cells (N-terminus), receptor recognition (central domain), and catalytic activity (C-terminus) (895). Different colicins are known to bind to different cell surface proteins, but little is known about the ways colicins subsequently cross the outer membrane. Colicin activity requires a transmembrane potential (539), and colicin remains accessible to exogenous trypsin when adsorbed to deenergized cells (920), indicating that energy may be required for outer membrane penetration. However, it is not clear how the energy potential across the cytoplasmic membrane could be coupled to the outer membrane, although such a role has been proposed for the TON B protein, which is required for the import of iron–siderophore complexes and vitamin B_{12} (all of which bind to cell surface receptors) and for the action of certain colicins (895). The possibility that colicins penetrate into cells at sites where the inner and outer membranes are juxtaposed (Chapter IV) has been raised on several occasions, but experiments directly addressing this aspect of colicin action have not yet been reported.

Pore-forming colicins have highly hydrophobic C-termini, which form the transmembrane channel. Isolated C-termini have almost the same pore-forming characteristics as do full-length colicins *in vitro* but are totally devoid of *in vivo* activity (221,693). All other colicins lack substantial regions of hydrophobicity, making it likely that they are transported through the outer membrane, and in some cases through the cytoplasmic membrane, through water-filled channels. Low pH has been reported to favor the penetration of both pore-forming and nuclease colicins into lipid membranes *in vitro* (222,291,848), apparently as a result of increased hydrophobicity, as is the case with toxins which act in eukaryotic cells (see Section 2 above). It is not clear whether the observed effects of low pH are relevant for the import of colicins *in vivo* because *E. coli* is not known to have an acidic compartment, although the transmembrane potential may cause the same effect by protonating amino acids in the colicin polypeptide (1037).

The N-terminus of the translocation domains of many of the characterized colicins is particularly rich in glycine residues, giving them considerable structural flexibility. Whether this feature is important for colicin uptake remains to be determined, but a significant reduction in the number of glycine residues in the pore-forming colicin N did not appear to affects its *in vivo* activity (896). Nevertheless, the effects of a mutation affecting the sequence of this part of the nuclease-type colicin E3 underscore the importance of this domain in colicin uptake (290).

A number of *E. coli* mutations which resist the action of one or more colicins have been described. Many of these mutations affect the production of several cell envelope components including lipopolysaccharide and proteins (895), making it difficult to determine whether there is a cause-and-effect relationship between their absence and the loss of colicin sensitivity. However, this concern overshadows a more important aspect of this class of mutations, namely the possibility that they affect both membrane biogenesis and colicin uptake. Further characterization of the products of the wild-type genes should provide useful insights into the mechanisms of colicin uptake.

Further reading

Goldstein, J. L., Brown, M. S., Anderson, R. E. W., Russel, D. W., and Schneider, W. (1985). Receptor-mediated endocytosis: Concepts emerging from the LDL receptor system. *Annu. Rev. Cell Biol.* **1,** 1–39.

Neville, D. M., Jr., and Hudson, T. H. (1986). Transmembrane transport of diphtheria toxin, related toxins and colicins. *Annu. Rev. Biochem.* **55,** 195–224.

Steinman, R. M., Mellman, I. S., Miller, W. A., and Cohn, Z. A. (1983). Endocytosis and the recycling of plasma membrane. *J. Cell Biol.* **96,** 1–27.

CHAPTER IX

Applications of protein targeting

So far we have been almost entirely concerned with the fundamental aspects of protein targeting. In this final chapter, we turn our attention to the potential applications of protein targeting. In fact, the separation of fundamental and applied aspects of protein targeting is somewhat artificial since much of the information described in preceding chapters was derived directly or indirectly from applied or medical research. Furthermore, the potential applications of fundamental research on protein targeting has provided a tremendous stimulus for continued interest in this area of research. The chapter is divided into three sections dealing with potential applications of protein export and secretion, organelle targeting, and protein import, respectively.

A. Applications of protein export and secretion

It was only shortly after the development of systems for cloning and expressing heterologous genes in bacteria, and the discovery of at least partial similarity between the early stages of protein secretion in bacteria and multicellular eukaryotes, that several groups became interested in the possibility of producing novel secretory proteins in bacteria. The aim of these tests was to take advantage of the considerable advantages of bacteria for the large-scale, economic production of proteins, principally those of medical and veterinary interest. These studies were carried out at a time of increasing interest in the high-level production of proteins of bacterial origin, particularly those already used in the biotechnology, food, and chemical industries. Initially, attempts to improve the efficiency of protein production, thereby reducing production costs, were essentially based on the classical strain improvement techniques then widely used in other aspects of applied microbiology. Subsequently, it became possible to use molecular genetic techniques to the same end. Shortly thereafter came the development of simple eukaryotes for the production of heterologous proteins, and improvements in the large-scale cultivation of animal cells have opened up new vistas for the production of hormones and other proteins. Each individual system (bacteria, simple eukaryotes, and animal cell cultures) has its own particular advantages and disadvantages, as discussed in the following sections (Table IX.1). Another source of the current interest in protein secretion by bacteria is that many secreted proteins (toxins and degradative enzymes) play a important role in bacterial pathogenicity. It is hoped that a greater understanding of the mechanisms of protein secretion, particularly by Gram-negative bacteria, may lead to new ways of controlling toxigenicity.

TABLE IX.1. Advantages and disadvantages of producing heterologous eukaryotic proteins in different cell cultures

Organism	Location	Advantages[a]	Disadvantages[a]
Bacteria		Simple expression vectors; easy to grow; high cell yields	Plasmids unstable; restrictions on cell types; no posttranslational modification
	Cytoplasm	Simple genetic constructs	Protein toxic to cells; proteins aggregate; loss of activity; proteolysis; no processing (SP)
	Periplasm	Reduced proteolysis; released in concentrated form; correct processing (SP); high yields	Export inefficient or easily saturated
	Outer membrane	Surface-exposed epitopes (live vaccines)	Export inefficient or easily saturated
	Secreted	Relatively pure; reduced proteolysis; correct processing (SP); high yields	Large volume, dilute product; export inefficient or saturated
Yeast	Secreted	Secretion pathway similar to other eukaryotes; correct processing (SP); posttranslational modification; easy to grow	Plasmids unstable; size restriction; incorrect posttranslational modification
Mammalian	Secreted	Correct processing and maturation	Stringent growth requirements; restrictions on cell types

[a] SP, signal peptide.

1. Export and secretion of heterologous proteins by bacteria

The main advantages of bacteria for the production of heterologous proteins are their flexibility and the ease with which they can be grown. Simple plasmid expression vectors are available for the most commonly used strains of bacteria (*E. coli, Pseudomonas aeruginosa, B. subtilis,* and other species of *Bacilli, Streptomyces,* and *Staphylococcus*). The efficiency of transformation by plasmid DNA is relatively low, but plasmids can be maintained, in pilot studies at least, by growing the bacteria in media containing the antibiotic to which the plasmid confers resistance. This strategy is often not applicable to large-scale cultures in industrial fermentors, in which case the cost of the antibiotic may become prohibitive, and their use in the preparation of most proteins may be prohibited for pharmacological reasons. Unfortunately, many recombinant bacterial plasmids are unstable and are lost during prolonged growth of the bacteria in the absence of selection. Several laboratories are now investigating methods of ensuring plasmid retention in the absence of external selection.

Most species of bacteria grow to high cell densities in relatively crude, inexpensive growth media. Strains with exotic growth requirements are avoided, as are strains which cannot tolerate the high cell densities and low oxygen tension which are characteristics of most industrial fermentors. Some strains cannot be used for the industrial production of heterologous proteins because they produce toxins.

Proteins produced by bacteria can be directed to one of five different locations, four of which are important for the production of heterologous proteins (Table IX.1). The disadvantages of allowing heterologous proteins to accumulate in the cytoplasm include (i) degradation by "housekeeping" proteases specifically designed to remove abnormal proteins; (ii) folding into inactive conformations or aggregation; (iii) and absence of posttranslational modification or processing. Furthermore, they may be difficult to purify because of the massive amounts of other proteins present in soluble cell extracts. Some of these disadvantages can be overcome by exporting the proteins into the periplasm (Gram-negative bacteria) or into the medium. However, the protein export machinery is often unable to cope with the large amounts of protein which must be produced to make the system economical, and the proteins consequently aggregate in the cytoplasm (see examples in Table IX.2). Considerable effort is currently being directed toward developing methods for denaturing and refolding these proteins. Nonetheless, heterologous proteins can accumulate in very large amounts when exported to the periplasm (61,985,1103).

TABLE IX.2. Examples of attempts to use *E. coli* K-12 to export or secrete heterologous proteins of commercial interest

Protein	Signal peptide[a]	Result	Reference
Insulin	BLA or insulin	Periplasmic, low yields	1105
Hirudin	PHO A	Periplasmic, 1 mg/ml	247
Human growth hormone	OMP A	Periplasmic, 15 μg/ml	61
	PHO A	Periplasmic, 2 μg/ml	806
Human β-endorphin	OMP F	Secreted 1 μg/ml	764
Insulin-like growth factor	Protein A	"Secreted"[b]	2
Pig secretin	BLA	Partially processed	1091
Wheat α-amylase	BLA	Periplasmic	355
Salmon growth hormone (SGH)	SGH	Cytoplasmic aggregates, 15% cell protein	1031
Chymosin	Chymosin	Cytoplasmic aggregates, 20% of cell protein	1021
Ig F_c chimera	OMP A, PHO A	Periplasmic, 5% periplasmic protein	1057
Ig F_{ab} chimera	PEL B	"Secreted"[b], 2 μg/ml	80
Superoxide dismutase	OMP A	Mainly periplasmic, 7–20% total cell protein	1104

[a] OMP A, *E. coli* outer membrane protein OMP A; OMP F, *E. coli* outer membrane protein OMP F; PHO A, *E. coli* periplasmic alkaline phosphatase; BLA, pBR322 plasmid-encoded β-lactamase; PEL B, *Erwinia carotovora* pectate lyase B.
[b] Probably by partial lysis.

The main advantage of exporting heterologous proteins to the periplasm is that they can be concentrated as part of the cell mass and then released by permeabilizing the outer membrane, which usually leads to insignificant release of cytoplasmic and membrane proteins, thereby facilitating purification. Bacteria generally secrete very few proteins; extracellular products are therefore present in the medium in a relatively pure form, but harvesting and further purification is hampered by the fact that the proteins are usually very dilute. This disadvantage may be overcome by using affinity adsorption chromatography (see below).

In considering the possible applications of protein export and secretion by bacteria, it is important to remember two facts. The first is that normally cytoplasmic proteins will almost invariably be unable to cross the bacterial cytoplasmic membrane if they are fed into the secretory pathway; i.e., a signal peptide is not sufficient to lead a normally cytoplasmic protein out of the cytoplasm (see Section III.A.3). The second is that bacteria are unable to glycosylate proteins. Since most eukaryotic secretory proteins are glycosylated, and glycosylation has a major effect on antigenic properties, eukaryotic proteins produced by bacteria will be antigenically different from those produced by eukaryotic cells. This is particularly important for proteins such as hormones or growth factors which, when produced by bacteria, would be recognized as "foreign" by the mammalian immune response system. This may limit the applications of heterologous protein export and secretion by bacteria to hormones and growth factors which are not glycosylated and to other, less profitable proteins which are not used in human or animal therapy.

The extensively characterized Gram-negative bacterium *E. coli* is generally preferred for studies on molecular genetics and often serves as intermediate host during the construction of plasmids carrying genes which will eventually be expressed elsewhere, either in other bacteria or in eukaryotic cells. *E. coli* strain K-12, the most extensively characterized of all bacteria, suffers from the major disadvantage that it does not have an endogenous system for secreting proteins into the medium. However, *E. coli* may have certain advantages which are only now being fully appreciated (see below).

Gram-positive bacteria are the source of most of the microbial enzymes currently used in food and drug industries (e.g., amylases, proteases, and lipases). These enzymes are secreted into the medium (see Chapter IV). Gram-positive bacteria, and in particular *B. subtilis* and related bacteria, have therefore been extensively characterized with respect to their potential for secreting heterologous proteins.

Some examples of the use of bacteria to produce heterologous proteins are listed in Tables IX.2 and IX.3. In most cases, heterologous proteins of eukaryotic origin are produced as hybrid polypeptides containing a bacterial signal peptide. This avoids possible complications caused by incompatibility of eukaryotic signal peptides with the prokaryotic protein export machinery (see Section III.A.2). The exceptions in Table IX.2 are insulin,

TABLE IX.3. Examples of the use of Gram-positive bacteria to secrete heterologous proteins of commercial interest

Protein	Bacterium	Signal peptide	Secreted	Yields	Reference
Human interferon $\alpha 2$	*B. subtilis*	α-Amylase	+	Low	844
	S. lividans	Interferon	−	Good	841
Mouse interferon β	*B. subtilis*	α-Amylase	+	Low	1036
Insulin-like growth factor	*S. aureus*	Protein A	+	Moderate	788
Human serum albumin	*B. subtilis*	α-Amylase or *N*-protease	+[a]	Good	998
Interleukin 1B	*S. lividans*	β-Galactosidase	+	Moderate	635a

[a] Retained by cell wall, release upon conversion to protoplasts.

chymosin, and salmon growth hormone, all of which are inefficiently exported to the periplasmic space in *E. coli*. Three of the hybrid proteins listed in Table IX.2 are reported to be secreted into the growth medium. The OMP F–β-endorphin hybrid is relatively short and may leak across the outer membrane from the periplasmic space (764). The situation with the protein A–insulin-like growth hormone and PEL B–IgF_{ab} hybrids is somewhat different because other periplasmic proteins are released at the same time. This hybrid may therefore directly or indirectly destabilize the outer membrane (2).

Gram-positive bacteria have been extensively used for the expression of genes coding for normally extracellular enzymes from other Gram-positive bacteria (295,721,1039,1138). These enzymes are generally resistant to the extracellular proteases which appear to be produced by all Gram-positive bacteria. Proteins of eukaryotic origin are highly sensitive to these proteases. Even mutations which block the production of the principal extracellular proteases do not completely stabilize secreted heterologous proteins produced by *B. subtilis* (295,1165a).

Increasing attention is now being paid to the ability of Gram-positive bacteria other than *Bacillus* to secrete proteins. Of particular interest is *Streptomyces*, which differs from all other bacteria tested so far in its ability to secrete normally cytoplasmic proteins such as *E. coli* β-galactosidase or galactokinase when they are fused to a *Streptomyces* signal peptide (653a; see Section III.A.3). The reason for this unusual and potentially useful property remains to be determined. Among the intriguing possibilities are a tighter coupling of secretory protein biosynthesis and secretion, a more efficient system for inhibiting premature folding of secretory protein precursors in the cytoplasm, or an unprecedented mechanism for the denaturation of prefolded secretory proteins as they are translocated through the cytoplasmic membrane.

Three other aspects of protein export and secretion by bacteria are also receiving increasing attention. The first is the use of the secretory pathway of Gram-negative bacteria to insert foreign epitopes into outer membrane proteins (6,153,329). These epitopes may be exposed on the surface of the bacteria if they are inserted in a position corresponding to a normally extracellular loop of the polypeptide and will therefore be immunogenic (Fig. IV.1). This approach is now being used to develop live vaccine strains, which substantially reduce the cost of vaccine preparation (154). Another approach to the same problem is illustrated by work by de la Cruz *et al.* (203), who cloned DNA coding for a foreign epitope into gene *III* of *E. coli* phage f1. Protein III is first inserted into the cytoplasmic membrane of infected cells and then assembled into the f1 phage coat. Phages are released by an unknown nonlytic process resembling secretion, and the cells continue to produce phage over long periods. Extracellular phages could be harvested relatively easily and were found to have the epitope on their surface and to be immunogenic.

The second area of increasing interest is in the use of secretory pathway-independent protein secretion to secrete heterologous proteins into the medium. The use of *E. coli* α-hemolysin seems most promising in this respect; preliminary studies have already shown that a small C-terminal fragment of α-hemolysin contains all of the information required for its secretion into the medium (677). It remains to be seen whether this property can be used to allow *E. coli* to secrete heterologous proteins as hybrids from which the α-hemolysin secretion signal could be subsequently excised as described below. This system may be more suitable than the signal peptide pathway for secreting normally cytoplasmic proteins.

The third novel application of protein export and secretion by bacteria is the use of the protein A secretion vehicles developed by Uhlèn and his colleagues. Protein A, a secreted protein produced by *Staphylococcus aureus,* has a high affinity for class G immunoglobulins. Hybrid proteins containing all or part of the IgG-binding domain of protein A can therefore be purified by affinity chromatography on an IgG affinity adsorption column. Uhlèn and his colleagues (659) have used this feature, together with the fact that protein A is made as a precursor with an efficient signal peptide, to devise methods for exporting or secreting hybrid polypeptides containing heterologous sequences in Gram-negative and Gram-positive bacteria, respectively. Initial pilot studies using β-galactosidase and alkaline phosphatase as reporter proteins were subsequently developed to study the production of a protein A–insulin-like growth factor hybrid (659). Other binding proteins of microbial origin are also being developed for use in secretion or affinity purification vehicles (65). In all of these cases, it may be important to separate the two parts of the secreted hybrid polypeptide. A number of agents cleave polypeptides at specific sites (Table IX.4). Gene fusions can be specifically engineered so that these sites occur at the joint between the two polypeptide sequences. This feature can then be used in a two-step affinity chromatography purification process (729), as illustrated in Fig. IX.1.

2. Secretion of heterologous proteins by yeasts

Yeasts offer several significant advantages over bacteria for the commercial production of proteins of eukaryotic origin, the principal one being that the secretory pathway is similar to that in more complex eukaryotes. Foreign secretory proteins will therefore be subject to extensive post-

TABLE IX.4. Potential cleavage sites permitting specific splitting of hybrid polypeptides

Site[a]	Agent	Site	Agent
Met ↔ X	Cyanogen bromide	Gly-Asp-Asp-Asp ↔ X	Enterokinase
Asp ↔ Pro	Formic acid	Ile-Glu-Gly-Arg ↔ X	Factor X
Asn ↔ Gly	Hydroxylamine	Pro-Val ↔ Gly-Pro	Collagenase
Trp ↔ X	Bromosuccinimide	Lys ↔ X	Pseudomonase Protease Ps-1
Cys ↔ X	Nitro-thiocyano-benzoate	Pro-Phe-Arg ↔ X	Plasma kallekrein
Ala Gly Ala Ser-X-Gly ↔ X Leu Ser Val Ile	Signal peptidase	Thr-Pro-Pro ↔ Thr-Pro-Ser-Pro ↔ Ser-Thr-Pro-Pro ↔ Thr ↔ Pro-Ser-Pro-Ser	IgA proteases
Gly-Pro-Arg ↔ X	Thrombin	Arg-Lys ↔ X	Yeast KEX2 protease

[a] X, Almost any amino acid; ↔, cleavage site.

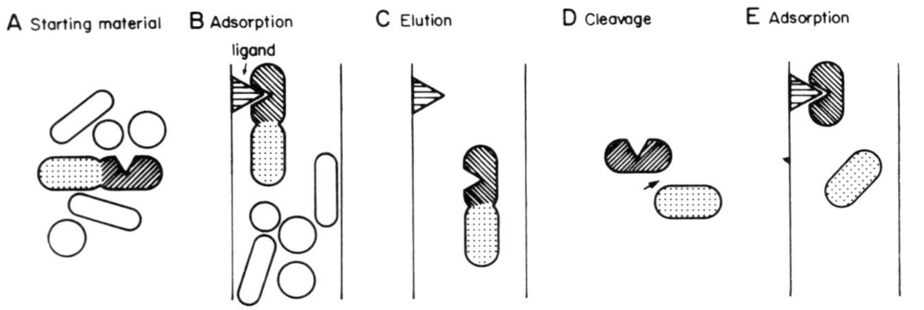

Fig. IX.1. Adsorption column chromatographic purification of secretory proteins from hybrid polypeptides. (A) The starting material (from the periplasmic space or medium) contains a hybrid protein which includes a "carrier" segment carrying an affinity recognition site, together with a mixture of other proteins. (B) The mixture is applied to an affinity adsorption column, which is then washed to eliminate all unbound proteins. The hybrid protein remains specifically bound to the column. (C) The hybrid protein is eluted (with high salt concentration, protein denaturant, or competing ligand). (D) The hybrid protein is cleaved with a reagent (Table IX.4) recognizing a specific site between the two parts of the molecule. (E) The mixture is then reapplied to the same column (after renaturation if appropriate). The carrier protein remains bound to the column, whereas the protein of interest is specifically present in the effluent.

translational modification, and may be structurally similar, if not identical, to the proteins secreted by the original producing cells. One disadvantage of yeasts is that secretory proteins are highly mannosylated, although this can be prevented by using mannosylation-defective mutants. Little is known about the location and identity of yeast Golgi enzymes involved in secretory protein glycosylation, but it may eventually be possible to engineer yeast strains genetically to perform the same glycosylation reactions as other eukaryotes. Yeasts, like bacteria, are able to grow to high densities in relatively crude media.

Some examples of the use of the yeast *S. cerevisiae* for the production of heterologous proteins are listed in Table IX.5. In most cases, the cloned gene was expressed under the control of an efficient yeast promoter, and the 5′ end encoding the natural signal peptide was sometimes replaced by DNA coding for a yeast signal peptide. Fusion to the signal peptides and mature segments of yeast prepro-α factor or killer toxin directly after a natural proteolytic processing site (Section V.D.4) offers the additional advantage that foreign proteins can be secreted without additional "carrier"-derived sequences. Proteins synthesized with their own or yeast signal peptides are generally efficiently processed and secreted into the medium, although some [e.g., immunoglobulins (1203)] form aggregates in the cytoplasm, and others [e.g., α-interferon (1241)] are trapped in the periplasmic space between the cytoplasmic membrane

TABLE IX.5. Secretion of heterologous proteins of commercial interest by *S. cerevisiae*

Protein	Signal peptide	Result	Reference
β-Endorphin	Prepro-α factor	Secreted	1241
Calcitonin	Prepro-α factor	Secreted	1241
α-Interferon	Prepro-α factor	Periplasmic	85,1241
Somatostatin	Prepro-α factor	Secreted	396
Epidermal growth factor	Prepro-α factor	Secreted	116
Egg lysozyme	Lysozyme	Secreted	798
Interleukin 1β	Killer toxin	Secreted	38
Insulin	Prepro-α factor	Secreted	1111
Trypsin inhibitor	Trypsin inhibitor	Secreted	527
Hirudin	Prepro-α factor	Secreted	650
Immunoglobulins	Immunoglobulins	Secreted[a]	1203
Cellulase (bacterial)	Killer toxin	Secreted	1058
Cellobiohydrolases (fungal)	Cellobiohydrolase	Secreted	859
α-Amylase (pancreatic)	Prepro-α factor	Secreted	26
α-Amylase (wheat)	α-Amylase	Secreted	973
α-Amylase (salivary)	α-Amylase	Secreted	767
α-Amylase (bacterial)	α-Amylase	Secreted	976

[a] Inefficient; mainly in cytoplasmic inclusion bodies.

and the cell wall. The latter situation is probably due to the large size of these polypeptides, which restricts their diffusion through the cell wall. The endogenous secreted proteins invertase (293) and acid phosphatase are also trapped by the cell wall rather than secreted into the medium. Periplasmic proteins can be released by chitinase treatment, which may offer advantages for protein purification. Relatively high protein yields have been reported [e.g., 5×10^5 molecules per cell for human salivary α-amylase (767), 1% of total cell protein for hirudin (650), and 0.2 μg/10^7 cells for somatostatin (396)].

The fermentation capacity of the yeast *S. cerevisiae* would be considerably enhanced if it were able to utilize carbohydrate polymers such as cellulose or starch. There is therefore considerable interest in the cloning and expression of genes for enzymes such as cellulases and amylases in yeast. Several examples of the production and secretion of these enzymes by recombinant yeasts are listed in Table IX.5. Other yeasts and fungi may also be suitable for the industrial production of secreted proteins (204,412).

3. Secretion of heterologous proteins by animal cell cultures

Several features of animal cell lines such as Chinese hamster ovary cells make them ideal for producing proteins for use in human or animal therapy. The protein products of heterologous mammalian genes expressed in these cell lines are more likely to be folded, modified, and secreted in the same way as they are in the parent cells than they are in yeasts. This avoids the risk that the products, although biologically active, will be antigenically different from the authentic product. The use of animal cell cultures was not feasible until techniques for growing large-volume cultures (up to 10,000 liters) were developed, and vectors for the stable expression of cloned genes became available. Even now, the costs of protein production are very high because expensive, serum- and protein-free, semi-defined media must be used. The technique is therefore only economically feasible for the production of proteins with a high market value (240,363,548,559–561,1169,1234). An increasingly attractive alternative to mammalian cell cultures is the use of baculovirus vectors to express foreign genes in insects. Again, many products of heterologous genes are antigenically identical to the authentic product produced by mammalian cells, and protein yields are similar to those obtained in serum-supplemented mammalian cell cultures (661).

Another recent development in the use of mammalian cell cultures has been the use of viral surface glycoprotein proteins to localize foreign sequences to the cell surface. An excellent example of this is provided by the recent work of Uijawa *et al.* (1130), who replaced the C-terminus of the type II membrane glycoprotein G of respiratory syncytial virus with the repeating epitope of an antigen expressed by *Plasmodium falciparum* circumsporozoites. The inserted epitope was exposed on the surface of cells expressing the gene fusion. The eventual hope is that recombinant viruses of this type may be used as live vaccines.

B. Applications of organelle targeting

Relatively little attention has been focused on the potential applications of protein targeting to organelles. One area of research which is being increasingly developed, however, is the targeting of proteins into chloroplasts of transgenic plants. Several studies illustrate the significant practical advantages of this technique in the development of herbicide-resistant plants. In these studies, a genetic transformation technique based on a plasmid found in a bacterium, *Agrobacterium tumefaciens,* which causes tumor growth on infected plants, was used to introduce a gene conferring herbicide resistance into plants. In one case, the gene concerned was of bacterial origin but was fused to DNA coding for a transit peptide. The protein was therefore targeted into the chloroplast and conferred herbicide resistance upon the plant (171). In another example, a chloroplast gene coding for a quinone-binding membrane protein which confers sensitivity to the herbicide abrazine was cloned, mutated so that it no longer bound the herbicide, fused to DNA coding for a transit peptide, and reintroduced into plants. The modified protein was imported into the chloroplasts, incorporated into photosynthetic membranes, and conferred abrazine resistance (167). These techniques allow the development of crop plant strains which are resistant to specific herbicides. The latter can then be used to kill all other plants growing on the same land.

The potential applications of other aspects of organelle targeting remain to be determined. One exciting possibility is the imposed regulation of genes integrated into chromosomes of eukaryotic cells by exogenous regulators. This may be done by fusing the gene for a simple regulator such as a bacterial repressor protein to a karyophilic signal so that its ability to recognize target DNA sequences is unaffected. This hybrid protein could therefore regulate the expression of specifically engineered nuclear genes. Another intriguing possibility is the packaging of heterologous proteins into yeast peroxisomes, which might facilitate their purification from whole cell lysates. Further characterization of peroxisomal routing signals (Section VI.D) and their ability to transport cytosolic proteins into the peroxisome may lead to the development of a potentially useful technique for segregating nonsecretable proteins from bulk cytosolic proteins.

C. Applications of protein import and endocytosis

As discussed in Chapter VIII, a wide variety of proteins can be imported into eukaryotic cells by endocytosis, including toxins. One potential application of this phenomenon is to target toxins to one particular class of cells, such as tumor cells. This approach became feasible when it was demonstrated that the receptor recognition domain of certain toxins could be replaced by other sequences without affecting their catalytic activity (502). These hybrid polypeptides are not toxic *in vivo,* unless the additional segment contains sequences recognized by cell surface receptors. This opens up the possibility of cross-linking a toxin polypeptide lacking its receptor recognition domain to antibodies specific for surface proteins present on tumor cells (immunotoxins), or of creating hybrid proteins carrying specific receptor recognition domains (155,665).

Further reading

Bigelis, R., and Das, R. C. (1988). Secretion research in industrial mycology. *In* "Protein Traffic and Organelle Biogenesis" (R. C. Das and P. W. Robbins, eds.), pp. 771–810. Academic Press, San Diego, California.

References

1. Abeijon, C., and Hirschberg, C. B. (1988). *Proc. Natl. Acad. Sci. U.S.A.* **85,** 1010–1014.
2. Abrahmsen, L. *et al.* (1986). *Nucleic Acids Res.* **14,** 7487–7500.
3. Achstetter, T. *et al.* (1988). *J. Biol. Chem.* **263,** 11711–11717.
4. Adams, G. A., and Rose, J. K. (1985). *Cell (Cambridge, Mass.)* **41,** 1007–1015.

4a. Addison, R. *et al.* (1988). *J. Biol. Chem.* **263,** 14281–14287.

5. Adler, L. A., and Arvidson, S. (1988). *J. Bacteriol.* **170,** 5337–5343.
6. Agterberg, M. *et al.* (1987). *Gene* **59,** 145–150.
7. Ahle, S. *et al.* (1988). *EMBO J.* **7,** 919–929.
8. Ainger, K. J., and Meyer, D. I. (1986). *EMBO J.* **5,** 951–955.
9. Ajioka, R. S., and Kaplan, J. (1986). *Proc. Natl. Acad. Sci. U.S.A.* **83,** 6445–6449.
10. Akiyama, T., and Ito, K. (1987). *EMBO J.* **6,** 3465–3470.
11. Alderson, J. W., and Miller, P. E. (1985). *Science* **228,** 993–996.
12. Allison, D. S., and Schatz, G. (1986). *Proc. Natl. Acad. Sci. U.S.A.* **83,** 9011–9015.
13. Allison, D. S., and Young, E. T. (1988). *Mol. Cell. Biol.* **8,** 1915–1922.
14. Allured, V. S. *et al.* (1987). *Proc. Natl. Acad. Sci. U.S.A.* **84,** 1320–1324.
15. Ammerer, G. *et al.* (1986). *Mol. Cell. Biol.* **6,** 2490–2499.
16. Anderson, D. J. *et al.* (1983). *Proc. Natl. Acad. Sci. U.S.A.* **80,** 7249–7253.
17. Anderson, R. G. W. *et al.* (1982). *J. Cell Biol.* **93,** 523–531.
18. Anderson, R. G. W., and Orci, L. (1988). *J. Cell Biol.* **106,** 539–543.
19. Anderson, R. G. W. *et al.* (1984). *Proc. Natl. Acad. Sci. U.S.A.* **81,** 4838–4842.
20. Andrews, D. W. *et al.* (1985). *Proc. Natl. Acad. Sci. U.S.A.* **82,** 785–789.
21. Andrews, D. W. *et al.* (1987). *EMBO J.* **6,** 3471–3477.
22. Andro, T. *et al.* (1984). *J. Bacteriol.* **160,** 1199–1203.
23. Arakawa, H. *et al.* (1987). *EMBO J.* **6,** 1361–1366.
24. Argan, C. *et al.* (1983). *J. Biol. Chem.* **258,** 6667–6670.
25. Armeither, H. *et al.* (1984). *Cell (Cambridge, Mass.)* **39,** 99–109.
26. Astofoli-Filho, S. *et al.* (1986). *Bio Technology* **4,** 311–315.
27. Atkinson, P. H., and Lee, J. P. (1984). *J. Cell Biol.* **98,** 2245–2249.
28. Audigier, Y. *et al.* (1987). *Proc. Natl. Acad. Sci. U.S.A.* **84,** 5783–5787.
29. Austen, B. M. (1979). *FEBS Lett.* **103,** 308–312.
30. Austen, B. M. *et al.* (1984). *Biochem. J.* **224,** 317–325.
31. Avran, P., and Castle, J. D. (1987). *J. Cell Biol.* **104,** 243–252.
32. Bacalloa, R. *et al.* (1986). *J. Biol. Chem.* **261,** 12907–12910.
33. Baeuerle, P. A., and Huttner, W. B. (1987). *J. Cell Biol.* **105,** 2655–2664.
34. Baga, M. *et al.* (1987). *Cell (Cambridge, Mass.)* **49,** 241–251.
35. Baker, A., and Schatz, G. (1987). *Proc. Natl. Acad. Sci. U.S.A.* **84,** 3117–3121.

35a. Baker, D. *et al.* (1988). *Cell (Cambridge, Mass.)* **54,** 335–344.

36. Baker, K. *et al.* (1987). *J. Mol. Biol.* **198,** 693–703.

36a. Baker, R. K., and Lively, M. O. (1987). *Biochemistry* **26,** 8561–8567.

37. Bakker, E. P., and Randall, L. L. (1984). *EMBO J.* **3,** 893–890.
38. Balcardi, C. *et al.* (1987). *EMBO J.* **6,** 229–234.
39. Balch, W. E. *et al.* (1984). *Cell (Cambridge, Mass.)* **39,** 405–416.
40. Balch, W. E. *et al.* (1984). *Cell (Cambridge, Mass.)* **39,** 525–536.
41. Balch, W. E. *et al.* (1986). *J. Biol. Chem.* **261,** 14681–14689.
41a. Bangs, J. D. *et al.* (1988). *J. Biol. Chem.* **263,** 17697–17705.
42. Bankaitis, V. A., and Bassford, P. J., Jr. (1985). *J. Bacteriol.* **161,** 169–178.
43. Bankaitis, V. A. *et al.* (1984). *Cell (Cambridge, Mass.)* **37,** 248–252.
44. Bankaitis, V. A. *et al.* (1986). *Proc. Natl. Acad. Sci. U.S.A.* **83,** 9075–9079.
45. Baraclough, R., and Ellis, R. J. (1980). *Biochim. Biophys. Acta* **608,** 19–31.
46. Barbus, J. A. *et al.* (1985). *FEBS Lett.* **188,** 73–76.
47. Barlow, D. J. *et al.* (1986). *Nature (London)* **322,** 747–748.
48. Barr, R. *et al.* (1984). *J. Biol. Chem.* **259,** 14064–14067.
49. Barriocanal, J. G. *et al.* (1986). *J. Biol. Chem.* **261,** 16755–16763.
50. Bartles, J. R. *et al.* (1987). *J. Cell Biol.* **105,** 1241–1251.
51. Bassüner, R. *et al.* (1983). *Eur. J. Biochem.* **133,** 321–326.
52. Batenburg, A. M. *et al.* (1988). *J. Biol. Chem.* **263,** 4202–4207.
52a. Batenburg, A. M. *et al.* (1988). *Biochemistry* **27,** 5678–5685.
53. Bathurst, K. *et al.* (1987). *Science* **235,** 348–349.
54. Bause, E., and Lehle, I. (1979). *Eur. J. Biochem.* **101,** 531–540.
55. Bayer, E. A., and Lamed, R. (1986). *J. Bacteriol.* **167,** 828–836.
56. Bayer, E. A. *et al.* (1985). *J. Bacteriol.* **163,** 552–559.
57. Bayer, M. E. (1979). *In* "Bacterial Outer Membranes: Biogenesis and Functions" (M. Inouye, ed.), pp. 167–202. Wiley, New York.
58. Bayer, M. H., and Bayer, M. E. (1985). *J. Bacteriol.* **162,** 50–54.
59. Bayer, M. H. *et al.* (1982). *J. Bacteriol.* **149,** 758–767.
60. Beck, E., and Bremer, E. (1980). *Nucleic Acids Res.* **8,** 3011–3024.
61. Becker, G. W., and Hsung, H. M. (1986). *FEBS Lett.* **204,** 145–150.
62. Beckers, C. J. M. *et al.* (1987). *Cell (Cambridge, Mass.)* **50,** 525–534.
63. Beckwith, J., and Ferro-Novick, S. (1986). *Curr. Top. Microbiol. Immunol.* **125,** 5–28.
64. Bedouelle, H. *et al.* (1980). *Nature (London)* **285,** 78–81.
65. Bedouelle, H., and Duplay, P. (1988). *Eur. J. Biochem.* **171,** 541–549.
66. Bedwell, D. M. *et al.* (1987). *Mol. Cell. Biol.* **7,** 4038–4047.
67. Begg, K. J. (1978). *J. Bacteriol.* **135,** 307–310.
68. Béguin, P. *et al.* (1985). *J. Bacteriol.* **162,** 102–105.
69. Bellion, E., and Goodman, J. M. (1987). *Cell (Cambridge, Mass.)* **48,** 165–173.
70. Beltzer, J. P. *et al.* (1988). *J. Biol. Chem.* **263,** 368–374.
71. Bendzko, P. *et al.* (1982). *Eur. J. Biochem.* **123,** 121–126.
72. Bennett, F. C. *et al.* (1983). *J. Cell Biol.* **97,** 1566–1572.
73. Bennett, J. (1977). *Nature (London)* **269,** 344–346.
74. Benson, S. A., and Silhavy, T. J. (1983). *Cell (Cambridge, Mass.)* **32,** 1325–1335.
75. Benson, S. A. *et al.* (1984). *Proc. Natl. Acad. Sci. U.S.A.* **81,** 3830–3834.
76. Benson, S. A. *et al.* (1985). *Annu. Rev. Biochem.* **54,** 101–134.
77. Benson, S. A. *et al.* (1987). *J. Bacteriol.* **169,** 4686–4691.
78. Berger, M., and Schmidt, M. F. G. (1986). *J. Biol. Chem.* **261,** 14912–14918.
79. Bergman, J. E., and Singer, S. J. (1983). *J. Cell Biol.* **97,** 1777–1787.
80. Better, M. *et al.* (1988). *Science* **240,** 1041–1043.
80a. Bever, R. A., and Iglewski, B. H. (1988). *J. Bacteriol.* **170,** 4309–4314.
81. Beveridge, T. J., and Davies, J. A. (1983). *J. Bacteriol.* **156,** 846–858.
82. Bielinska, M. *et al.* (1979). *Proc. Natl. Acad. Sci. U.S.A.* **76,** 6152–6156.

83. Birbeck, M. S. C., and Mercer, E. H. (1961). *Nature* (*London*) **189,** 558–560.
84. Bird, P. *et al.* (1987). *J. Cell Biol.* **105,** 2905–2914.
85. Bitter, G. A. *et al.* (1984). *Proc. Natl. Acad. Sci. U.S.A.* **81,** 5330–5334.
86. Blachley-Dyson, E., and Stevens, T. H. (1987). *J. Cell Biol.* **104,** 1183–1191.
87. Blasland, A. J. (1986). *J. Biol. Chem.* **261,** 12723–12732.
88. Blobel, G. (1980). *Proc. Natl. Acad. Sci. U.S.A.* **77,** 1496–1500.
89. Blobel, G. (1985). *Proc. Natl. Acad. Sci. U.S.A.* **82,** 8527–8531.
90. Blobel, G., and Dobberstein, B. (1975). *J. Cell Biol.* **67,** 835–851.
91. Blobel, G., and Dobberstein, B. (1975). *J. Cell Biol.* **67,** 852–862.
92. Blobel, G., and Sabatini, D. (1970). *J. Cell Biol.* **67,** 835–851.
93. Blobel, G., and Sabatini, D. (1971). *Biomembranes* **2,** 193–195.
93a. Block, M. R. *et al.* (1988). *Proc. Natl. Acad. Sci. U.S.A.* **85,** 7852–7856.
94. Boeke, J. *et al.* (1980). *J. Mol. Biol.* **144,** 103–116.
95. Böhni, P. C. *et al.* (1983). *J. Biol. Chem.* **258,** 4937–4943.
96. Böhni, P. C. *et al.* (1988). *J. Cell Biol.* **106,** 1035–1042.
97. Bole, D. G. *et al.* (1986). *J. Cell Biol.* **102,** 1558–1566.
97a. Bolla, J-M. *et al.* (1988). *EMBO J.* **7,** 3595–3599.
98. Bonner, W. D. (1975). *J. Cell Biol.* **64,** 421–430.
99. Bonner, W. D. (1975). *J. Cell Biol.* **64,** 431–437.
100. Bonnerot, C. *et al.* (1987). *Proc. Natl. Acad. Sci. U.S.A.* **84,** 6795–6799.
101. Bordier, C. (1981). *J. Biol. Chem.* **256,** 1604–1607.
102. Borgese, N., and Meldolesi, E. (1980). *J. Cell Biol.* **85,** 501–515.
103. Borgese, N. *et al.* (1974). *J. Mol. Biol.* **88,** 559–580.
104. Bos, T. J. *et al.* (1984). *Proc. Natl. Acad. Sci. U.S.A.* **81,** 2327–2331.
105. Bosch, D., and Tommassen, J. (1987). *Mol. Gen. Genet.* **208,** 485–489.
106. Bosch, D. *et al.* (1986). *J. Mol. Biol.* **189,** 449–455.
107. Boulay, F. *et al.* (1987). *EMBO J.* **6,** 2643–2650.
108. Boulay, F. *et al.* (1988). *J. Cell Biol.* **106,** 629–639.
109. Bourne, H. R. (1988). *Cell* (*Cambridge, Mass.*) **53,** 669–671.
110. Boutry, M., and Chua, N.-H. (1984). *EMBO J.* **4,** 2159–2105.
111. Boutry, M. *et al.* (1987). *Nature* (*London*) **328,** 340–342.
112. Boyd, A. E., III *et al.* (1982). *J. Cell Biol.* **92,** 425–434.
113. Boyd, D. *et al.* (1987). *Proc. Natl. Acad. Sci. U.S.A.* **84,** 8525–8529.
114. Brada, D., and Schekman, R. (1988). *J. Bacteriol.* **170,** 2775–2783.
115. Braell, W. A. (1987). *Proc. Natl. Acad. Sci. U.S.A.* **84,** 1137–1141.
116. Brake, A. J. *et al.* (1984). *Proc. Natl. Acad. Sci. U.S.A.* **81,** 4642–4646.
117. Brands, R. *et al.* (1985). *J. Cell Biol.* **101,** 1724–1732.
118. Brass, J. M. *et al.* (1986). *J. Bacteriol.* **165,** 787–794.
119. Braulke, T. *et al.* (1987). *J. Cell Biol.* **104,** 1736–1742.
120. Braun-Breton, C. *et al.* (1988). *Nature* (*London*) **332,** 457–459.
121. Breckenbridge, L. J., and Almers, W. (1987). *Proc. Natl. Acad. Sci. U.S.A.* **84,** 1945–1949.
122. Bretscher, M. S. (1984). *Science* **244,** 681–686.
123. Bretscher, M. S. *et al.* (1980). *Proc. Natl. Acad. Sci. U.S.A.* **77,** 4156–4159.
124. Brickman, E. R. *et al.* (1984). *Mol. Gen. Genet.* **196,** 24–27.
125. Briggs, M. S., and Gierasch, L. M. (1984). *Biochemistry* **23,** 3111–3114.
126. Briggs, M. S., and Gierasch, L. M. (1986). *Adv. Protein Res.* **32,** 109–180.
127. Briggs, M. S. *et al.* (1985). *Science* **228,** 1096–1099.
128. Briggs, M. S. *et al.* (1986). *Science* **233,** 206–208.
129. Broeck, G. van den *et al.* (1985). *Nature* (*London*) **313,** 358–363.
130. Brown, W. G. *et al.* (1984). *Mol. Gen. Genet.* **197,** 351–357.

131. Brown, W. J., and Farquhar, M. G. (1984). *Cell (Cambridge, Mass.)* **36,** 295–307.
132. Brown, W. J., and Goldstein, J. L. (1976). *Cell (Cambridge, Mass.)* **9,** 663–674.
133. Brown, W. J. *et al.* (1983). *Cell (Cambridge, Mass.)* **32,** 633–667.
134. Brown, W. J. *et al.* (1986). *J. Cell Biol.* **103,** 1235–1247.
135. Burgess, T. L. *et al.* (1987). *J. Cell Biol.* **105,** 659–668.
136. Burgess, T. L., and Kelly, R. B. (1987). *Annu. Rev. Cell Biol.* **3,** 243–293.
137. Burglin, T. R., and De Robertis, E. M. (1987). *EMBO J.* **6,** 2617–2625.
138. Burgoyne, R. B. (1987). *Nature (London)* **328,** 112–113.
139. Busson-Mabillot, S. *et al.* (1982). *J. Cell Biol.* **95,** 105–117.
139a. Cabelli, R. J. *et al.* (1988). *Cell (Cambridge, Mass.)* **55,** 683–692.
140. Caplan, M. J. *et al.* (1985). *J. Cell Biol.* **105,** 183a.
141. Caplan, M. J. *et al.* (1986). *Cell (Cambridge, Mass.)* **46,** 623–631.
142. Caplan, M. J. *et al.* (1978). *Nature (London)* **329,** 632–635.
143. Caras, I. W. *et al.* (1987). *Science* **238,** 1280–1283.
144. Carde, J.-P. *et al.* (1982). *Biol. Cell.* **44,** 315–324.
145. Carlson, M., and Botstein, D. (1982). *Cell (Cambridge, Mass.)* **28,** 145–154.
146. Carlson, M. *et al.* (1983). *Mol. Cell. Biol.* **3,** 439–447.
147. Castilho, J. M. (1984). *J. Bacteriol.* **158,** 488–495.
148. Caulfield, M. P. *et al.* (1984). *Proc. Natl. Acad. Sci. U.S.A.* **81,** 7772–7776.
149. Caulfield, M. P. *et al.* (1985). *Proc. Natl. Acad. Sci. U.S.A.* **82,** 4031–4035.
150. Caulfield, M. P. *et al.* (1985). *Proc. Natl. Acad. Sci. U.S.A.* **82,** 10953–10956.
151. Ceriotti, A., and Colman, A. (1988). *EMBO J.* **7,** 633–638.
152. Cerretti, P. D. *et al.* (1983). *Nucleic Acids Res.* **11,** 2599–2619.
153. Charbit, A. *et al.* (1986). *EMBO J.* **5,** 3029–3037.
154. Charbit, A. *et al.* (1987). *J. Immunol.* **139,** 1658–1664.
155. Chaudhary, V. K. *et al.* (1987). *Proc. Natl. Acad. Sci. U.S.A.* **84,** 4538–4542.
156. Cheetham, R. D. *et al.* (1971). *J. Cell Biol.* **49,** 899–905.
157. Chen, L., and Tai, P.-C. (1985). *Proc. Natl. Acad. Sci. U.S.A.* **82,** 4384–4388.
158. Chen, L., and Tai, P.-C. (1986). *J. Bacteriol.* **168,** 828–832.
159. Chen, L., and Tai, P.-C. (1987). *Nature (London)* **328,** 164–166.
160. Chen, L. and Tai, P.-C. (1987). *J. Bacteriol.* **169,** 2373–2379.
161. Chen, L. *et al.* (1985). *J. Bacteriol.* **161,** 973–980.
162. Chen, L. *et al.* (1987). *J. Biol. Chem.* **262,** 1427–1429.
163. Chen, W.-J., and Douglas, M. G. (1987). *J. Biol. Chem.* **262,** 15598–15604.
164. Chen, W.-J., and Douglas, M. G. (1987). *J. Biol. Chem.* **262,** 15605–15609.
165. Chen, W.-J., and Douglas, M. G. (1987). *Cell (Cambridge, Mass.)* **49,** 651–658.
166. Chen, W.-J., and Douglas, M. G. (1988). *J. Biol. Chem.* **263,** 4997–5000.
167. Cheung, A. Y. *et al.* (1988). *Proc. Natl. Acad. Sci. U.S.A.* **85,** 391–395.
168. Chia, C. P., and Arntlein, C. J. (1986). *J. Cell Biol.* **103,** 725–731.
169. Chicheportiche, Y. *et al.* (1984). *J. Cell Biol.* **99,** 2200–2210.
170. Chioppa, G. della, and Kishmore, G. M. (1988). *EMBO J.* **7,** 1299–1305.
171. Chioppa, G. della *et al.* (1987). *Bio/Technology* **5,** 579–584.
172. Chirico, W. J. *et al.* (1988). *Nature (London)* **322,** 805–810.
173. Chou, P. Y., and Fasman, G. D. (1978). *Biochemistry* **13,** 222–245.
174. Chu, T. W. *et al.* (1987). *J. Biol. Chem.* **262,** 15759–15764.
175. Chvatchko, Y. *et al.* (1986). *Cell (Cambridge, Mass.)* **46,** 355–364.
176. Cieplak, W. *et al.* (1987). *J. Biol. Chem.* **262,** 13246–13253.
177. Citi, S. *et al.* (1988). *Nature (London)* **333,** 272–276.
178. Clayton, C. E. (1987). *J. Cell Biol.* **105,** 2649–2654.
179. Clément, J.-M., and Hofnung, M. (1981). *Cell (Cambridge, Mass.)* **27,** 507–514.
180. Cline, K. *et al.* (1985). *J. Biol. Chem.* **260,** 3691–3696.

181. Cline, K. (1986). *J. Biol. Chem.* **261,** 14804–14810.
182. Cole, S. T., and Maldener, M. (1986). *FEMS Microbiol. Lett.* **33,** 133–136.
183. Cole, S. T. *et al.* (1985). *Mol. Gen. Genet.* **198,** 465–472.
184. Colledge, W. H. *et al.* (1986). *Mol. Cell. Biol.* **6,** 4136–4139.
185. Collier, D. N. *et al.* (1988). *Cell (Cambridge, Mass.)* **53,** 273–283.
186. Collombatte, M. *et al.* (1986). *J. Biol. Chem.* **261,** 3030–3035.
187. Comb, M. *et al.* (1982). *Nature (London)* **295,** 663–666.
188. Comb, M. *et al.* (1985). *EMBO J.* **4,** 3115–3122.
189. Conboy, J. G. *et al.* (1982). *Biochem. Biophys. Res. Commun.* **105,** 1–7.
190. Connelly, T., and Gilmore, R. (1986). *J. Cell Biol.* **103,** 2253–2261.
191. Conzelmann, A. *et al.* (1986). *EMBO J.* **5,** 3291–3296.
191a. Conzelmann, A. *et al.* (1988). *EMBO J.* **7,** 2233–2240.
192. Cooke, W. R. *et al.* (1986). *J. Bacteriol.* **168,** 1430–1438.
193. Copeland, C. S. *et al.* (1986). *J. Cell Biol.* **103,** 1179–1191.
194. Copeland, C. S. *et al.* (1988). *Cell (Cambridge, Mass.)* **53,** 197–209.
195. Cornet, P. *et al.* (1983). *Bio/Technology* **1,** 589–594.
196. Cover, W. H. *et al.* (1987). *J. Bacteriol.* **169,** 1794–1800.
197. Crick, E. M. *et al.* (1988). *J. Bacteriol.* **170,** 2005–2011.
198. Crimaudo, C. *et al.* (1986). *EMBO J.* **6,** 75–82.
199. Critchley, D. R. *et al.* (1981). *J. Biol. Chem.* **256,** 8724–8731.
200. Crooke, E., and Wickner, W. (1987). *Proc. Natl. Acad. Sci. U.S.A.* **84,** 5216–5220.
200a. Crooke, E. *et al.* (1988). *EMBO J.* **7,** 1831–1835.
200b. Crooke, E. *et al.* (1988). *Cell (Cambridge, Mass.)* **54,** 1003–1011.
201. Cross, G. A. M. (1987). *Cell (Cambridge, Mass.)* **48,** 179–181.
202. Crowelsmith, I. *et al.* (1981). *Eur. J. Biochem.* **113,** 375–380.
203. Cruz, V. F. de la *et al.* (1981). *J. Biol. Chem.* **263,** 4318–4322.
204. Cullen, D. *et al.* (1987). *Bio/Technology* **5,** 369–375.
205. Cullis, P. R., and de Kruijff, B. (1979). *Biochim. Biophys. Acta* **559,** 399–420.
206. Cutler, D. F., and Garoff, H. (1986). *J. Cell Biol.* **102,** 889–901.
207. Cutler, D. F., and Garoff, H. (1986). *J. Cell Biol.* **102,** 902–910.
208. Dahms, N. M. *et al.* (1987). *Cell (Cambridge, Mass.)* **50,** 181–192.
209. Dalbey, R. E., and Wickner, W. (1985). *J. Biol. Chem.* **260,** 15925–15931.
210. Dalbey, R. E., and Wickner, W. (1986). *J. Biol. Chem.* **261,** 13844–13850.
211. Dalbey, R. E., and Wickner, W. (1987). *Science* **235,** 783–787.
212. Dalbey, R. E., and Wickner, W. (1988). *J. Biol. Chem.* **263,** 404–408.
213. Dalbey, R. E. *et al.* (1987). *J. Biol. Chem.* **262,** 13241–13245.
214. Daniels, C. J. *et al.* (1981). *Proc. Natl. Acad. Sci. U.S.A.* **78,** 5396–5400.
215. Date, T. (1983). *J. Bacteriol.* **154,** 76–83.
216. Date, T. *et al.* (1980). *Proc. Natl. Acad. Sci. U.S.A.* **77,** 4669–4673.
217. Daum, G. *et al.* (1982). *J. Biol. Chem.* **257,** 13028–13033.
218. Daum, G. *et al.* (1982). *J. Biol. Chem.* **257,** 13075–13080.
218a. Dautry-Varsat, A. *et al.* (1983). *Proc. Natl. Acad. Sci. U.S.A.* **80,** 2258–2262.
219. Davey, J. *et al.* (1985). *Cell (Cambridge, Mass.)* **40,** 667–675.
220. Davey, J. *et al.* (1985). *Cell (Cambridge, Mass.)* **43,** 643–652.
221. Davidson, V. L. *et al.* (1984). *J. Biol. Chem.* **259,** 594–600.
222. Davidson, V. L. *et al.* (1985). *Proc. Natl. Acad. Sci. U.S.A.* **82,** 1386–1390.
223. Davis, C. G. *et al.* (1986). *Cell (Cambridge, Mass.)* **45,** 15–24.
224. Davis, C. G. *et al.* (1987). *J. Biol. Chem.* **262,** 4075–4082.
225. Davis, L. I., and Blobel, G. (1986). *Cell (Cambridge, Mass.)* **45,** 699–709.
226. Davis, N. G., and Model, P. (1985). *Cell (Cambridge, Mass.)* **41,** 607–614.
227. Davis, N. G. *et al.* (1985). *J. Mol. Biol.* **181,** 111–121.

228. Reference deleted.
229. De Robertis, E. M. (1983). *Cell (Cambridge, Mass.)* **32,** 1021–1025.
230. De Robertis, E. M. *et al.* (1978). *Nature (London)* **272,** 254–256.
231. Desaymard, C. *et al.* (1986). *EMBO J.* **5,** 1383–1388.
232. Deschenes, R. J., and Broach, J. R. (1987). *Mol. Cell Biol.* **7,** 2344–2351.
233. Deshaies, R. J., and Schekman, R. (1987). *J. Cell Biol.* **105,** 633–645.
234. Deshaies, R. J. *et al.* (1988). *Nature (London)* **322,** 800–805.
235. Deutscher, S. L. *et al.* (1983). *Proc. Natl. Acad. Sci. U.S.A.* **80,** 3938–3942.
236. Dev, I. K., and Ray, P. H. (1984). *J. Biol. Chem.* **259,** 11114–11120.
236a. Dev, I. K. *et al.* (1985). *J. Biol. Chem.* **260,** 5891–5894.
237. Diaz, R. *et al.* (1988). *J. Biol. Chem.* **263,** 6093–6100.
238. Dierstein, R., and Wickner, W. (1985). *J. Biol. Chem.* **260,** 15919–15924.
239. Diesenhofer, J. *et al.* (1985). *Nature (London)* **318,** 618–624.
240. Dijkmans, R. *et al.* (1987). *J. Biol. Chem.* **262,** 2528–2535.
241. Dingwall, C., and Laskey, R. A. (1986). *Annu. Rev. Cell Biol.* **2,** 367–390.
241a. Dingwall, C. *et al.* (1982). *Cell (Cambridge, Mass.)* **30,** 449–458.
242. Dingwall, C. *et al.* (1988). *J. Cell Biol.* **107,** 841–849.
242a. DiPaola, M., and Maxfield, F. R. (1980). *J. Biol. Chem.* **259,** 9163–9173.
243. DiRienzo, J. M., and Inouye, M. (1979). *Cell (Cambridge, Mass.)* **17,** 155–161.
244. Distel, B. *et al.* (1987). *EMBO J.* **6,** 3111–3116.
245. Dmochowska, A. *et al.* (1987). *Cell (Cambridge, Mass.)* **50,** 573–584.
246. Dobberstein, B. *et al.* (1977). *Proc. Natl. Acad. Sci. U.S.A.* **74,** 1082–1085.
247. Dodt, J. *et al.* (1986). *FEBS Lett.* **202,** 373–377.
248. Doherty, A. *et al.* (1986). *J. Bacteriol.* **166,** 1072–1082.
249. Dolci, E. D., and Palade, G. E. (1985). *J. Biol. Chem.* **260,** 10728–10735.
250. Doms, R. W. *et al.* (1987). *J. Cell Biol.* **105,** 1957–1969.
251. Dorner, A. J. *et al.* (1987). *J. Cell Biol.* **105,** 2665–2674.
252. Dorset, D. L. *et al.* (1983). *J. Mol. Biol.* **165,** 701–710.
253. Douce, R. *et al.* (1982). *Subcell. Biochem.* **10,** 1–84.
254. Douglas, C. M. *et al.* (1987). *J. Bacteriol.* **169,** 4962–4966.
255. Douglas, M. G. *et al.* (1984). *Proc. Natl. Acad. Sci. U.S.A.* **81,** 3983–3987.
256. Downham, W. *et al.* (1985). *EMBO J.* **4,** 179–184.
257. Dowsey, S. J. *et al.* (1985). *J. Cell Biol.* **101,** 19–27.
258. Dowsey, S. J. *et al.* (1986). *J. Cell Biol.* **103,** 53a.
259. Doyle, C. *et al.* (1986). *J. Cell Biol.* **103,** 1193–1204.
260. Dragsten, P. R. *et al.* (1981). *Nature (London)* **294,** 718–722.
261. Drews, B. van *et al.* (1988). *J. Cell Biol.* **106,** 253–267.
262. Drickamer, K. *et al.* (1984). *J. Biol. Chem.* **259,** 770–778.
263. Drummond, D. *et al.* (1985). *J. Cell Biol.* **100,** 1148–1156.
263a. Duffaud, G., and Inouye, M. (1988). *J. Biol. Chem.* **263,** 10224–10228.
264. Dunn, W. A., and Hubbard, L. A. (1984). *J. Cell Biol.* **98,** 2148–2159.
265. Dunphy, W. G., and Rothman, J. E. (1983). *J. Cell Biol.* **97,** 270–275.
266. Dunphy, W. G., and Rothman, J. E. (1985). *Cell (Cambridge, Mass.)* **42,** 13–21.
267. Dunphy, W. G. *et al.* (1985). *Cell (Cambridge, Mass.)* **40,** 463–472.
268. Dunphy, W. G. *et al.* (1986). *Proc. Natl. Acad. Sci. U.S.A.* **83,** 1622–1626.
269. Dworetzki, S. I., and Feldherr, C. M. (1988). *J. Cell Biol.* **106,** 575–584.
269a. Dworetzki, S. I. *et al.* (1988). *J. Cell Biol.* **107,** 1279–1287.
269b. Eakle, K. A. *et al.* (1988). *Mol. Cell. Biol.* **8,** 4098–4109.
270. Ebina, Y. *et al.* (1985). *Cell (Cambridge, Mass.)* **40,** 747–758.
271. Eble, B. E. *et al.* (1987). *Mol. Cell. Biol.* **7,** 3591–3601.
272. Edelman, A. *et al.* (1987). *Mol. Microbiol.* **1,** 101–106.

273. Eilers, M., and Schatz, G. (1986). *Nature (London)* **322,** 228–232.
274. Eilers, M. *et al.* (1987). *EMBO J.* **6,** 1073–1077.
275. Eilers, M. *et al.* (1988). *EMBO J.* **7,** 1139–1149.
276. Eisenberg, D. *et al.* (1984). *J. Mol. Biol.* **179,** 125–142.
277. Emr, S. D., and Bassford, P. J., Jr. (1982). *J. Biol. Chem.* **257,** 5852–5860.
278. Emr, S. D. *et al.* (1981). *Cell (Cambridge, Mass.)* **23,** 79–88.
279. Emr, S. D. *et al.* (1983). *Proc. Natl. Acad. Sci. U.S.A.* **80,** 7080–7084.
280. Emr, S. D. *et al.* (1984). *J. Cell Biol.* **102,** 523–533.
281. Emr, S. D. *et al.* (1984). *Mol. Cell Biol.* **4,** 2347–2355.
282. Emr, S. D. *et al.* (1986). *J. Cell Biol.* **102,** 523–533.
283. Endo, T., and Schatz, G. (1988). *EMBO J.* **7,** 1153–1158.
284. Enequist, H. G. *et al.* (1981). *Eur. J. Biochem.* **116,** 227–233.
285. Enfert, C. d', and Pugsley, A. P. (1987). *Mol. Microbiol.* **1,** 159–168.
286. Enfert, C. d' *et al.* (1987). *Mol. Microbiol.* **1,** 107–116.
287. Enfert, C. d' *et al.* (1987). *EMBO J.* **6,** 3531–3538.
288. Engelman, D. M., and Steitz, T. A. (1981). *Cell (Cambridge, Mass.)* **23,** 411–422.
289. Epand, R. M. *et al.* (1986). *J. Biol. Chem.* **261,** 10017–10020.
290. Escuyer, V., and Mock, M. (1987). *Mol. Microbiol.* **1,** 82–85.
291. Escuyer, V. *et al.* (1986). *J. Biol. Chem.* **261,** 10891–10898.
292. Esmon, B. P. *et al.* (1981). *Cell (Cambridge, Mass.)* **25,** 451–460.
293. Esmon, P. C. *et al.* (1987). *J. Biol. Chem.* **262,** 4387–4394.
294. Evans, E. A. *et al.* (1986). *Proc. Natl. Acad. Sci. U.S.A.* **83,** 581–585, 996.
295. Fahnenstock, S. R., and Fisher, K. E. (1986). *J. Bacteriol.* **165,** 796–804.
296. Fairbanks, G. *et al.* (1971). *Biochemistry* **10,** 2606–2617.
297. Fandl, J. P., and Tai, P.-C. (1987). *Proc. Natl. Acad. Sci. U.S.A.* **87,** 7448–7462.
298. Farabakhsh, Z. T. *et al.* (1987). *J. Biol. Chem.* **262,** 2256–2261.
299. Farquhar, M. G. (1985). *Annu. Rev. Cell Biol.* **1,** 447–488.
300. Farquhar, M. G., and Palade, G. E. (1981). *J. Cell Biol.* **91,** 77s–103s.
301. Faust, P. L. *et al.* (1987). *J. Cell Biol.* **105,** 1937–1945.
302. Faust, P. L. *et al.* (1977). *J. Cell Biol.* **105,** 1947–1955.
303. Fecycz, I. T., and Blobel, G. (1987). *Proc. Natl. Acad. Sci. U.S.A.* **87,** 3723–3727.
304. Fecycz, I. T., and Campbell, J. N. (1985). *Eur. J. Biochem.* **146,** 35–42.
305. Feldherr, C. M., and Ogburn, J. A. (1980). *J. Cell Biol.* **87,** 589–593.
306. Feldherr, C. M., *et al.* (1984). *J. Cell Biol.* **99,** 2216–2222.
307. Fellems, R. *et al.* (1975). *J. Cell Biol.* **65,** 1–14.
308. Felmlee, T. *et al.* (1985). *J. Bacteriol.* **163,** 88–93.
309. Felmlee, T. *et al.* (1985). *J. Bacteriol.* **163,** 94–105.
310. Ferguson, M. A. J., and Cross, G. A. M. (1984). *J. Biol. Chem.* **259,** 3011–3015.
311. Ferguson, M. A. J. *et al.* (1986). *J. Biol. Chem.* **261,** 356–362.
312. Fernandez, J. M. *et al.* (1984). *Nature (London)* **312,** 453–455.
313. Figura, K. von, and Haslik, A. (1986). *Annu. Rev. Biochem.* **55,** 167–193.
314. Fikes, J. D., and Bassford P. J., Jr. (1987). *J. Bacteriol.* **169,** 2352–2359.
315. Fikes, J. D. *et al.* (1987). *J. Bacteriol.* **169,** 2345–2351.
316. Filson, A. J. *et al.* (1985). *J. Biol. Chem.* **260,** 3164–3172.
317. Finlay, D. R. *et al.* (1987). *J. Cell Biol.* **104,** 189–200.
317a. Fisher, J. M. *et al.* (1988). *Cell (Cambridge, Mass.)* **54,** 813–822.
318. Fisher, P. A. (1987). *Cell (Cambridge, Mass.)* **48,** 175–176.
319. Fishman, J. B., and Fine, R. E. (1987). *Cell (Cambridge, Mass.)* **48,** 157–164.
320. Fitting, T., and Kabat, D. (1982). *J. Biol. Chem.* **257,** 14011–14017.
321. Flügge, I., and Hinz, G. (1986). *Eur. J. Biochem.* **160,** 563–570.
322. Forbes, D. J. *et al.* (1983). *Cell (Cambridge, Mass.)* **34,** 13–23.

323. Franke, W. W. *et al.* (1981). *J. Cell Biol.* **91,** 39S–50S.
324. Fraser, J. H., and Bruce, B. J. (1978). *Proc. Natl. Acad. Sci. U.S.A.* **75,** 3936–3940.
325. Freedman, R. B. (1984). *Trends Biochem. Sci.* **11,** 438–441.
326. Freedman, R. B. (1987). *Nature (London)* **329,** 196–197.
327. Freitag, H. *et al.* (1982). *Eur. J. Biochem.* **126,** 197–202.
328. Freudl, R. *et al.* (1985). *EMBO J.* **4,** 3593–3598.
329. Freudl, R. *et al.* (1986). *J. Mol. Biol.* **188,** 491–494.
330. Freudl, R. *et al.* (1986). *J. Biol. Chem.* **261,** 11355–11361.
331. Freudl, R. *et al.* (1988). *J. Biol. Chem.* **263,** 344–349.
331a. Freudl, R. *et al.* (1988). *J. Biol. Chem.* **263,** 17084–17091.
331b. Friederich, E. *et al.* (1988). *J. Cell Biol.* **107,** 1655–1667.
332. Friedlander, M., and Blobel, G. (1985). *Nature (London)* **318,** 338–343.
333. Friend, D. S. (1969). *J. Cell Biol.* **41,** 269–279.
334. Fries, E., and Rothman, J. E. (1980). *Proc. Natl. Acad. Sci. U.S.A.* **77,** 3870–3874.
335. Fries, E. *et al.* (1984). *EMBO J.* **3,** 147–152.
336. Fritz, L. C. *et al.* (1986). *Proc. Natl. Acad. Sci. U.S.A.* **83,** 4114–4118.
337. Fritz, L. C. *et al.* (1987). *J. Biol. Chem.* **262,** 12409–12412.
338. Fujiki, Y. *et al.* (1984). *Proc. Natl. Acad. Sci. U.S.A.* **81,** 7127–7131.
339. Fujimoto, Y. *et al.* (1984). *J. Biochem. (Tokyo)* **96,** 1125–1131.
340. Fuller, S. D., and Simons, K. (1986). *J. Cell Biol.* **103,** 1767–1779.
341. Fumagalli, G., and Zanani, A.(1985). *J. Cell Biol.* **100,** 2019–2024.
342. Futerman, A. H. (1985). *Biochem. Biophys. Res. Commun.* **129,** 312–317.
343. Gabel, C. A., and Kornfeld, S. (1984). *J. Cell Biol.* **99,** 296–305.
344. Gahmberg, N. *et al.* (1986). *EMBO J.* **5,** 3111–3118.
345. Garavito, M. *et al.* (1983). *J. Mol. Biol.* **164,** 313–328.
346. Garcia, P. D., and Walters, P. (1988). *J. Cell Biol.* **106,** 1043–1048.
347. Garcia, P. D. *et al.* (1987). *J. Biol. Chem.* **262,** 9463–9468.
348. Garcia, P. D. *et al.* (1988). *J. Cell Biol.* **106,** 1093–1104.
349. Gardel, C. *et al.* (1987). *J. Bacteriol.* **169,** 1286–1290.
350. Garnier, J. *et al.* (1978). *J. Mol. Biol.* **120,** 97–120.
351. Gascuel, O., and Danchin, A.(1986). *J. Mol. Evol.* **24,** 130–142.
352. Gasser, S. M., and Schatz, G. (1983). *J. Biol. Chem.* **258,** 3427–3430.
353. Gasser, S. M. *et al.* (1982). *J. Biol. Chem.* **257,** 13034–13041.
354. Gasser, S. M. *et al.* (1982). *Proc. Natl. Acad. Sci. U.S.A.* **79,** 267–271.
355. Gatenby, A. A. *et al.* (1986). *Gene* **45,** 11–18.
356. Gatenby, A. A. *et al.* (1988). *EMBO J.* **7,** 1307–1314.
357. Gearing, D. P., and Nagley, P. (1986). *EMBO J* **5,** 3651–3655.
358. Geli, V. *et al.* (1988). *Proc. Natl. Acad. Sci. U.S.A.* **85,** 689–693.
359. Geller, B. L., and Wickner, W. (1985). *J. Biol. Chem.* **260,** 13281–13286.
360. Geller, B. L. *et al.* (1986). *Proc. Natl. Acad. Sci. U.S.A.* **83,** 4219–4222.
361. Gentz, R. *et al.* (1988). *J. Bacteriol.* **170,** 2212–2220.
362. Gerace, L. *et al.* (1982). *J. Cell Biol.* **95,** 826–837.
363. Gestry, L. E. *et al.* (1987). *Mol. Cell. Biol.* **7,** 3148–3427.
364. Gething, M.-J. *et al.* (1986). *Cell (Cambridge, Mass.)* **46,** 939–950.
365. Geuze, H. J. *et al.* (1984). *J. Cell Biol.* **98,** 2047–2054.
366. Geuze, H. J. *et al.* (1984). *Cell (Cambridge, Mass.)* **37,** 195–204.
367. Geuze, H. J. *et al.* (1987). *J. Cell Biol.* **104,** 1715–1723.
368. Ghersa, D. *et al.* (1986). *J. Biol. Chem.* **261,** 7969–7974.
369. Ghrayeb, J., and Inouye, M. (1984). *J. Biol. Chem.* **259,** 463–467.
370. Giddings, T. H., and Staehelin, L. A. (1980). *J. Cell Biol.* **85,** 145–152.

371. Gillespie, L. L. (1987). *J. Biol. Chem.* **262,** 7939–7942.
372. Gillespie, L. L. *et al.* (1985). *J. Biol. Chem.* **260,** 16045–16048.
373. Gilmore, R., and Blobel, G. (1983). *Cell (Cambridge, Mass.)* **35,** 677–685.
374. Gilmore, R., and Blobel, G. (1985). *Cell (Cambridge, Mass.)* **42,** 497–505.
375. Gilmore, R. *et al.* (1982). *J. Cell Biol.* **95,** 463–469.
376. Gilmore, R. *et al.* (1982). *J. Cell Biol.* **95,** 470–477.
377. Goebel, W. *et al.* (1984). *In* "Plasmids in Bacteria" (D. R. Helinski *et al.*, eds.), pp. 791–805. Plenum, New York.
378. Goetz, F. *et al.* (1985). *Nucleic Acids Res.* **13,** 5895–5903.
379. Goldberg, D. E., and Kornfeld, S. (1981). *J. Biol. Chem.* **256,** 13060–13067.
380. Goldberg, D. E., and Kornfeld, S. (1983). *J. Biol. Chem.* **258,** 3159–3165.
381. Goldberg, J. B., and Ohman, D. E. (1987). *J. Bacteriol.* **169,** 4532–4539.
382. Goldfarb, D. S. *et al.* (1986). *Nature (London)* **322,** 641–644.
383. Goldmacher, V. S. *et al.* (1987). *J. Biol. Chem.* **262,** 3205–3209.
384. Goldman, B. M., and Blobel, G. (1978). *Proc. Natl. Acad. Sci. U.S.A.* **75,** 5060–5070.
385. Goldstein, J. L. *et al.* (1979). *Nature (London)* **279,** 679–685.
386. Goldstein, J. L. *et al.* (1985). *Annu. Rev. Cell Biol.* **1,** 1–39.
387. Gonzalez, A. *et al.* (1987). *Proc. Natl. Acad. Sci. U.S.A.* **84,** 3738–3742.
388. Goodman, J. M. *et al.* (1984). *J. Biol. Chem.* **259,** 8485–8493.
389. Gotschlich, E. C. *et al.* (1987). *Proc. Natl. Acad. Sci. U.S.A.* **84,** 8135–8139.
390. Gottlied, T. A. *et al.* (1986). *Proc. Natl. Acad. Sci. U.S.A.* **83,** 2100–2104.
391. Goud, B. *et al.* (1988). *Cell (Cambridge, Mass.)* **53,** 753–768.
392. Gould, A. R. *et al.* (1974). *J. Bacteriol.* **122,** 34–40.
393. Gould, S. J. *et al.* (1987). *J. Cell Biol.* **105,** 2923–2931.
393a. Gould, S. J. (1988). *J. Cell Biol.* **107,** 897–905.
394. Gray, G. L. *et al.* (1984). *Proc. Natl. Acad. Sci. U.S.A.* **81,** 2645–2649.
395. Green, R., and Shields, D. (1984). *J. Cell Biol.* **99,** 97–104.
396. Green, R. *et al.* (1986). *J. Biol. Chem.* **261,** 7558–7565.
397. Green, S. A. *et al.* (1987). *J. Cell Biol.* **105,** 1227–1240.
398. Griffiths, G., and Simons, K. (1986). *Science* **234,** 438–443.
399. Griffiths, G. *et al.* (1982). *J. Cell Biol.* **95,** 781–792.
400. Griffiths, G. *et al.* (1983). *J. Cell Biol.* **96,** 835–850.
401. Griffiths, G. *et al.* (1985). *J. Cell Biol.* **101,** 949–964.
402. Griffiths, G. *et al.* (1988). *Cell (Cambridge, Mass.)* **52,** 329–344.
403. Gritz, L. *et al.* (1985). *Mol. Cell. Biol.* **5,** 3435–3442.
404. Gruenberg, J. E., and Howell, K. E. (1986). *EMBO J.* **5,** 3091–3101.
405. Gruenberg, J. E., and Howell, K. E. (1987). *Proc. Natl. Acad. Sci. U.S.A.* **84,** 5758–5762.
406. Grundy, F. J. *et al.* (1978). *J. Bacteriol.* **169,** 4442–4450.
406a. Guan, J. L. *et al.* (1988). *Mol. Cell. Biol.* **8,** 2869–2874.
407. Guan, T. *et al.* (1985). *J. Bacteriol.* **164,** 107–113.
408. Guiard, B. (1985). *EMBO J.* **4,** 3265–3272.
409. Gumbiner, B. (1987). *Am. J. Physiol.* **253,** C749–C758.
410. Gumbiner, B., and Louvard, D. (1985). *Trends Biochem. Sci.* **10,** 435–438.
411. Gurdon, J. B. (1970). *Proc. R. Soc. London, Ser. B* **176,** 303–314.
412. Gwynne, D. I. *et al.* (1987). *Bio/Technology* **5,** 713–719.
413. Haas, I. G., and Wabl, M. (1983). *Nature (London)* **306,** 387–389.
414. Habener, J. F. *et al.* (1978). *Proc. Natl. Acad. Sci. U.S.A.* **75,** 2626–2620.
415. Habener, J. F. *et al.* (1980). *J. Biol. Chem.* **254,** 10596–10599.

416. Hall, M. N. *et al.* (1984). *Cell (Cambridge, Mass.)* **36,** 1057–1065.
417. Hankte, K., and Braun, V. (1973). *Eur. J. Biochem.* **34,** 284–296.
418. Hannover, J. A. *et al.* (1980). *J. Biol. Chem.* **255,** 6713–6716.
419. Hansen, W. *et al.* (1986). *Cell (Cambridge, Mass.)* **45,** 397–406.
420. Hansen, W., and Walter, P. (1988). *J. Cell Biol.* **106,** 1075–1081.
421. Reference deleted.
422. Harick-Ort, V. *et al.* (1987). *J. Biol. Chem.* **104,** 855–863.
423. Harrison, T. M. *et al.* (1974). *Eur. J. Biochem.* **47,** 613–620.
424. Hartl, F.-U. *et al.* (1986). *Cell (Cambridge, Mass.)* **47,** 939–951.
425. Hartl, F.-U. *et al.* (1987). *Cell (Cambridge, Mass.)* **51,** 1027–1037.
426. Hase, T. *et al.* (1984). *EMBO J.* **3,** 3157–3164.
427. Hase, T. *et al.* (1986). *FEMS Lett.* **197,** 199–203.
428. Haselbeck, A., and Schekman, R. (1986). *Proc. Natl. Acad. Sci. U.S.A.* **83,** 2017–2021.
429. Haselbeck, A., and Tanner, W. (1983). *FEBS Lett.* **158,** 335–338.
430. Hawlitschek, G. *et al.* (1988). *Cell (Cambridge, Mass.)* **53,** 795–806.
431. Hayashi, S. *et al.* (1984). *J. Biol. Chem.* **259,** 10448–10454.
432. Hayashi, S. *et al.* (1985). *J. Biol. Chem.* **260,** 5753–5759.
433. Heijne, G. von (1983). *Eur. J. Biochem.* **133,** 17–21.
434. Heijne, G. von (1985). *J. Mol. Biol.* **184,** 99–105.
435. Heijne, G. von (1986). *J. Mol. Biol.* **192,** 287–290.
436. Heihne, G. von (1986). *Nucleic Acids Res.* **14,** 4683–4690.
437. Heijne, G. von (1986). *EMBO J.* **5,** 3021–3027.
438. Heijne, G. von (1986). *J. Mol. Biol.* **189,** 239–242.
439. Heijne, G. von (1986). *EMBO J.* **5,** 1335–1342.
440. Heihne, G. von (1986). *FEBS Lett.* **198,** 1–4.
441. Heijne, G. von, and Blomberg, C. (1981). *Eur. J. Biochem.* **97,** 175–181.
441a. Heijne, G. von, and Gavel, I. (1988). *Eur. J. Biochem.* **174,** 671–678.
442. Heijne,, G. von *et al.* (1988). *Proc. Natl. Acad. Sci. U.S.A.* **85,** 3363–3366.
443. Helenius, A. *et al.* (1983). *Trends Biochem. Sci.* **8,** 245–250.
444. Hellerman, J. G. (1984). *Proc. Natl. Acad. Sci. U.S.A.* **81,** 5340–5344.
445. Hemmingsen, S. M. *et al.* (1988). *Nature (London)* **333,** 330–335.
446. Hengge-Aronius, R., and Boos, W. (1986). *J. Bacteriol.* **167,** 462–466.
447. Hennig, B. *et al.* (1983). *Proc. Natl. Acad. Sci. U.S.A.* **80,** 4963–4967.
448. Herrin, D., and Michaels, A. (1985). *FEBS Lett.* **184,** 90–95.
449. Herzog, V., and Reggio, H. (1980). *Eur. J. Cell Biol.* **21,** 141–150.
450. Hicks, S. J. *et al.* (1969). *Science* **164,** 584–585.
450a. Hiebert, S. N., and Lamb, R. A. (1988). *J. Cell Biol.* **107,** 865–876.
451. Hijikato, M. *et al.* (1987). *J. Biol. Chem.* **262,** 8151–8158.
452. Hirschberg, C. B., and Snider, M. D. (1987). *Annu. Rev. Biochem.* **56,** 63–87.
453. Hirst, T. R., and Holmgren, J. (1987). *Proc. Natl. Acad. Sci. U.S.A.* **84,** 7418–7422.
454. Hirst, T. R., and Holmgren, J. (1987). *J. Bacteriol.* **169,** 1037–1045.
455. Hirst, T. R. *et al.* (1983). *J. Bacteriol.* **153,** 21–26.
456. Hirst, T. R. *et al.* (1984). *Proc. Natl. Acad. Sci. U.S.A.* **81,** 7752–7756.
457. Hisami, Y. *et al.* (1986). *J. Biol. Chem.* **261,** 11398–11405.
458. Hobot, J. A. *et al.* (1984). *J. Bacteriol.* **160,** 143–152.
459. Hoffman, C., and Wright, A. (1985). *Proc. Natl. Acad. Sci. U.S.A.* **82,** 5107–5110.
460. Hoffman, K. E., and Gilmore, R. (1988). *J. Biol. Chem.* **263,** 4381–4385.
461. Hoflak, B., and Kornfeld, S. (1985). *Proc. Natl. Acad. Sci. U.S.A.* **82,** 4428–4432.
462. Holcomb, C. L. *et al.* (1988). *J. Cell Biol.* **106,** 641–648.
463. Hold, G. D. *et al.* (1987). *J. Cell Biol.* **104,** 1157–1164.

464. Holmes, R. K. *et al.* (1975). *J. Clin. Invest.* **55,** 551–560.
465. Homans, S. W. *et al.* (1988). *Nature (London)* **333,** 269–272.
466. Homma, M. *et al.* (1983). *J. Bacteriol.* **154,** 413–418.
467. Hopkins, C. R. (1983). *Cell (Cambridge, Mass.)* **35,** 321–330.
468. Hopkins, C. R., and Trowbridge, I. S. (1983). *J. Cell Biol.* **97,** 508–521.
469. Hopper, A. K. *et al.* (1982). *Cell (Cambridge, Mass.)* **28,** 543–550.
470. Horiuchi, S. *et al.* (1983). *J. Bacteriol.* **154,** 1215–1221.
471. Horiuchi, S. *et al.* (1983). *Proc. Natl. Acad. Sci. U.S.A.* **80,** 3287–3291.
472. Hortin, G., and Boime, I. (1980). *Proc. Natl. Acad. Sci. U.S.A.* **77,** 1356–1360.
473. Hortin, G., and Boime, I. (1981). *Cell (Cambridge, Mass.)* **24,** 453–461.
474. Hortin, G. *et al.* (1986). *Biochem. Biophys. Res. Commun.* **141,** 326–333.
475. Hortsch, M., and Meyer, D. I. (1985). *Eur. J. Biochem.* **150,** 559–564.
476. Hortsch, M., and Meyer, D. I. (1986). *Int. Rev. Cytol.* **102,** 215–242.
477. Hortsch, M. *et al.* (1985). *J. Biol. Chem.* **260,** 9137–9145.
478. Hortsch, M. *et al.* (1985). *Eur. J. Cell Biol.* **38,** 271–279.
479. Hortsch, M. *et al.* (1986). *J. Cell Biol.* **103,** 241–253.
480. Hortsch, M., and Meyer, D. I. (1988). *Biochem. Biophys. Res. Commun.* **150,** 111–117.
481. Horwich, A. L. *et al.* (1985). *EMBO J.* **4,** 1129–1135.
482. Horwich, A. L. *et al.* (1985). *Proc. Natl. Acad. Sci. U.S.A.* **82,** 4930–4933.
483. Horwich, A. L. *et al.* (1986). *Cell (Cambridge, Mass.)* **44,** 451–459.
484. Horwich, A. L. *et al.* (1987). *J. Cell Biol.* **105,** 669–677.
485. Howard, S. P., and Buckley, J. P. (1983). *J. Bacteriol.* **154,** 413–418.
486. Howard, S. P., and Buckley, J. P. (1985). *J. Bacteriol.* **161,** 1118–1124.
487. Howard, S. P., and Buckley, J. P. (1986). *Mol. Gen. Genet.* **204,** 289–295.
488. Hudson, T. H. *et al.* (1989). *J. Biol. Chem.* **263,** 4773–4781.
489. Huet, C. *et al.* (1987). *J. Cell Biol.* **105,** 345–357.
490. Huffaker, T. V., and Robbins, P. W. (1983). *Proc. Natl. Acad. Sci. U.S.A.* **80,** 7466–7470.
491. Hunter, T. (1984). *Nature (London)* **311,** 414–416.
492. Hurt, E. C., and Schatz, G. (1987). *Nature (London)* **325,** 499–503.
493. Hurt, E. C. *et al.* (1984). *EMBO J.* **3,** 3149–3156.
494. Hurt, E. C. *et al.* (1985). *EMBO J.* **4,** 2062–2068.
495. Hurt, E. C. *et al.* (1985). *EMBO J.* **4,** 3509–3518.
496. Hurt, E. C. *et al.* (1986). *EMBO J.* **5,** 1343–1350.
497. Hurt, E. C. *et al.* (1986). *J. Biol. Chem.* **261,** 11440–11443.
498. Hurt, E. C. *et al.* (1987). *J. Biol. Chem.* **262,** 1420–1424.
499. Hussain, M. *et al.* (1980). *J. Biol. Chem.* **255,** 3707–3712.
500. Hussain, M. *et al.* (1982). *Eur. J. Biochem.* **129,** 233–239.
501. Huttner, W. B. (1982). *Nature (London)* **299,** 273–276.
502. Hwang, J. *et al.* (1987). *Cell (Cambridge, Mass.)* **48,** 129–136.
503. Ibrahami, I. (1987). *J. Cell Biol.* **104,** 61–66.
504. Ibrahami, I. (1987). *J. Cell Biol.* **105,** 1555–1560.
505. Ibrahami, I., and Gentz, R. (1987). *J. Biol. Chem.* **262,** 10189–10194.
506. Ibrahami, I. *et al.* (1986). *Eur. J. Biochem.* **155,** 571–576.
507. Ichihara, S. *et al.* (1981). *J. Biol. Chem.* **256,** 3125–3129.
508. Ichihara, S. *et al.* (1986). *J. Biol. Chem.* **261,** 9405–9411.
509. Iida, A. *et al.* (1985). *EMBO J.* **4,** 1875–1880.
510. Iino, T. *et al.* (1987). *J. Biol. Chem.* **262,** 7412–7417.
511. Ikeda, T. *et al.* (1983). *J. Bacteriol.* **153,** 506–510.
512. Ikemura, H. *et al.* (1987). *J. Biol. Chem.* **262,** 7859–7864.

513. Ikemura, Y. *et al.* (1976). *Biochem. Biophys. Res. Commun.* **72,** 319–326.
513a. Imamoto-Sonobe, N. *et al.* (1988). *Proc. Natl. Acad. Sci. U.S.A.* **85,** 3426–3430.
514. Imanaka, T. *et al.* (1987). *J. Biol. Chem.* **105,** 2915–2922.
515. Innerarity, T. L. *et al.* (1984). *J. Biol. Chem.* **259,** 7261–7267.
516. Innis, M. A. *et al.* (1984). *Proc. Natl. Acad. Sci. U.S.A.* **81,** 3708–3712.
517. Inouye, M., and Halegoua, S. (1980). *CRC Crit. Rev. Biochem.* **7,** 339–371.
518. Inouye, S. *et al.* (1982). *Proc. Natl. Acad. Sci. U.S.A.* **79,** 3438–3441.
519. Inouye, S. *et al.* (1986). *J. Biol. Chem.* **261,** 10970–10973.
520. Inukai, M. (1978). *J. Antibiot.* **31,** 421–425.
521. Inukai, M. *et al.* (1979). *J. Bacteriol.* **140,** 1098–1101.
521a. Iocopetta, B. J. *et al.* (1988). *Cell (Cambridge, Mass.)* **54,** 485–489.
522. Ishidate, K. *et al.* (1986). *J. Biol. Chem.* **261,** 428–443.
523. Ito, K. *et al.* (1981). *Cell (Cambridge, Mass.)* **24,** 707–717.
524. Ito, K. (1983). *Cell (Cambridge, Mass.)* **32,** 789–797.
525. Ito, K. *et al.* (1986). *J. Bacteriol.* **167,** 201–204.
526. Izui, K. *et al.* (1980). *Biochemistry* **19,** 1182–1186.
527. Izumoto, Y. *et al.* (1987). *Gene* **59,** 151–159.
528. Jackowski, S., and Rock, C. O. (1986). *J. Biol. Chem.* **261,** 11328–11333.
529. Jackson, M. E., and Pratt, J. M. (1987). *Mol. Microbiol.* **1,** 23–28.
530. Jackson, M. E. *et al.* (1986). *J. Mol. Biol.* **189,** 477–486.
531. Jackson, R. C. (1983). *In* "Methods in Enzymology" (S. Fleischer and B. Fleischer, eds.), Vol. 96, pp. 784–794. Academic Press, New York.
532. Jackson, R. C., and Blobel, G. (1977). *Proc. Natl. Acad. Sci. U.S.A.* **74,** 5598–5602.
533. Jamieson, J. D., and Palade, G. E. (1968). *J. Cell Biol.* **39,** 589–603.
534. Jamieson, J. D., and Palade, G. E. (1971). *J. Cell Biol.* **48,** 503–522.
535. Jannatipour, M. *et al.* (1978). *J. Bacteriol.* **169,** 3785–3791.
536. Jenness, D. D., and Spatnak, P. (1986). *Cell (Cambridge, Mass.)* **46,** 345–353.
537. Jenness, D. D. *et al.* (1983). *Cell (Cambridge, Mass.)* **35,** 521–529.
538. Jenness, D. D. *et al.* (1986). *Mol. Cell. Biol.* **35,** 521–529.
538a. Jensen, R. E., and Yaffe, M. P. (1988). *EMBO J.* **7,** 3863–3871.
539. Jetten, A. M., and Jetten, R. E. R. (1975). *Biochim. Biophys. Acta* **387,** 12–22.
540. Johnson, D. C., and Spear, D. G. (1983). *Cell (Cambridge, Mass.)* **32,** 987–997.
541. Johnson, L. M. *et al.* (1987). *Cell (Cambridge, Mass.)* **48,** 875–885.
542. Josephson, L.-G., and Randall, L. L. (1981). *J. Biol. Chem.* **256,** 2504–2507.
543. Josephson, L.-G., and Randall, L. L. (1981). *Cell (Cambridge, Mass.)* **25,** 151–157.
544. Julius, D. *et al.* (1983). *Cell (Cambridge, Mass.)* **32,** 839–852.
545. Julius, D. *et al.* (1984). *Cell (Cambridge, Mass.)* **36,** 309–316.
546. Julius, D. *et al.* (1984). *Cell (Cambridge, Mass.)* **37,** 1075–1089.
547. Kaderbhai, M. A. *et al.* (1988). *FEBS Lett.* **232,** 313–316.
548. Kaetzel, D. M. *et al.* (1985). *Proc. Natl. Acad. Sci. U.S.A.* **82,** 7280–7283.
549. Kagan, B. L. *et al.* (1981). *Proc. Natl. Acad. Sci. U.S.A.* **78,** 4950–4954.
550. Kaiser, C. A., and Botstein, D. (1986). *Mol. Cell. Biol.* **6,** 2382–2391.
551. Kaiser, C. A. *et al.* (1987). *Science* **235,** 312–317.
552. Kalderon, D. *et al.* (1984). *Nature (London)* **311,** 33–38.
553. Kalderon, D. *et al.* (1984). *Cell (Cambridge, Mass.)* **39,** 499–509.
554. Kalkkinen, N. *et al.* (1986). *FEBS Lett.* **200,** 18–22.
554a. Kalousek, F. *et al.* (1988). *Proc. Natl. Acad. Sci. U.S.A.* **85,** 7536–7540.
555. Kamps, M. P. *et al.* (1986). *Cell (Cambridge, Mass.)* **45,** 105–112.
556. Karlin-Neumann, G. A., and Tobin, E. M. (1986). *EMBO J.* **5,** 9–13.
557. Kassenbrock, C. K. *et al.* (1988). *Nature (London)* **333,** 90–93.
558. Reference deleted.

559. Kaufman, R. J. *et al.* (1982). *J. Mol. Biol.* **159,** 601–621.
560. Kaufman, R. J. *et al.* (1986). *J. Biol. Chem.* **261,** 9622–9628
561. Kaufman, R. J. *et al.* (1988). *J. Biol. Chem.* **263,** 6352–6362.
562. Kawazu, T. *et al.* (1987). *J. Bacteriol.* **169,** 1564–1570.
563. Reference deleted.
564. Keen, J. H. *et al.* (1982). *Proc. Natl. Acad. Sci. U.S.A.* **79,** 2912–2916.
565. Keen, N. T. *et al.* (1984). *J. Bacteriol.* **159,** 825–831.
566. Keller, G.-A. *et al.* (1987). *Proc. Natl. Acad. Sci. U.S.A.* **84,** 3264–3268.
567. Kelly, R. B. (1985). *Science* **230,** 25–32.
568. Kendall, D. *et al.* (1986). *Nature (London)* **321,** 706–708.
569. Keng, T. *et al.* (1986). *Mol. Cell. Biol.* **6,** 355–364.
570. Kirwin, P. M. *et al.* (1987). *J. Biol. Chem.* **262,** 16386–16390.
571. Kleene, R. *et al.* (1987). *EMBO J.* **6,** 2627–2633.
572. Klionsky, D. J., *et al.* (1988). *Mol. Cell. Biol.* **8,** 2105–2116.
572a. Klose, M. *et al.* (1988). *J. Biol. Chem.* **263,** 13291–13296.
572b. Klose, M. *et al.* (1988). *J. Biol. Chem.* **263,** 13297–13302.
573. Kolata, G. (1985). *Science* **229,** 850.
574. Kondor-Koch, C. *et al.* (1985). *Cell (Cambridge, Mass.)* **43,** 297–306.
574a. Kontinen, V. P., and Sarvas, M. (1988). *J. Gen. Microbiol.* **134,** 2333–2344.
575. Koren, R. *et al.* (1983). *Proc. Natl. Acad. Sci. U.S.A.* **80,** 7205–7209.
576. Kornacker, M. G. *et al.* (1988). *In* "Membrane Biogenesis" (J. A. F. Op den Kamp, ed.), pp. 429–438. Springer-Verlag, Berlin.
577. Kornfeld, R., and Kornfeld, S. (1985). *Annu. Rev. Biochem.* **54,** 631–664.
577a. Koronakis, V. *et al.* (1988). *Mol. Gen. Genet.* **213,** 551–555.
578. Korteland, J. *et al.* (1985). *Eur. J. Biochem.* **152,** 691–697.
579. Koshland, D., and Botstein, D. (1982). *Cell (Cambridge, Mass.)* **30,** 903–914.
580. Kotwal, G. J., and Ghosh, H. P. (1984). *J. Biol. Chem.* **259,** 4699–4701.
581. Kotwal, G. J. *et al.* (1986). *J. Biol. Chem.* **261,** 8936–8943.
582. Kovacevic, S. *et al.* (1985). *J. Bacteriol.* **162,** 521–528.
583. Kozutsumi, Y. *et al.* (1988). *Nature (London)* **332,** 462–464.
584. Krakow, J. L. *et al.* (1986). *J. Biol. Chem.* **261,** 12147–12153.
585. Kreibach, G. *et al.* (1978). *J. Cell Biol.* **77,** 464–487.
586. Kreibach, G. *et al.* (1978). *J. Cell Biol.* **77,** 488–506.
587. Kreibach, G. *et al.* (1982). *J. Cell Biol.* **93,** 111–121.
588. Kreibach, G. *et al.* (1983). *In* "Methods in Enzymology" (S. Fleischer and B. Flesicher, eds.), Vol. 96, pp. 520–530. Academic Press, New York.
589. Kreis, T. E., and Lodish, H. F. (1986). *Cell (Cambridge, Mass.)* **46,** 929–937.
590. Kreis, T. E. (1986). *EMBO J.* **5,** 931–941.
591. Krieg, U. C. *et al.* (1986). *Proc. Natl. Acad. Sci. U.S.A.* **83,** 8604–8608.
592. Krieger, M. *et al.* (1981). *J. Mol. Biol.* **150,** 167–184.
593. Krieger, M. *et al.* (1983). *Proc. Natl. Acad. Sci. U.S.A.* **80,** 5607–5611.
594. Krupp, M. N., and Lane, M. D. (1982). *J. Biol. Chem.* **257,** 1372–1377.
595. Kruppa, J., and Sabatini, D. (1977). *J. Cell Biol.* **74,** 414–427.
596. Kuhn, A. (1987). *Science* **238,** 1413–1415.
597. Kuhn, A., and Wickner, W. (1985). *J. Biol. Chem.* **260,** 15914–15918.
598. Kukurazinska, M. A. *et al.* (1987). *Annu. Rev. Biochem.* **56,** 915–944.
599. Kumamoto, S., and Beckwith, J. (1983). *J. Bacteriol.* **154,** 253–260.
600. Kumar, G. *et al.* (1986). *J. Cell Biol.* **103,** 767–776.
601. Kuntz, M. *et al.* (1986). *Mol. Gen. Genet.* **205,** 454–460.
602. Kurjan, J., and Herskowitz, I. (1982). *Cell (Cambridge, Mass.)* **30,** 933–943.
603. Kurzchalia, T. V. *et al.* (1986). *Nature (London)* **320,** 634–636.

604. Kvist, S. *et al.* (1982). *Cell (Cambridge, Mass.)* **29,** 61–69.
605. Kyte, J., and Doolittle, R. (1982). *J. Mol. Biol.* **157,** 105–152.
606. Lai, J. S. *et al.* (1981). *J. Bacteriol.* **145,** 657–660.
607. Lambrecht, B., and Schmidt, M. F. G. (1986). *FEBS Lett.* **202,** 127–132.
608. Lamed, R. *et al.* (1983). *J. Bacteriol.* **156,** 828–836.
609. Lamppa, G. K., and Abad, M. S. (1987). *J. Cell Biol.* **105,** 2642–2648.
610. Lanford, R. E., and Butel, J. S. (1984). *Cell (Cambridge, Mass.)* **37,** 801–813.
611. Lanford, R. E. *et al.* (1986). *Cell (Cambridge, Mass.)* **46,** 575–582.
612. Lang, L. *et al.* (1984). *J. Biol. Chem.* **259,** 14663–14667.
613. Lauffer, L. *et al.* (1985). *Nature (London)* **318,** 334–338.
613a. Lazarovits, J., and Roth, J. M. (1988). *Cell (Cambridge, Mass.)* **53,** 743–752.
614. Lazarow, P. B., and Fujuki, Y. (1985). *Annu. Rev. Cell Biol.* **1,** 489–530.
614a. Lazzarino, D. A., and Gabel, C. A. (1988). *J. Biol. Chem.* **263,** 10118–10126.
615. Ledford, D. E., and Davis, D. F. (1983). *J. Biol. Chem.* **258,** 3304–3308.
616. Lee, C. A., and Beckwith, J. (1986). *J. Bacteriol.* **166,** 878–883.
617. Lee, C. A. *et al.* (1985). *J. Bacteriol.* **161,** 1156–1161.
618. Lee, N. *et al.* (1983). *J. Bacteriol.* **155,** 407–411.
619. Lee, R. W. H., and Huttner, W. B. (1985). *Proc. Natl. Acad. Sci. U.S.A.* **82,** 6143–6147.
620. Lee, S. Y. *et al.* (1971). *J. Cell Biol.* **49,** 683–691.
621. Lehle, L., and Bause, E. (1984). *Biochim. Biophys. Acta* **799,** 246–251.
622. Lehnardt, S. *et al.* (1987). *J. Biol. Chem.* **262,** 1716–1719.
623. Lehner, C. F. *et al.* (1986). *Proc. Natl. Acad. Sci. U.S.A.* **83,** 2096–2099.
624. Lehrman, M. A. *et al.* (1985). *Cell (Cambridge, Mass.)* **41,** 735–743.
625. Lehrman, M. A. *et al.* (1985). *Science* **227,** 140–146.
626. Lemansky, P. *et al.* (1987). *J. Cell Biol.* **104,** 1743–1749.
627. Lemmon, S. K., and Jones, E. W. (1987). *Science* **238,** 504–509.
628. Lemos-Chiarandini, C. de *et al.* (1978). *J. Cell Biol.* **104,** 209–219.
629. Lepault, J. *et al.* (1988). *EMBO J.* **7,** 261–268.
630. Lewis, M. J. *et al.* (1985). *J. Biol. Chem.* **260,** 6926–6931.
631. Ley, P. van der *et al.* (1985). *Eur. J. Biochem.* **147,** 401–407.
632. Ley, P. van der *et al.* (1986). *J. Biol. Chem.* **261,** 1222–1225.
633. Ley, P. van der *et al.* (1987). *Mol. Gen. Genet.* **209,** 585–591.
633a. Li, P. *et al.* (1988). *Proc. Natl. Acad. Sci. U.S.A.* **85,** 7685–7689.
634. Li, W.-Y. *et al.* (1982). *J. Biol. Chem.* **257,** 5136–5142.
635. Li, Y. *et al.* (1981). *Eur. J. Biochem.* **116,** 199–205.
635a. Lichenstein, H. *et al.* (1988). *J. Bacteriol.* **170,** 3924–3929.
636. Liebke, H. H. (1987). *J. Bacteriol.* **169,** 1174–1181.
636a. Lill, R. *et al.* (1988). *Cell (Cambridge, Mass.)* **54,** 1013–1018.
637. Lindberg, F. *et al.* (1987). *Nature (London)* **328,** 84–87.
638. Lingappa, V. R. *et al.* (1984). *Proc. Natl. Acad. Sci. U.S.A.* **81,** 456–460.
639. Lingelbach, K. R. *et al.* (1986). *Eur. J. Biochem.* **161,** 19–23.
640. Lipp, J., and Dobberstein, B. (1986). *Cell (Cambridge, Mass.)* **46,** 1103–1112.
641. Lipp, J., and Dobberstein, B. (1986). *J. Cell Biol.* **102,** 2169–2175.
642. Lipp, J. *et al.* (1987). *J. Biol. Chem.* **262,** 1680–1684.
642a. Lipponcott-Schwartz, J. *et al.* (1988). *Cell (Cambridge, Mass.)* **54,** 209–222.
643. Liss, L. R., and Oliver, D. B. (1986). *J. Biol. Chem.* **261,** 2299–2303.
644. Liss, L. R. *et al.* (1985). *J. Bacteriol.* **164,** 925–928.
645. Lobel, P. *et al.* (1988). *J. Biol. Chem.* **263,** 2563–2570.
646. Lodish, H. F. (1988). *J. Biol. Chem.* **263,** 2107–2110.
647. Lodish, H. F., and Kong, N. (1984). *J. Cell Biol.* **98,** 1720–1729.

648. Lodish, H. F. *et al.* (1983). *Nature (London)* **304,** 80–83.
649. Lodish, H. F. *et al.* (1987). *J. Cell Biol.* **104,** 221–230.
650. Loison, G. *et al.* (1988). *Bio/Technology* **6,** 72–77.
651. Loon, A. P. G. M. van, and Schatz, G. (1987). *EMBO J.* **6,** 2440–2448.
652. Loon, A. P. G. M. van, and Young, E. T. (1986). *EMBO J.* **5,** 161–165.
653. Loon, A. P. G. M. van *et al.* (1986). *Cell (Cambridge, Mass.)* **44,** 801–812.
654. Loon, A. P. G. M. van *et al.* (1987). *EMBO J.* **6,** 2433–2439.
655. Lory, S. *et al.* (1983). *J. Bacteriol.* **156,** 695–702.
656. Louvard, P. (1980). *Proc. Natl. Acad. Sci. U.S.A.* **77,** 4132–4136.
657. Low, M. G., and Kincade, P. W. (1985). *Nature (London)* **318,** 62–64.
658. Low, M. G. *et al.* (1986). *Trends Biochem. Sci.* **11,** 212–215.
659. Löwenadler, I. B. *et al.* (1986). *EMBO J.* **5,** 2393–2398.
660. Lubben, T. H., and Keegstra, K. (1986). *Proc. Natl. Acad. Sci. U.S.A.* **83,** 5502–5506.
661. Luckow, V. A., and Summers, M. D. (1988). *Bio/Technology* **6,** 47–55.
662. Lucocq, J. M., and Warren, G. (1987). *EMBO J.* **6,** 3239–3246.
663. Lucocq, J. M. *et al.* (1984). *J. Cell Biol.* **99,** 354a.
664. Lunn, C., and Inouye, M. (1987). *J. Biol. Chem.* **262,** 8318–8324.
665. Lurberboum-Galski, H. (1988). *Proc. Natl. Acad. Sci. U.S.A.* **85,** 1922–1926.
666. Lusis, A. J. *et al.* (1976). *J. Biol. Chem.* **251,** 7753–7760.
667. Lyons, R. H. *et al.* (1987). *Mol. Cell. Biol.* **7,** 2451–2456.
668. Maccecchini, M.-L. *et al.* (1979). *Proc. Natl. Acad. Sci.U.S.A.* **76,** 343–347.
669. MacDonald, R. G. *et al.* (1988). *Science* **239,** 1134–1137.
670. Machamer, C. E., and Rose, J. K. (1987). *J. Cell Biol.* **105,** 503–519.
671. Machamer, C. E., and Rose, J. K. (1988). *J. Biol. Chem.* **263,** 5955–5960.
672. Machamer, C. E. *et al.* (1985). *Mol. Cell. Biol.* **5,** 3074–3083.
673. MacIntyre, S. *et al.* (1987). *J. Biol. Chem.* **262,** 8416–8422.
674. Mackin, R. B., and Noe, B. D. (1987). *J. Biol. Chem.* **262,** 6453–6456.
675. Mackman, N. *et al.* (1985). *Mol. Gen. Genet.* **201,** 529–536.
676. Mackman, N. *et al.* (1986). *Curr. Top. Microbiol. Immunol.* **125,** 159–182.
677. Mackman, N. *et al.* (1987). *EMBO J.* **6,** 2835–2841.
678. Madsen, P. *et al.* (1986). *J. Cell Biol.* **103,** 2083–2089.
678a. Maeba, P. Y. (1986). *J. Bacteriol.* **166,** 644–650.
679. Magee, A. I., and Courtheidge, S. A. (1985). *EMBO J.* **4,** 1137–1144.
680. Maher, P. A., and Singer, S. J. (1986). *Proc. Natl. Acad. Sci. U.S.A.* **83,** 9001–9005.
681. Mains, P. E., and Sibley, C. H. (1983). *J. Biol. Chem.* **258,** 5027–5033.
682. Majzoub, J. A. *et al.* (1980). *J. Biol. Chem.* **255,** 11478–11483.
683. Makarow, M. (1985). *EMBO J.* **4,** 1861–1866.
684. Makarow, M. (1985). *EMBO J.* **4,** 1855–1860.
685. Makarow, M. (1988). *EMBO J.* **7,** 1475–1482.
685a. Malhorta, V. *et al.* (1988). *Cell (Cambridge, Mass.)* **54,** 221–227.
686. Mallay, C. *et al.* (1982). *Proc. Natl. Acad. Sci. U.S.A.* **79,** 2260–2263.
687. Mandel, G., and Wickner, W. (1979). *Proc. Natl. Acad. Sci. U.S.A.* **76,** 256–240.
688. Manoil, C., and Beckwith, J. (1985). *Proc. Natl. Acad. Sci. U.S.A.* **82,** 8129–8133.
689. Manoil, C., and Beckwith, J. (1986). *Science* **23,** 1403–1408.
690. Marcantonio, E. E. *et al.* (1984). *J. Cell Biol.* **99,** 2254–2259.
691. Markoff, L. *et al.* (1984). *Mol. Cell. Biol.* **4,** 8–16.
692. Marrs, C. F. *et al.* (1985). *J. Bacteriol.* **163,** 132–139.
693. Martinez, M. C. *et al.* (1983). *EMBO J.* **2,** 1501–1507.
694. Marty-Mazars, D. *et al.* (1983). *J. Bacteriol.* **154,** 1381–1388.
695. Matlin, K. S., and Simons, K. (1984). *J. Cell Biol.* **99,** 2131–2139.

695a. Matoba, S. *et al.* (1988). *Mol. Cell. Biol.* **8,** 4904–4916.
696. Matoo, A. K., and Edelman, M. (1987). *Proc. Natl. Acad. Sci. U.S.A.* **84,** 1497–1501.
697. Matsura, S. *et al.* (1978). *J. Cell Biol.* **78,** 503–519.
698. Mattaj, I. W., and De Robertis, E. M. (1985). *Cell (Cambridge, Mass.)* **40,** 111–118.
699. McAda, P., and Douglas, M. G. (1982). *J. Biol. Chem.* **257,** 3177–3182.
700. McClelland, A., *et al.* (1984). *Cell (Cambridge, Mass.)* **39,** 267–274.
701. McMullin, T. W., and Hallbcrg, R. L. (1988). *Mol. Cell. Biol.* **8,** 374–380.
702. McQueen, N. L. *et al.* (1986). *Proc. Natl. Acad. Sci. U.S.A.* **83,** 9318–9322.
703. McQueen, N. L. *et al.* (1987). *J. Biol. Chem.* **262,** 16233–16240.
704. Medda, S. *et al.* (1987). *Cell (Cambridge, Mass.)* **50,** 301–310.
705. Meek, R. L. *et al.* (1982). *J. Biol. Chem.* **257,** 12245–12251.
706. Meer, G. van (1986). *Trends Biochem. Sci.* **11,** 194–195.
707. Meer, G. van and Simons, K. (1986). *EMBO J.* **5,** 1455–1464.
708. Meer, G. van *et al.* (1986). *Nature (London)* **322,** 639–641.
709. Meer, G. van *et al.* (1987). *J. Cell Biol.* **105,** 1623–1635.
710. Melançon, P. *et al.* (1987). *Cell (Cambridge, Mass.)* **51,** 1053–1062.
711. Melcher, B. *et al.* (1982). *J. Biol. Chem.* **257,** 11203–11206.
712. Mellman, I. *et al.* (1986). *Annn. Rev. Biochem.* **55,** 663–700.
712a. Messner, P., and Sleytr, U.B. (1988). *FEBS Lett.* **228,** 317–320.
713. Metcalf, M., and Holland, I. B. (1981). *FEMS Microbiol. Lett.* **7,** 111–114.
714. Meyer, D. I. (1985). *EMBO J.* **4,** 2031–2033.
715. Meyer, D. I., and Dobberstein, B. (1980). *J. Cell Biol.* **87,** 498–502.
716. Meyer, D. I., and Dobberstein, B. (1980). *J. Cell Biol.* **87,** 503–508.
717. Meyer, D. I. *et al.* (1982). *Nature (London)* **297,** 647–650.
718. Meyer, T. *et al.* (1982). *Cell (Cambridge, Mass.)* **30,** 45–52.
719. Mézes, P. S. F. *et al.* (1985). *J. Biol. Chem.* **260,** 1218–1223.
720. Michaelis, S. *et al.* (1983). *J. Bacteriol.* **154,** 356–365.
721. Miller, J. R. *et al.* (1987). *J. Bacteriol.* **169,** 3508–3414.
722. Milstein, C. *et al.* (1972). *Nature (London) New Biol.* **239,** 117–120.
723. Minami, Y. *et al.* (1987). *Proc. Natl. Acad. Sci. U.S.A.* **84,** 2688–2692.
724. Mishkind, M. L. *et al.* (1985). *J. Cell Biol.* **100,** 226–233.
725. Miura, S. *et al.* (1984). *J. Biol. Chem.* **259,** 6397–6402.
726. Miyakawa, T. *et al.* (1987). *J. Bacteriol.* **169,** 1626–1631.
727. Mizuno, T. *et al.* (1983). *J. Biol. Chem.* **258,** 6932–6940.
728. Mizuno, T. *et al.* (1987). *Mol. Gen. Genet.* **207,** 217–223.
729. Moks, T. *et al.* (1987). *Bio/Technology* **5,** 379–382.
730. Montesano, R. *et al.* (1982). *Nature (London)* **296,** 651–653.
731. Moore, H.-P., and Kelly, R. B. (1986). *Nature (London)* **321,** 443–446.
732. Moore, H.-P., *et al.* (1983). *Cell (Cambridge, Mass.)* **35,** 351–358.
733. Moore, H.-P. *et al.* (1983). *Nature (London)* **302,** 434–436.
734. Moore, K. E., and Miura, S. (1987). *J. Biol. Chem.* **262,** 8806–8813.
735. Moreland, R. B. *et al.* (1985). *Proc. Natl. Acad. Sci. U.S.A.* **82,** 6561–6565.
736. Moreland, R. B. *et al.* (1987). *Mol. Cell. Biol.* **7,** 4048–4057.
737. Moreno, F. *et al.* (1980). *Nature (London)* **286,** 356–359.
738. Morgan, D. O. *et al.* (1987). *Nature (London)* **329,** 301–307.
739. Mori, M. *et al.* (1985). *Eur. J. Biochem.* **149,** 181–186.
740. Morona, R. *et al.* (1984). *J. Bacteriol.* **159,** 570–582.
741. Mostov, K. E., and Deitcher, D. L. (1986). *Cell (Cambridge, Mass.)* **46,** 613–621.
742. Mostov, K. E. *et al.* (1984). *Nature (London)* **308,** 37–43.

743. Mostov, K. E. *et al.* (1987). *J. Cell Biol.* **105,** 2031–2036.
744. Mroz, E. A., and Lechene, C. (1986). *Science* **232,** 871–872.
745. Muekler, M., and Lodish, H. F. (1986). *Cell (Cambridge, Mass.)* **44,** 629–637.
746. Muekler, M., and Lodish, H. F. (1986). *Nature (London)* **322,** 549–552.
747. Reference deleted.
748. Müller, G., and Zimmermann, R. (1987). *EMBO J.* **6,** 2099–2107.
749. Müller, G., and Zimmermann, R. (1988). *EMBO J.* **7,** 639–648.
750. Müller, M., and Blobel, G. (1984). *Proc. Natl. Acad. Sci. U.S.A.* **81,** 7421–7425.
751. Müller, M., and Blobel, G. (1984). *Proc. Natl. Acad. Sci. U.S.A.* **81,** 7737–7741.
752. Reference deleted.
753. Müller, M. *et al.* (1982). *J. Biol. Chem.* **257,** 11860–11863.
754. Müller, M. *et al.* (1987). *EMBO J.* **6,** 3855–3861.
755. Mumford, R. *et al.* (1980). *J. Biol. Chem.* **255,** 2227–2230.
756. Mundy, D. I., and Strittmatter, W. J. (1985). *Cell (Cambridge, Mass.)* **40,** 645–656.
757. Munro, S., and Pelham, H. R. B. (1984). *EMBO J.* **3,** 3087–3093.
758. Munro, S., and Pelham, H. R. B. (1986). *Cell (Cambridge, Mass.)* **46,** 291–300.
759. Munro, S., and Pelham, H. R. B. (1987). *Cell (Cambridge, Mass.)* **48,** 899–907.
759a. Murakomi, H. *et al.* (1988). *J. Cell Biol.* **107,** 2051–2057.
760. Murén, E. M., and Randall, L. L. (1985). *J. Bacteriol.* **164,** 712–716.
761. Murphy, R. F. *et al.* (1984). *J. Cell Biol.* **98,** 1757–1762.
762. Musgrove, J. E. *et al.* (1987). *Eur. J. Biochem.* **163,** 529–534.
763. Musil, L. S., and Baeziger, J. U. (1987). *J. Cell Biol.* **104,** 1725–1733.
764. Nagahari, K. *et al.* (1985). *EMBO J.* **4,** 3589–3592.
765. Nagaraj, R. (1984). *FEBS Lett.* **165,** 79–82.
766. Nakamura, M. *et al.* (1983). *Mol. Gen. Genet.* **191,** 1–9.
767. Nakamura, Y. *et al.* (1986). *Gene* **50,** 239–245.
767a. Nakano, A. *et al.* (1988). *J. Cell Biol.* **107,** 851–863.
767b. Nakayama, N. *et al.* (1985). *EMBO J.* **4,** 2643–2648.
768. Natata, A. *et al.* (1987). *In* "Phosphate Metabolism and Cellular Regulation in Microorganisms." pp. 139–141. Am. Soc. Microbiol., Washington, D.C.
768a. Neefjes, J. J. *et al.* (1988). *J. Cell Biol.* **107,** 79–87.
769. Neer, E. J., and Clapham, D. E. (1988). *Nature (London)* **333,** 129–134.
770. Nelson, M. J., and Veshnock, P. J. (1987). *Nature (London)* **328,** 533–536.
771. Ness, S. A., and Weiss, R. L. (1987). *Proc. Natl. Acad. Sci. U.S.A.* **84,** 6692–6696.
772. Neugebauer, K. *et al.* (1981). *Nucleic Acids Res.* **9,** 2577–2588.
773. Neuhaus, J.-M. *et al.* (1982). *Ann. Microbiol (Paris)* **133A,** 27–32.
774. Neville, D. M., Jr., and Hudson, T. H. (1986). *Annu. Rev. Biochem.* **55,** 195–224.
775. Newman, A. P., and Ferro-Novick, S. (1987). *J. Cell Biol.* **105,** 1587–1594.
776. Newmeyer, D. D., and Forbes, D. J. (1988). *Cell (Cambridge, Mass.)* **52,** 641–654.
777. Newmeyer, D. D. *et al.* (1986). *Cell (Cambridge, Mass.)* **45,** 500–510.
778. Newmeyer, D. D. *et al.* (1986). *J. Cell Biol.* **103,** 2091–2102.
779. Newport, J. W., and Forbes, D. J. (1987). *Annu. Rev. Biochem.* **56,** 535–565.
780. Nguyen, M. *et al.* (1987). *J. Cell Biol.* **104,** 1193–1198.
780a. Nguyen, M. *et al.* (1988). *J. Cell Biol.* **106,** 1499–1505.
781. Nicaud, J.-M. *et al.* (1986). *FEBS Lett.* **187,** 339–344.
782. Nicholson, D. W. *et al.* (1987). *Eur. J. Biochem.* **164,** 147–157.
783. Nielsen, J. B. K., and Lampen, J. O. (1982). *J. Bacteriol.* **152,** 315–322.
784. Nielsen, J. B. K., and Lampen, J. O. (1983). *Biochemistry* **22,** 4652–4656.
785. Nielsen, J. B. K. *et al.* (1981). *Proc. Natl. Acad. Sci. U.S.A.* **78,** 3511–3515.
786. Niemann, H. *et al.* (1982). *EMBO J.* **1,** 1499–1504.

787. Nikaido, H., and Wu, H. C. (1984). *Proc. Natl. Acad. Sci. U.S.A.* **81,** 1048–1052.
788. Nilsson, B. *et al.* (1985). *Nucleic Acids Res.* **13,** 1151–1162.
789. Noe, B. D., and Moran, M. N. (1984). *J. Cell Biol.* **99,** 418–424.
790. Noe, B. D. *et al.* (1986). *J. Cell Biol.* **103,** 1205–1211.
791. Nogami, T. *et al.* (1985). *J. Bacteriol.* **164,** 797–801.
792. Novak, P. *et al.* (1986). *J. Biol. Chem.* **261,** 420–427.
793. Novick, P., and Botstein, D. (1985). *Cell (Cambridge, Mass.)* **40,** 405–416.
794. Novick, P., and Schekman, R. (1979). *Proc. Natl. Acad. Sci. U.S.A.* **76,** 1858–1862.
795. Novick, P. *et al.* (1980). *Cell (Cambridge, Mass.)* **21,** 205–215.
796. Novikoff, P. M. (1983). *Proc. Natl. Acad. Sci. U.S.A.* **80,** 4364–4368.
797. Nowack, D. D. *et al.* (1987). *Proc. Natl. Acad. Sci. U.S.A.* **84,** 6089–6102.
798. Oberto, J., and Davidson, J. (1986). *Gene* **40,** 57–65.
799. Octave, J. N. *et al.* (1983). *Trends Biochem. Sci.* **8,** 217–220.
800. Ohashi, A. *et al.* (1982). *J. Biol. Chem.* **257,** 13042–13047.
801. Ohba, M., and Schatz, G. (1987). *EMBO J.* **6,** 2109–2115.
802. Ohba, M., and Schatz, G. (1987). *EMBO J.* **6,** 2117–2122.
803. Ohta, S., and Schatz, G. (1984). *EMBO J.* **3,** 651–657.
804. Ohta, S. *et al.* (1987). *FEBS Lett.* **226,** 171–175.
805. Ojakian, G. K. *et al.* (1977). *J. Cell Biol.* **72,** 530–551.
806. Oka, T. *et al.* (1985). *Proc. Natl. Acad. Sci. U.S.A.* **82,** 7212–7216.
807. Okhuma, S., and Poole, B. (1978). *Proc. Natl. Acad. Sci. U.S.A.* **75,** 3377–3381.
808. Oliver, D. B. (1985). *J. Bacteriol.* **161,** 285–291.
809. Oliver, D. B., and Beckwith, J. (1981). *Cell (Cambridge, Mass.)* **25,** 765–772.
810. Oliver, D. B., and Beckwith, J. (1982). *Cell (Cambridge, Mass.)* **30,** 311–319.
811. Oliver, D. B., and Beckwith, J. (1982). *J. Bacteriol.* **150,** 686–691.
812. Olsen, E. M., and Spizz, G. (1986). *J. Biol. Chem.* **261,** 2458–2466.
813. Ono, N., and Tuboi, S. (1987). *Eur. J. Biochem.* **168,** 509–514.
814. Ono, N., and Tuboi, S. (1988). *J. Biol. Chem.* **263,** 3188–3193.
815. Ono, N. *et al.* (1985). *J. Biol. Chem.* **260,** 3402–3407.
816. Orci, L. (1985). *Cell (Cambridge, Mass.)* **42,** 671–681.
817. Orci, L. *et al.* (1981). *Proc. Natl. Acad. Sci. U.S.A.* **78,** 293–297.
818. Orci, L. *et al.* (1984). *Cell (Cambridge, Mass.)* **39,** 39–47.
819. Orci, L. *et al.* (1984). *Proc. Natl. Acad. Sci. U.S.A.* **81,** 6743–6746.
820. Orci, L. *et al.* (1986). *Cell (Cambridge, Mass.)* **46,** 171–184.
821. Orci, L. *et al.* (1986). *J. Biol. Chem.* **103,** 2273–2281.
822. Orci, *et al.* (1986). *EMBO J.* **5,** 2097–2101.
823. Orci, L. *et al.* (1987). *Nature (London)* **326,** 77–79.
824. Orci, L. *et al.* (1987). *Cell (Cambridge, Mass.)* **51,** 1039–1051.
825. Orci, L. *et al.* (1987). *Cell (Cambridge, Mass.)* **49,** 865–868.
826. Oshima, A. *et al.* (1988). *J. Biol. Chem.* **263,** 2553–2562.
827. Osinga, K. A. *et al.* (1985). *EMBO J.* **4,** 3811–3817.
828. Overbeeke, N. *et al.* (1983). *J. Mol. Biol.* **163,** 513–532.
829. Owada, M., and Neufeld, E. F. (1982). *Biochem. Biophys. Res. Commun.* **105,** 814–820.
830. Oxender D. L. *et al.* (1980). *Proc. Natl. Acad. Sci. U.S.A.* **77,** 2005–2009.
831. Oxender, D. L. *et al.* (1984). *In* "Microbiology 1984" (D. Schlessinger, ed.), pp. 4–7. Am. Soc. Microbiol. Washington, D.C.
832. Paabo, S. *et al.* (1987). *Cell (Cambridge, Mass.)* **50,** 311–317.
833. Pacaud, M. (1982). *J. Biol. Chem.* **257,** 4333–4339.
834. Padden, C. J., and Hartley, R. W. (1986). *Gene* **40,** 231–239.
835. Paddington, L. *et al.* (1987). *Proc. Natl. Acad. Sci. U.S.A.* **84,** 2756–2760.

836. Pagès, J. M., and Lazdunski, C. (1981). *FEMS Microbiol. Lett.* **12,** 65–69.
837. Pain, D., and Blobel, G. (1987). *Proc. Natl. Acad. Sci. U.S.A.* **84,** 3288–3292.
838. Pain, D. *et al.* (1988). *Nature (London)* **331,** 232–237.
839. Paine, P. L. (1975). *J. Cell Biol.* **66,** 652–657.
840. Palade, G. (1975). *Science* **189,** 347–358.
841. Palido, D. *et al.* (1986). *Gene* **45,** 167–174.
842. Palva, E. T., and Randall, L. L. (1978). *J. Bacteriol.* **133,** 279–286.
843. Palva, I. *et al.* (1981). *Gene* **15,** 43–51.
844. Palva, I. *et al.* (1983). *Gene* **22,** 229–235.
844a. Papini, E. *et al.* (1988). *EMBO J.* **7,** 3353–3359.
845. Park, K. K. *et al.* (1987). *Proc. Natl. Acad. Sci. U.S.A.* **84,** 6462–6466.
846. Park, S. *et al.* (1988). *Science* **239,** 1033–1035.
847. Parodi, A. J. *et al.* (1984). *J. Biol. Chem.* **259,** 6351–6357.
848. Pattus, F. *et al.* (1983). *Biochemistry* **22,** 5698–5703.
849. Patzak, A., and Winkler, H. (1986). *J. Cell Biol.* **102,** 510–515.
850. Patzelt, C. P. *et al.* (1978). *Proc. Natl. Acad. Sci. U.S.A.* **75,** 1260–1264.
851. Paul, A. V. *et al.* (1987). *Proc. Natl. Acad. Sci. U.S.A.* **84,** 7827–7831.
852. Payne, C. G., and Schekman, R. (1985). *Science* **230,** 1009–1018.
853. Payne, G. S. *et al.* (1987). *Mol Cell. Biol.* **7,** 3888–3898.
853a. Payne, G. S. *et al.* (1988). *J. Cell Biol.* **106,** 1453–1461.
854. Pearce, B. M. F. (1985). *EMBO J.* **4,** 2457–2460.
855. Pearce, B. M. F. (1987). *EMBO J.* **6,** 2507–2512.
855a. Pearce, B. M. F. (1988). *EMBO J.* **7,** 3331–3336.
856. Pearson, G. D. N., and Mekalanos, J. J. (1982). *Proc. Natl. Acad. Sci. U.S.A.* **79,** 2976–2980.
857. Pelham, H. R. B. (1986). *Cell (Cambridge, Mass.)* **46,** 959–961.
858. Pelham, H. R. B. (1988). *EMBO J.* **7,** 913–918.
858a. Pelham, H. R. B. (1988). *EMBO J.* **7,** 1757–1762.
859. Pentillä, M. F. *et al.* (1988). *Gene* **63,** 103–112.
860. Perera, E. *et al.* (1986). *Science* **232,** 348–352.
861. Perlman, D., and Halvorson, H. O. (1983). *J. Mol. Biol.* **167,** 391–409.
862. Perlman, D. *et al.* (1982). *Proc. Natl. Acad. Sci. U.S.A.* **79,** 781–785.
863. Perlman, D. *et al.* (1986). *Proc. Natl. Acad. Sci. U.S.A.* **83,** 5033–5037.
864. Perumal, N. D., and Minkley, E. G., Jr. (1984). *J. Biol. Chem.* **259,** 5357–5360.
865. Pfaller, R., and Neupert, W. (1987). *EMBO J.* **6,** 2635–2642.
866. Pfaller, R. *et al.* (1985). *J. Biol. Chem.* **260,** 8188–8193.
866a. Pfaller, R. *et al.* (1988). *J. Cell Biol.* **107,** 2483–2490.
867. Pfanner, N., and Neupert, W. (1985). *EMBO J.* **4,** 2818–2825.
868. Pfanner, N., and Neupert, W. (1987). *J. Biol. Chem.* **262,** 7328–7336.
869. Pfanner, N. *et al.* (1987). *Cell (Cambridge, Mass.)* **48,** 815–823.
870. Pfanner, N. *et al.* (1987). *EMBO J.* **6,** 3449–3454.
871. Pfanner, N. *et al.* (1988). *J. Biol. Chem.* **263,** 4049–4051.
872. Pfeffer, S. R., and Rothman, J. E. (1987). *Annu. Rev. Biochem.* **56,** 829–852.
873. Pfeiffer, S. *et al.* (1985). *J. Biol. Chem.* **101,** 470–476.
874. Pfisterer, J. *et al.* (1982). *Eur. J. Biochem.* **126,** 143–148.
875. Picard, D., and Yamamoto, K. R. (1987). *EMBO J.* **6,** 3333–3340.
876. Pilgrim, D., and Young, E. T. (1987). *Mol. Cell Biol.* **7,** 294–304.
877. Pines, O. *et al.* (1988). *Mol. Microbiol.* **2,** 209–217.
878. Pinto, M. *et al.* (1983). *Biol. Cell.* 47, 323–330.
879. Plückthun, A., and Knowles, J. R. (1987). *J. Biol. Chem.* **262,** 3951–3957.
879a. Plückthun, A., and Pfitzinger, I. (1988). *J. Biol. Chem.* **263,** 14315–14322.

880. Pohler, J. *et al.* (1986). *Mol. Gen. Genet.* **205,** 501–506.
881. Pohler, J. *et al.* (1987). *Nature (London)* **325,** 458–462.
881a. Pollack, R. A. *et al.* (1988). *EMBO J.* **7,** 3493–3500.
882. Pollitt, N. S., and Inouye, M. (1988). *J. Bacteriol.* **170,** 2051–2055.
883. Pollitt, N. S. *et al.* (1985). *In* "Microbiology 1985" (D. Schlessinger, ed.), pp. 308–311. Am. Soc. Microbiol., Washington, D.C.
883a. Poritz, M. A. *et al.* (1988). *Proc. Natl. Acad. Sci. U.S.A.* **85,** 4315–4319.
883b. Poritz, M. A. *et al.* (1988). *Cell (Cambridge, Mass.)* **55,** 4–6.
883c. Poruchynsky, M. S., and Atkinson, P. H. (1988). *J. Cell Biol.* **107,** 1697–1707.
884. Poruchynsky, M. S. (1985). *J. Cell. Biol.* **101,** 2199–2209.
885. Power, S. D. *et al.* (1986). *Proc. Natl. Acad. Sci. U.S.A.* **83,** 3896–3100.
886. Powers, S. *et al.* (1986). *Cell (Cambridge, Mass.)* **47,** 413–422.
887. Pratje, E., and Guiard, B. (1986). *EMBO J.* **5,** 1313–1317.
888. Pratt, J. M. *et al.* (1986). *EMBO J.* **5,** 2399–2405.
889. Pratt, R. E. *et al.* (1988). *J. Biol. Chem.* **263,** 3137–3141.
890. Prehn, S. *et al.* (1980). *Eur. J. Biochem.* **107,** 185–195.
891. Prehn, S. *et al.* (1981). *FEBS Lett.* **123,** 79–84.
892. Prehn, S. *et al.* (1987). *EMBO J.* **6,** 2093–2097.
893. Preston, R. A. *et al.* (1987). *J. Cell Biol.* **105,** 1981–1987.
894. Prynes, R. *et al.* (1986). *EMBO J.* **5,** 2179–2190.
895. Pugsley, A. P. (1984). *Microbiol. Sci.* **1,** 168–175, 203–205.
896. Pugsley, A. P. (1987). *Mol. Microbiol.* **1,** 317–325.
897. Pugsley, A. P. (1988). *In* "Protein Transfer and Organelle Biogenesis" (R. Das and P. A. Robbins, eds.), pp. 607–652. Academic Press, San Diego, California.
898. Pugsley, A. P. (1988). *In* "Membrane Biogenesis" (J. A. F. Op den Kamp, ed.), pp. 399–417. Springer-Verlag, Berlin.
899. Pugsley, A. P. (1988). *Mol. Gen. Genet.* **211,** 335–341.
900. Pugsley, A. P., and Cole, S. T. (1987). *J. Gen. Microbiol.* **133,** 2411–2420.
901. Pugsley, A. P., and Schwartz, M. (1983). *J. Bacteriol.* **156,** 109–114.
902. Pugsley, A. P., and Schwartz, M. (1984). *EMBO J.* **3,** 2393–2397.
903. Pugsley, A. P., and Schwartz, M. (1985). *FEMS Microbiol. Rev.* **32,** 3–38.
904. Pugsley, A. P. *et al.* (1986). *J. Bacteriol.* **166,** 1083–1088.
905. Quinn, P. *et al.* (1983). *J. Cell Biol.* **96,** 851–856.
906. Rachubinski, R. A. *et al.* (1984). *J. Cell Biol.* **99,** 2241–2246.
907. Randall, L. L. (1983). *Cell (Cambridge, Mass.)* **33,** 231–240.
908. Randall, L. L., and Hardy, S. J. S. (1977). *Eur. J. Biochem.* **75,** 43–53.
909. Randall, L. L., and Hardy, S. J. S. (1986). *Cell (Cambridge, Mass.)* **46,** 921–928.
910. Randall, L. L. *et al.* (1987). *Annu. Rev. Microbiol.* **41,** 507–541.
911. Ranelli, D. M. *et al.* (1985). *Proc. Natl. Acad. Sci. U.S.A.* **82,** 5850–5854.
911a. Rasmussen, B. A., and Bassford, P. J. Jr., (1984). *J. Bacteriol.* **160,** 612–617.
912. Rasmussen, B. A., and Bassford, P. J. Jr., (1985). *J. Bacteriol.* **161,** 258–264.
913. Ray, P. *et al.* (1986). *Curr. Top. Microbiol. Immunol.* **125,** 75–102.
914. Redman, C. M. (1969). *J. Biol. Chem.* **244,** 4308–4315.
915. Reid, G., and Schatz, G. (1982). *J. Biol. Chem.* **257,** 13056–13061.
916. Reid, G., and Schatz, G. (1982). *J. Biol. Chem.* **257,** 13062–13067.
917. Reid, G., and Schatz, G. (1982). *J. Biol. Chem.* **257,** 13068–13074.
917a. Reid, J. *et al.* (1988). *J. Biol. Chem.* **263,** 7753–7759.
918. Reiss, B. *et al.* (1987). *Mol. Gen. Genet.* **209,** 116–121.
919. Reitman, M. L., and Kornfeld, S. (1981). *J. Biol. Chem.* **256,** 11977–11980.
920. Reynolds, B. L., and Reeves, P. R. (1963). *Biochem. Biophys. Res. Commun.* **11,** 140–145.

921. Rhodes, C., and Halban, P. A. (1987). *J. Cell Biol.* **105,** 143–153.
922. Ribes, V. *et al.* (1988). *EMBO J.* **7,** 231–237.
923. Richardson, W. D. *et al.* (1986). *Cell (Cambridge, Mass.)* **44,** 77–85.
924. Richardson, W. D. *et al.* (1988). *Cell (Cambridge, Mass.)* **52,** 655–664.
925. Rietveld, A. *et al.* (1985). *J. Biol. Chem.* **261,** 3846–3856.
926. Riezman, H. (1985). *Cell (Cambridge, Mass.)* **40,** 1001–1009.
927. Riezman, H. *et al.* (1983). *EMBO J.* **2,** 1113–1118.
928. Riggs, P. D. *et al.* (1988). *Genetics* **118,** 571–579.
929. Rindler, M. J. *et al.* (1985). *J. Cell Biol.* **100,** 136–151.
930. Rindler, M. J. *et al.* (1987). *J. Cell Biol.* **104,** 231–241.
930a. Rindler, M. J. *et al.* (1988). *J. Cell Biol.* **106,** 471–479.
931. Rine, J. *et al.* (1983). *Proc. Natl. Acad. Sci. U.S.A.* **80,** 6750–8754.
931a. Rizzolo, L. J., and Kornfeld, R. (1988). *J. Biol. Chem.* **263,** 9520–9525.
932. Roa, M., and Blobel, G. (1983). *Proc. Natl. Acad. Sci. U.S.A.* **80,** 6872–6876.
933. Roberts, B. L. *et al.* (1987). *Cell (Cambridge, Mass.)* **50,** 465–475.
934. Roberts, L. M., and Lord, J. M. (1981). *Eur. J. Biochem.* **119,** 43–49.
935. Roberts, W. L. *et al.* (1987). *Proc. Natl. Acad. Sci. U.S.A.* **84,** 7817–7821.
936. Robinson, A. *et al.* (1987). *Biochem. J.* **242,** 767–777.
937. Robinson, C., and Ellis, R. J. (1984). *Eur. J. Biochem.* **142,** 337–342.
938. Robinson, C., and Ellis, R. J. (1984). *Eur. J. Biochem.* **142,** 343–346.
939. Robinson, C., and Ellis, R. J. (1985). *Eur. J. Biochem.* **152,** 67–73.
939a. Robinson, J. S. *et al.* (1988). *Mol. Cell. Biol.* **8,** 4936–4958.
940. Robson, L. M., and Chambliss, G. H. (1986). *J. Bacteriol.* **165,** 612–619.
941. Rocque, W. J. *et al.* (1987). *J. Bacteriol.* **169,** 4003–4010.
942. Rodriguez-Boulan, E. J., and Pendergast, M. (1980). *Cell (Cambridge, Mass.)* **20,** 45–54.
943. Rodriquez-Boulan, E. J., and Sabatini, D. D. (1978). *Proc. Natl. Acad. Sci. U.S.A.* **75,** 5071–5075.
944. Rogalski, A. A. (1984). *J. Cell Biol.* **99,** 1101–1109.
945. Roggenkamp, R. *et al.* (1984). *Mol. Gen. Genet.* **194,** 489–493.
946. Roggenkamp, R. *et al.* (1985). *J. Biol. Chem.* **260,** 1508–1512.
947. Roise, D., and Schatz, G. (1988). *Cell (Cambridge, Mass.)* **263,** 4509–4511.
948. Roise, D. *et al.* (1986). *EMBO J.* **5,** 1327–1334.
949. Roise, D. *et al.* (1988). *EMBO J.* **7,** 649–653.
949a. Rollo, E. E., and Oliver, D. B. (1988). *J. Bacteriol.* **170,** 3281–3282.
950. Rosbach, M., and Penman, S. (1971). *J. Mol. Biol.* **59,** 227–241.
951. Rose, J. K. *et al.* (1984). *Proc. Natl. Acad. Sci. U.S.A.* **81,** 2050–2054.
952. Rosenbusch, J. P. (1974). *J. Biol. Chem.* **249,** 8019–8029.
953. Rosenfeld, M. G. *et al.* (1982). *J. Cell Biol.* **93,** 135–143.
954. Rosenfeld, M. G. *et al.* (1984). *J. Cell Biol.* **99,** 1076–1082.
955. Roth, J., and Berger, E. G. (1982). *J. Cell Biol.* **93,** 223–229.
956. Roth, J. *et al.* (1978). *J. Histochem. Cytochem.* **26,** 1974–1981.
957. Roth, J. *et al.* (1985). *Cell (Cambridge, Mass.)* **43,** 287–285.
958. Roth, M. G. *et al.* (1986). *J. Cell Biol.* **102,** 1271–1283.
959. Roth, M. G. *et al.* (1987). *J. Cell Biol.* **104,** 769–782.
960. Roth, T. F., and Porter, K. R. (1964). *J. Cell Biol.* **20,** 313–332.
961. Rothblatt, J. A., and Meyer, D. I. (1986). *Cell (Cambridge, Mass.)* **44,** 619–628.
962. Rothblatt, J. A., and Meyer, D. I. (1986). *EMBO J.* **5,** 1031–1036.
963. Rothblatt, J. A. *et al.* (1987). *EMBO J.* **6,** 3455–3463.
964. Rothenberger, S. *et al.* (1987). *Cell (Cambridge, Mass.)* **49,** 423–431.
965. Rothman, J. E. (1981). *Science* **213,** 1212–1220.

966. Rothman, J. E. (1987). *J. Biol. Chem.* **262,** 12502–12510.
967. Rothman, J. E., and Schmid, S. L. (1986). *Cell (Cambridge, Mass.)* **46,** 5–9.
968. Rothman, J. E. *et al.* (1979). *Cell (Cambridge, Mass.)* **15,** 1447–1454.
969. Rothman, J. E. *et al.* (1984). *J. Cell Biol.* **99,** 248–259.
970. Rothman, J. E. *et al.* (1984). *J. Cell Biol.* **99,** 260–271.
971. Rothman, J. H., and Stevens, T. H. (1986). *Cell (Cambridge, Mass.)* **47,** 1041–1051.
972. Rothman, J. H. *et al.* (1986). *Proc. Natl. Acad. Sci. U.S.A.* **83,** 3248–3252.
972a. Rothman, R. E. *et al.* (1988). *J. Biol. Chem.* **263,** 10470–10480.
973. Rothstein, S. J. *et al.* (1984). *Nature (London)* **308,** 662–665.
974. Rottier, P. J. M. *et al.* (1987). *J. Biol. Chem.* **262,** 8889–8895.
975. Ruohola, H., and Ferro-Novick, S. (1987). *Proc. Natl. Acad. Sci. U.S.A.* **84,** 8468–8472.
976. Ruohonen, L. *et al.* (1987). *Gene* **59,** 161–170.
977. Russel, M., and Model, P. (1981). *Proc. Natl. Acad. Sci. U.S.A.* **78,** 1717–1721.
978. Ryan, J. P. *et al.* (1986). *J. Biol. Chem.* **261,** 3389–3395.
979. Ryter, A. *et al.* (1975). *J. Bacteriol.* **122,** 295–301.
980. Sabatini, D. D. *et al.* (1966). *J. Mol. Biol.* **19,** 503–524.
981. Sachs, A. B. *et al.* (1986). *Cell (Cambridge, Mass.)* **45,** 827–835.
982. Sahagian, G. G. (1981). *Proc. Natl. Acad. Sci. U.S.A.* **78,** 4289–4293.
983. Sakagami, Y. *et al.* (1981). *Science* **212,** 1525–1527.
984. Sakaguchi, M. *et al.* (1987). *EMBO J.* **6,** 2425–2431.
985. Sako, T. (1985). *Eur. J. Biochem.* **149,** 551–563.
986. Salas, A. J. L. *et al.* (1986). *J. Cell Biol.* **102,** 1853–1857.
987. Salminen, A., and Novick, P. J. (1987). *Cell (Cambridge, Mass.)* **49,** 527–538.
988. Salpeter, M. M., and Farquhar, M. G. (1981). *J. Cell Biol.* **91,** 240–246.
989. Salzman, N. H., and Neufeld, F. R. (1988). *J. Cell Biol.* **106,** 1083–1091.
990. Sandvig, K. *et al.* (1986). *J. Biol. Chem.* **261,** 11639–11644.
991. Sandvig, K. *et al.* (1987). *J. Cell Biol.* **105,** 679–689.
992. Santos, M. J. (1988). *J. Biol. Chem.* **263,** 10502–10509.
992a. Sanz, P., and Meyer, D. I. (1988). *EMBO J.* **7,** 3553–3557.
993. Sara, M., and Sleytr, U. B. (1987). *J. Bacteriol.* **169,** 4092–4098.
994. Saraste, K., and Hedman, K. (1983). *EMBO J.* **2,** 2001–2006.
995. Saraste, J. *et al.* (1986). *Proc. Natl. Acad. Sci. U.S.A.* **83,** 6425–6429.
996. Saraste, J. *et al.* (1987). *J. Cell Biol.* **105,** 2021–2029.
997. Sastry, P. A. *et al.* (1986). *J. Bacteriol.* **164,** 571–577.
998. Saunders, C. W. *et al.* (1987). *J. Bacteriol.* **169,** 2917–2925.
999. Sayre, R. T. *et al.* (1986). *Cell (Cambridge, Mass.)* **47,** 601–608.
1000. Schad, P. A., and Iglewski, B. H. (1988). *J. Bacteriol.* **170,** 2784–2789.
1001. Schad, P. A. *et al.* (1987). *J. Bacteriol.* **169,** 2691–2696.
1002. Schauer, I. *et al.* (1985). *J. Cell Biol.* **100,** 1664–1675.
1003. Schekman, R. (1985). *Annu. Rev. Cell Biol.* **1,** 115–143.
1004. Schenkman, S. *et al.* (1983). *J. Bacteriol.* **155,** 1382–1392.
1005. Schenkman, S. *et al.* (1984). *J. Biol. Chem.* **269,** 7570–7576.
1006. Schindler, M., and Jiung, L. W. (1986). *J. Cell Biol.* **102,** 859–862.
1007. Schlenstedt, G., and Zimmermann, R. (1987). *EMBO J.* **6,** 699–703.
1008. Schleyer, M., and Neupert, W. (1985). *Cell (Cambridge, Mass.)* **43,** 339–350.
1009. Schleyer, M. *et al.* (1982). *Eur. J. Biochem.* **125,** 109–116.
1010. Schlossman, D. A. *et al.* (1984). *J. Cell Biol.* **99,** 723–733.
1011. Schmidt, G. W., and Mishkind, M. L. (1983). *Proc. Natl. Acad. Sci. U.S.A.* **80,** 2632–2636.
1012. Schmidt, G. W., and Mishkind, M. L. (1986). *Annu. Rev. Biochem.* **55,** 579–612.

1013. Schmidt, G. W. *et al.* (1981). *J. Cell Biol.* **91,** 468–478.
1014. Schmidt, H. G. *et al.* (1988). *Cell (Cambridge, Mass.)* **53,** 635–647.
1015. Schmidt, M. F. G. (1984). *EMBO J.* **3,** 2295–2300.
1016. Schmidt, M. F. G., and Schlesinger, M. J. (1980). *J. Biol. Chem.* **255,** 3334–3339.
1017. Schmidt, M. F. G., and Schmidt, M. (1988). *In* "Membrane Biogenesis" (J.A.F. Op den Kamp, ed.). pp. 235–256. Springer-Verlag, Berlin.
1018. Schmidt, S. L. *et al.* (1988). *Cell (Cambridge, Mass.)* **52,** 73–83.
1019. Schnaitman, C. A. (1971). *J. Bacteriol.* **108,** 545–552.
1020. Schneider, C. *et al.* (1984). *Nature (London)* **311,** 675–678.
1021. Schoemaker, J. M. *et al.* (1985). *EMBO J.* **4,** 775–780.
1022. Schulze-Lohoff, E. *et al.* (1985). *J. Cell Biol.* **101,** 824–829.
1023. Schumacher, G. *et al.* (1986). *Nucleic Acids Res* **14,** 5713–5727.
1024. Schwaiger, M. *et al.* (1987). *J. Cell Biol.* **105,** 235–246.
1025. Schwaizer, H. *et al.* (1982). *Biochem. Biophys. Res. Commun.* **104,** 950–956.
1026. Schweizer, M. *et al.* (1978). *Eur. J. Biochem.* **82,** 211–217.
1027. Scoulica, E. *et al.* (1987). *Eur. J. Biochem.* **163,** 519–528.
1028. Sefton, B. M., and Buss, J. E. (1987). *J. Cell Biol.* **104,** 1449–1453.
1029. Sefton, B. M. *et al.* (1982). *Cell (Cambridge, Mass.)* **31,** 465–474.
1030. Segev, N. *et al.* (1988). *Cell (Cambridge, Mass.)* **52,** 915–924.
1031. Sekine, S. *et al.* (1985). *Proc. Natl. Acad. Sci. U.S.A.* **82,** 4306–4310.
1032. Shaltiel, L. S., and Wisneieski, B. J. (1984). *Proc. Natl. Acad. Sci. U.S.A.* **81,** 3336–3341.
1032a. Shaw, A. S. *et al.* (1988). *Proc. Natl. Acad. Sci. U.S.A.* **85,** 7592–7596.
1032b. Shelness, G. S., *et al.* (1988). *J. Biol. Chem.* **263,** 17063–17070.
1033. Shiba, K. *et al.* (1984). *EMBO J.* **3,** 631–635.
1034. Shiba, K. *et al.* (1986). *J. Bacteriol.* **166,** 849–856.
1035. Shinnar, A. E., and Kaiser, E. T. (1984). *J. Am. Chem. Soc.* **106,** 5006–5007.
1036. Shiroza, T. *et al.* (1985). *Gene* **34,** 1–8.
1037. Shiver, J. W. *et al.* (1987). *J. Biol. Chem.* **262,** 14273–14281.
1038. Shortle, D. (1983). *Gene* **22,** 181–189.
1039. Shibakov, M. (1986). *Eur. J. Biochem.* **155,** 577–581.
1040. Sidasubamanian, N., and Nayak, D. P. (1987). *Proc. Natl. Acad. Sci. U.S.A.* **84,** 1–5.
1041. Sidhu, R. S., and Bollon, A. P. (1987). *Gene* **54,** 175–184.
1042. Siegel, V., and Walter, P. (1985). *J. Cell Biol.* **100,** 1913–1921.
1043. Siegel, V., and Walter, P. (1988). *Cell (Cambridge, Mass.)* **52,** 39–49.
1044. Siegel, V., and Walter, P. (1988). *Proc. Natl. Acad. Sci. U.S.A.* **85,** 1801–1805.
1045. Silve, S. *et al.* (1987). *Mol. Cell. Biol.* **7,** 3306–3314.
1046. Silver, P. *et al.* (1981). *Cell (Cambridge, Mass.)* **25,** 341–345.
1047. Silver, P. A., and Hall, M. N. (1988). *In* "Protein Transfer and Organelle Biogenesis" (R. Das and P. A. Robbins, eds.), pp. 749–769. Academic Press, San Diego, California.
1048. Silver, P. A. *et al.* (1984). *Proc. Natl. Acad. Sci. U.S.A.* **81,** 5951–5954.
1049. Simon, K. *et al.* (1987). *J. Cell Biol.* **104,** 1165–1172.
1050. Simons, K, and Fuller, S. D. (1985). *Annu. Rev. Cell Biol.* **1,** 243–288.
1050a. Simons, K., and Van Meer, G. (1988). *Biochemistry* **27,** 6197–6202.
1051. Simons, K., and Virta, H. (1987). *EMBO J.* **6,** 2241–2247.
1052. Simons, K. *et al.* (1980). *J. Mol. Biol.* **126,** 673–690.
1053. Singer, S. J. *et al.* (1987). *Proc. Natl. Acad. Sci. U.S.A.* **84,** 1015–1019.
1054. Singer, S. J. *et al.* (1987). *Proc. Natl. Acad. Sci. U.S.A.* **84,** 1960–1964.
1055. Sjöström, M. *et al.* (1987). *EMBO J.* **6,** 823–831.

1056. Skerjanc, I. S. *et al.* (1987) *EMBO J.* **6,** 3117–3123.
1057. Skerra, A., and Plückthun, A. (1988). *Science* **240,** 1030–1040.
1058. Skipper, N. *et al.* (1985). *Science* **230,** 958–960.
1059. Smagala, C., and Douglas, M. G. (1988). *J. Biol. Chem.* **263,** 6783–6790.
1060. Small, G. M. *et al.* (1987). *Mol. Cell Biol.* **7,** 1848–1855.
1061. Small, G. M. *et al.* (1988). *EMBO J.* **7,** 1167–1173.
1062. Smeekens, S. *et al.* (1985). *Nature (London)* **317,** 456–458.
1063. Smeekens, S. *et al.* (1986). *Cell (Cambridge, Mass.)* **46,** 365–375.
1064. Smit, J., and Nikaido, H. (1978). *J. Bacteriol.* **135,** 687–702.
1065. Smith, R. A. *et al.* (1985). *Science* **229,** 1219–1224.
1066. Smith, W. P. *et al.* (1977). *Proc. Natl. Acad. Sci. U.S.A.* **74,** 2830–2834.
1067. Smith, W. P. *et al.* (1978). *Proc. Natl. Acad. Sci. U.S.A.* **75,** 814–817.
1068. Snider, M. D., and Rogers, D. C. (1986). *J. Cell Biol.* **103,** 265–275.
1069. Snow, C. M. *et al.* (1987). *J. Cell Biol.* **104,** 1143–1156.
1070. Söll, J., and Buchanan, B. B. (1983). *J. Biol. Chem.* **258,** 6686–6689.
1071. Spiess, M., and Lodish, H. F. (1986). *Cell (Cambridge, Mass.)* **44,** 177–185.
1072. Staehelin, L. A., and Arntzen, C. J. (1983). *J. Cell Biol.* **97,** 1327–1337.
1073. Stahl, M. L., and Ferrari, E. (1984). *J. Bacteriol.* **158,** 411–418.
1074. Stark, M. J. R., and Boyd, A. (1986). *EMBO J.* **5,** 1995–2002.
1075. Steeg, H. van *et al.* (1986). *EMBO J.* **5,** 3643–3650.
1076. Stein, B. S. and Sussman, H. H. (1986). *J. Biol. Chem.* **261,** 10319–10331.
1077. Stein, M. *et al.* (1987). *EMBO J.* **6,** 2677–2681.
1078. Steinman, R. M. *et al.* (1976). *J. Cell Biol.* **68,** 665–687.
1079. Steinman, R. M. *et al.* (1983). *J. Cell Biol.* **96,** 1–27.
1080. Stephens, E. B., and Compans, R. W. (1986). *Cell (Cambridge, Mass.)* **47,** 1053–1059.
1081. Steven, A. C. *et al.* (1977). *J. Cell Biol.* **72,** 292–301.
1082. Stevens, T. *et al.* (1982). *Cell (Cambridge, Mass.)* **30,** 439–448.
1083. Stevens, T. H. *et al.* (1986). *J. Cell Biol.* **102,** 1551–1557.
1084. Stirzaher, S. C. *et al.* (1987). *J. Cell Biol.* **105,** 2897–2903.
1085. Stochaj, V. *et al.* (1988). *J. Bacteriol.* **170,** 2639–2645.
1086. Stoynowski, I. *et al.* (1987). *Cell (Cambridge, Mass.)* **50,** 759–768.
1087. Streuli, C. H., and Griffin, B. E. (1987). *Nature (London)* **326,** 619–622.
1088. Strous, G. J. *et al.* (1983). *J. Cell Biol.* **97,** 1815–1822.
1089. Strous, G. J. *et al.* (1987). *J. Biol. Chem.* **262,** 3620–3625.
1090. Suissa, M., and Schatz, G. (1982). *J. Biol. Chem.* **257,** 13048–13055.
1091. Suzuki, M. *et al.* (1982). *Proc. Natl. Acad. Sci. U.S.A.* **79,** 2475–2479.
1092. Suzuki, T. *et al.* (1987). *J. Bacteriol.* **169,** 2523–2528.
1093. Swanson, J. A., and McNeil, P. L. (1988). *Science* **238,** 548–550.
1094. Swinkles, B. W. *et al.* (1988). *EMBO J.* **7,** 1159–1165.
1095. Szczesna-Skorupa, E. *et al.* (1987). *J. Biol. Chem.* **262,** 8896–8900.
1096. Szczesna-Skorupa, E. *et al.* (1988). *Proc. Natl. Acad. Sci. U.S.A.* **85,** 738–742.
1097. Sztul, E. S. *et al.* (1987). *J. Cell Biol.* **105,** 2631–2639.
1098. Taatjes, D. J. *et al.* (1988). *J. Biol. Chem.* **263,** 6302–6309.
1099. Tabe, L. *et al.* (1984). *J. Mol. Biol.* **180,** 645–666.
1100. Tague, B. W., and Chrispeels, M. J. (1987). *J. Cell Biol.* **105,** 1971–1979.
1101. Tajima, S. *et al.* (1986). *J. Cell. Biol.* **103,** 1167–1178.
1102. Takagi, M. *et al.* (1985). *J. Bacteriol.* **163,** 824–831.
1103. Takahara, M. *et al.* (1985). *J. Biol. Chem.* **260,** 2670–2674.
1104. Takahara, M. *et al.* (1988). *Bio/Technology* **6,** 195–198.

1104a. Takase, K. *et al.* (1988). *J. Biol. Chem.* **263,** 11548–11553.
1105. Talmadge, K. *et al.* (1980). *Proc. Natl. Acad. Sci. U.S.A.* **77,** 3369–3373.
1106. Tamm, L. K. (1986). *Biochemistry* **23,** 3232–3240.
1107. Tanner, W., and Lehle, L. (1987). *Biochim. Biophys. Acta* **906,** 81–99.
1108. Tartakoff, A. M. (1983). *Cell (Cambridge, Mass.)* **32,** 1026–1028.
1109. Tartakoff, A. M., and Vassalli, P. (1978). *J. Cell Biol.* **79,** 694–707.
1110. Tartakoff, A. M., and Vassalli, P. (1983). *J. Cell Biol.* **97,** 1243–1248.
1110a. Thom, J. R., and Randall, L. L. (1988). *J. Bacteriol.* **170,** 5654–5661.
1111. Thin, L. *et al.* (1986). *Proc. Natl. Acad. Sci. U.S.A.* **83,** 6766–6770.
1112. Tillmann, V. *et al.* (1986). *Eur. J. Biochem.* **162,** 635–642.
1113. Tokunaga, M. *et al.* (1982). *J. Biol. Chem.* **257,** 9922–9925.
1114. Tokunaga, M. *et al.* (1982). *Proc. Natl. Acad. Sci. U.S.A.* **79,** 2255–2259.
1115. Tokunaga, M. *et al.* (1983). *J. Biol. Chem.* **258,** 12101–12106.
1116. Tokunaga, M. *et al.* (1984). *J. Biol. Chem.* **259,** 3825–3830.
1117. Tokunaga, M. *et al.* (1985). *J. Biol. Chem.* **259,** 5610–5615.
1118. Tommassen, J., and deKroon, T. (1987). *FEBS Lett.* **221,** 226–230.
1119. Tommassen, J. *et al.* (1983). *EMBO J.* **2,** 1275–1279.
1120. Tommassen, J. *et al.* (1985). *EMBO J.* **4,** 1041–1047.
1121. Tommassen, J. *et al.* (1985). *EMBO J.* **4,** 1583–1587.
1122. Tooze, J., and Tooze, S. A. (1986). *J. Cell Biol.* **103,** 839–850.
1123. Tooze, J. *et al.* (1987). *J. Cell Biol.* **105,** 1215–1226.
1124. Towler, D. A. *et al.* (1987). *J. Biol. Chem.* **262,** 1030–1036.
1125. Tsuji, T. *et al.* (1985). *J. Biol. Chem.* **260,** 8552–8558.
1126. Tsukagoshi, N. *et al.* (1984). *Mol. Gen. Genet.* **193,** 58–63.
1127. Tsuneoka, M. *et al.* (1986). *J. Biol. Chem.* **261,** 1829–1834.
1128. Tufaro, F. *et al.* (1987). *J. Cell Biol.* **105,** 647–657.
1129. Turco, S. J., and Robbins, P. W. (1979). *J. Biol. Chem.* **254,** 4560–4567.
1130. Uijawa, S. *et al.* (1988). *Mol. Cell Biol.* **7,** 1709–1714.
1131. Ullrich, A. *et al.* (1984). *Nature (London)* **309,** 418–425.
1132. Ullu, E. *et al.* (1982). *Cell (Cambridge, Mass.)* **29,** 195–202.
1133. Unwin, P. N. T. (1977). *Nature (London)* **269,** 118–122.
1134. Unwin, P. N. T., and Milligan, R. A. (1982). *J. Cell Biol.* **93,** 63–75.
1135. Urban, J. *et al.* (1987). *J. Cell Biol.* **105,** 2735–2743.
1136. Vale, R. D. (1987). *Annu. Rev. Cell Biol.* **3,** 347–378.
1137. Valls, L. A. *et al.* (1987). *Cell (Cambridge, Mass.)* **48,** 887–897.
1138. Vasantha, N., and Thompson, L. D. (1986). *J. Bacteriol.* **165,** 837–842.
1139. Vasantha, N., and Thompson, L. D. (1987). *Gene* **49,** 23–28.
1140. van Veldhoven, P. P. *et al.* (1987). *J. Biol. Chem.* **262,** 4310–4318.
1141. van Venetië, R., and Vertleij, A. J. (1982). *Biochim. Biophys. Acta* **692,** 396–405.
1142. Vassarotti, A. *et al.* (1987). *EMBO J.* **6,** 705–711.
1143. Vega-Salas, D. E. *et al.* (1987). *J. Cell Biol.* **104,** 905–916.
1144. Verner, K., and Schatz, G. (1987). *EMBO J.* **6,** 2449–2456.
1145. Vestweber, D., and Schatz, G. (1988). *EMBO J.* **7,** 1147–1151.
1145a. Vestweber, D., and Schatz, G. (1988). *J. Cell Biol.* **107,** 2037–2043.
1146. Viebrock, A. *et al.* (1982). *EMBO J.* **1,** 565–571.
1147. Vierling, E. *et al.* (1988). *EMBO J.* **7,** 575–581.
1148. Vigers, G. P. A. *et al.* (1986). *EMBO J.* **5,** 2079–2085.
1149. Virschup, O. M., and Bennett, V. (1988). *J. Cell Biol.* **106,** 39–50.
1150. Vlasuk, C. P. *et al.* (1983). *J. Biol. Chem.* **256,** 7141–7148.
1151. Vogel, H., and Jähnig, F. (1985). *J. Mol. Biol.* **190,** 191–199.

1152. Voorhout, W. *et al.* (1988). *J. Gen. Microbiol.* **134,** 599–604.
1153. Vos-Scheperkeuter, G. H., and Witholt, B. (1984). *J. Mol. Biol.* **175,** 511–528.
1154. Vos-Scheperkeuter, G. H. *et al.* (1984). *J. Bacteriol.* **159,** 440–447.
1155. Vrije, T. de *et al.* (1987) *Biochim. Biophys. Acta.* **900,** 63–72.
1155a. Vrije, T. de *et al.* (1988). *Nature (London)* **334,** 173–175.
1156. Walk, R.-A., and Hock, B. (1981). *Biochem. Biophys. Res. Commun.* **81,** 636–643.
1157. Wall, D. A. *et al.* (1980). *Cell (Cambridge, Mass.)* **21,** 79–93.
1158. Wallace, B. A. (1986). *Proc. Natl. Acad. Sci. U.S.A.* **83,** 9423–9427.
1159. Walter, P. (1987). *Nature (London)* **328,** 763–764.
1160. Walter, P., and Blobel, G. (1980). *Proc. Natl. Acad. Sci. U.S.A.* **77,** 7112–7116.
1161. Walter, P., and Blobel, G. (1981). *J. Cell Biol.* **91,** 557–561.
1162. Walter, P., and Blobel, G. (1982). *Nature (London)* **229,** 691–698.
1163. Walter, P., and Blobel, G. (1983). *Cell (Cambridge, Mass.)* **34,** 525–533.
1164. Walter, P. *et al.* (1979). *Proc. Natl. Acad. Sci., U.S.A.* **76,** 1796–1799.
1165. Walworth, N. C., and Novick, P. J. (1987). *J. Cell Biol.* **105,** 163–174.
1165a. Wang, L.-F. *et al.* (1988). *Gene* **69,** 39–47.
1166. Warren, G. (1987). *Nature (London)* **327,** 17–18.
1167. Warren, G., and Dobberstein, B. (1978). *Nature (London)* **273,** 569–571.
1168. Warren, T. G., and Shields, D. (1984). *Cell (Cambridge, Mass.)* **39,** 547–555.
1169. Wasley, L. C. *et al.* (1987). *J. Biol. Chem.* **262,** 14766–14772.
1170. Wasmann, C. C. *et al.* (1986). *Mol. Gen. Genet.* **205,** 446–453.
1171. Watanabe, K., and Kuboi, S. (1982). *Eur. J. Biochem.* **123,** 587–592.
1171a. Watanabe, T. *et al.* (1988). *J. Bacteriol.* **170,** 4001–4007.
1172. Waters, M. G., and Blobel, G. (1986). *J. Cell Biol.* **102,** 1543–1550.
1173. Waters, M. G. *et al.* (1988). *J. Biol. Chem.* **263,** 6309–6214.
1174. Watson, M. E. E. (1984). *Nucleic Acids Res.* **12,** 4155–4174.
1175. Watts, C. *et al.* (1981). *Cell (Cambridge, Mass.)* **25,** 347–356.
1176. Watts, C. *et al.* (1983). *Proc. Natl. Acad. Sci. U.S.A.* **80,** 2809–2813.
1177. Wehland, J. *et al.* (1981). *Cell (Cambridge, Mass.)* **25,** 105–119.
1178. Welsh, J. D. *et al.* (1986). *Mol. Cell. Biol.* **6,** 2207–2212.
1179. Wen, D., and Schlesinger, M. J. (1984). *Mol. Cell Biol.* **4,** 688–694.
1180. Weng, Q. *et al.* (1988). *J. Bacteriol.* **170,** 126–131.
1180a. Wessels, H. P., and Spiess, M. (1988). *Cell (Cambridge, Mass.)* **55,** 61–70.
1181. Weyer, K. A. *et al.* (1987). *Biochemistry* **26,** 2909–2914.
1182. Wickner, W. (1979). *Annu. Rev. Biochem.* **48,** 23–45.
1183. Wickner, W. (1980). *Science* **210,** 861–868.
1184. Wickner, W. *et al.* (1987). *J. Bacteriol.* **169,** 3821–3822.
1185. Widnell, C. C. *et al.* (1982). *Cell (Cambridge, Mass.)* **28,** 61–70.
1186. Wiech, H. *et al.* (1987). *EMBO J.* **6,** 1011–1016.
1187. Wiedmann, M. *et al.* (1984). *Nature (London)* **309,** 637–639.
1188. Wiedmann, M. *et al.* (1987). *J. Cell Biol.* **104,** 201–208.
1189. Wiedmann, M. *et al.* (1987). *Nature (London)* **328,** 830–833.
1190. Wieland, F. T. *et al.* (1987). *Cell (Cambridge, Mass.)* **50,** 289–300.
1191. Willey, D. L. *et al.* (1984). *Cell (Cambridge, Mass.)* **36,** 555–562.
1192. Williams, D. B. *et al.* (1988). *J. Biol. Chem.* **263,** 4549–4560.
1193. Williams, M. A., and Lamb, R. A. (1987). *Mol. Cell. Biol.* **6,** 4317–4328.
1193a. Wilson, C. *et al.* (1988). *J. Cell Biol.* **107,** 69–77.
1194. Willumsen, B. M. *et al.* (1984). *Nature (London)* **310,** 583–586.
1195. Witte, C. *et al.* (1988) *EMBO J.* **7,** 1439–1447.
1196. Wold, F. (1981). *Annu. Rev. Biochem.* **50,** 783–814.
1197. Wold, F. (1986). *Trends Biochem. Sci.* **11,** 58–59.
1198. Wolfe, P. D. (1988). *J. Biol. Chem.* **263,** 6908–6915.

1199. Wolfe, P. D. *et al.* (1982). *J. Biol. Chem.* **257,** 7898–7902.
1200. Wolfe, P. D. *et al.* (1983). *J. Biol. Chem.* **258,** 12073–12080.
1201. Wolfe, P. D. *et al.* (1985). *J. Biol. Chem.* **260,** 1836–1841.
1202. Wong, S.-L., and Doi, R. H. (1986). *J. Biol. Chem.* **261,** 10176–10181.
1203. Wood, C. R. *et al.* (1985). *Nature (London)* **314,** 446–449.
1204. Woodman, P. G., and Edwardson, J. M. (1986). *J. Cell Biol.* **103,** 1829–1835.
1205. Woods, J. W. *et al.* (1986). *J. Cell Biol.* **103,** 277–286.
1206. Woolford, C. A. *et al.* (1986). *Mol. Cell. Biol.* **6,** 2500–2510.
1207. Wright, R. M. *et al.* (1986). *J. Biol. Chem.* **261,** 17183–17191.
1208. Wu, H. C., and Tokunaga, M. (1986). *Curr. Top. Microbiol. Immunol.* **125,** 127–158.
1209. Wychowski, C. *et al.* (1986). *EMBO J.* **5,** 2569–2576.
1210. Yaffe, M. P., and Schatz, G. (1984). *Proc. Natl. Acad. Sci. U.S.A.* **81,** 4819–4823.
1211. Yaffe, M. P. *et al.* (1985). *EMBO J.* **4,** 2069–2074.
1212. Reference deleted.
1213. Yamagata, H. *et al.* (1982). *J. Bacteriol.* **152,** 1163–1168.
1214. Yamaguchi, K. *et al.* (1988). *Cell (Cambridge, Mass.)* **53,** 423–432.
1215. Yamamoto, T. *et al.* (1984). *Cell (Cambridge, Mass.)* **34,** 27–38.
1216. Yamane, K. *et al.* (1987). *J. Biol. Chem.* **262,** 2358–2362.
1217. Yamane, K. *et al.* (1988). *J. Biol. Chem.* **263,** 5368–5372.
1218. Yanagida, N. *et al.* (1986). *J. Bacteriol.* **166,** 937–944.
1218a. Yang, M. *et al.* (1988). *EMBO J.* **7,** 3857–3862.
1219. Yang, M. Y. *et al.* (1984). *J. Bacteriol.* **160,** 15–21.
1220. Ye, R. D. *et al.* (1988). *J. Biol. Chem.* **263,** 4869–4875.
1221. Yeo, K.-T. *et al.* (1985). *J. Biol. Chem.* **260,** 7896–7902.
1222. Yewdell, J. W. *et al.* (1988). *Cell (Cambridge, Mass.)* **52,** 843–852.
1223. Yokota, S., and Fahimi, H. D. (1981). *Proc. Natl. Acad. Sci. U.S.A.* **78,** 4970–4974.
1224. Yong, P. Y. *et al.* (1988). *J. Biol. Chem.* **263,** 2585–2589.
1225. Yost, C. S. *et al.* (1983). *Cell (Cambridge, Mass.)* **34,** 759–766.
1226. Youle, R. J., and Colombatti, M. (1987). *J. Biol. Chem.* **262,** 4676–4682.
1227. Young, E. T., and Pilgrim, D. (1985). *Mol. Cell. Biol.* **5,** 3024–3034.
1228. Yu, F. *et al.* (1984). *FEBS Lett.* **173,** 264–268.
1229. Yu, F. *et al.* (1986). *J. Biol. Chem.* **261,** 2284–2288.
1230. Zanini, A. *et al.* (1980). *J. Cell Biol.* **86,** 260–272.
1231. Zastrow, M. von, and Castle, J. D. (1987). *J. Cell Biol.* **105,** 2675–2684.
1232. Zeller, R. *et al.* (1983) *Cell (Cambridge, Mass.)* **32,** 425–434.
1233. Zerial, M. *et al.* (1987). *Cell (Cambridge, Mass.)* **48,** 147–155.
1234. Zettmeissl, G. *et al.* (1987). *Bio/Technology* **5,** 720–725.
1234a. Zhao, L-J., and Padmanabhan, R. (1988). *Cell (Cambridge, Mass.)* **55,** 1005–1015.
1235. Zieg, J., and Simon, M. (1980). *Proc. Natl. Acad. Sci. U.S.A.* **77,** 4169–4200.
1236. Zilberstein, A. *et al.* (1980). *Cell (Cambridge, Mass.)* **21,** 417–427.
1236a. Zimmer, F. J. *et al.* (1988). *J. Cell Biol.* **106,** 1435–1444.
1237. Zimmermann, R., and Mollay, C. (1986). *J. Biol. Chem.* **261,** 12889–12895.
1238. Zimmermann, R., and Neupert, W. (1980). *Eur. J. Biochem.* **109,** 217–229.
1239. Zimmermann, R., and Neupert, W. (1980). *Eur. J. Biochem.* **112,** 225–233.
1240. Zimmermann, R. *et al.* (1981). *Eur. J. Biochem.* **116,** 455–460.
1240a. Zimmermann, R. *et al.* (1988). *EMBO J.* **7,** 2875–2880.
1241. Zsebo, K. M. *et al.* (1986). *J. Biol. Chem.* **261,** 5858–5865.
1242. Zuniga, M. C., and Hood, L. E. (1986). *J. Cell Biol.* **102,** 1–10.
1243. Zwizinski, C., and Wickner, W. (1980). *J. Biol. Chem.* **255,** 7973–7977.
1244. Zwizinski, C. *et al.* (1983). *J. Biol. Chem.* **258,** 13340–13346.
1245. Zwizinski, C. *et al.* (1983). *J. Biol. Chem.* **258,** 4071–4074.

Index

A

AAC, *see* ADP/ATP carrier
Acid phosphatase (eukaryotic)
  location of in Golgi, 148
  as reporter protein, 24
Acid phosphatase-β-galactosidase hybrids (in yeast), 52
Acid pH
  endocytosis and, 218
  lysosomal sorting and, 161
  polarized sorting and, 154
  regulated exocytosis and, 157
Actin, secretion in yeast and, 155
Acyl transferases
  location in eukaryotic cells, 110, 142
  protein fatty acylation and, 106, 110
Adenylate kinase, mitochondrial routing of, 179
ADP/ATP carrier, mitochondrial routing of, 179
Adrenocorticothyroid hormone, exocytosis of, 157
Aerolysin, activation of, 125
Affinity chromatography, purification of hybrid proteins and, 235
Aggregation, regulated secretion and, 157
Agrobacterium tumefaciens, transgenic plants and, 239
Alcohol dehydrogenase
  isoforms, 171
  prepeptide, 171
ALG7 gene product (yeast), glycosylation and, 129
Alkaline phosphatase (PHO A) (bacterial)
  processing of, 113
  as reporter protein, 26, 64
  signal peptide of, 57
  use of to determine membrane protein topology, 26
Alkaline phosphatase (eukaryotic), polarized exocytosis of, 152
Alpha factor
  endocytosis, 225
  heterologous protein secretion and, 235
  processing, 140
  receptor, 225
  secretion, 82
  signal peptide, 60
α-globin, as reporter protein, 24
α-glucosidases
  endoplasmic reticulum and, 128
  inhibition of, 137
α-hemolysin
  activation of, 125
  secretion of, 123
  use of as secretion vehicle, 235
α-isopropylmalate synthetase, mitochondrial routing of, 172
α-mannosidase
  endoplasmic reticulum and, 128
  sorting of to yeast vacuole membrane, 164
Alu sequences, in SRP 7S RNA, 74
Amide-linked fatty acids, bacterial lipoproteins and, 105
Amino acid analogs, use of to inactivate routing signals, 28, 175
8-Aminolevulase, prepeptide in β-galactosidase hybrid, 172
Aminopeptidase N, polarized secretion of, 152
Ammonium chloride, effect on polarized sorting, 154
Amphiphilic helix
  in membrane anchor, 103
  in prepeptides, 173, 175
Ankyrin
  fatty acylation of, 109
  possible role in cell polarity, 155

Antibodies
  endocytosis of, 221
  use of, 20, 120, 131, 159
Antitrypsin, transport to Golgi, 137
Apical membrane in polarized cells, 3, 38, 151
Apocytochrome *c*, mitochondrial routing of, 179, 190
Apolipoprotein
  endocytosis of, 221
  fatty acylation of, 109
  receptor, 212
Asialoglycoprotein signal sequence, 63
ATP, 42
  in bacterial protein export, 85, 89, 97
  in chloroplast protein import, 182, 187
  in endocytic invagination and fusion, 217
  in exocytosis, 150
  in mitochondrial protein import, 182, 187
  in nuclear assembly, 201
  in nuclear protein import, 208
  in peroxisomal protein import, 195
  in polarized sorting, 155
  in protein coat removal, 165
  in protein folding, 132
  in protein translocation, 82, 97, 99
  in protein transport through Golgi, 145
  in protein transport to Golgi, 137
  in toxin uptake, 217, 223
ATPase
  in ATP hydrolysis, 42
  in clathrin disassembly, 165
ATPase subunit 8, mitochondrial sorting of, 192
Autophosphorylation of chloroplast proteins, 183

B

Bacteriophages
  receptor and resistant mutants, 21, 116
  as vaccine carriers, 234
Basolateral membrane, 3, 38, 151
Bayer patches
  lipoprotein export and, 107
  outer membrane protein sorting and, 117
β-galactosidase (LAC Z)
  aborted export of, 50
  hybrid proteins, 26, 51, 172, 203
  mitochondrial routing of, 172
  nuclear targeting of, 203
  as reporter protein, 24, 26
  toxicity of, 30, 56
β-glucanase, retention of in ER, 136
β-lactamase (BLA)
  hybrid proteins, 26, 64
  fatty acylation of, 104
  release of, 113
  as reporter protein, 25
BiP
  as competence factor, 132
  protein folding and, 132
  retention of in ER, 134
BP, *see* Chaperonin

C

Ca^{2+}-influx in exocytosis, 159
Caco-2 cells, 39, 152
Candida, peroxisomes in, 195
Capsid protein (picornavirus)
  export of, 166
  myristoylation of, 166
Carbomyl phosphate synthase, import into mitochondria, 183
Carboxypeptidase Y (yeast)
  default secretion of, 163
  effect of heat shock protein on, 83
  propeptide of, 163
  signal peptide of, 61
  vacuolar sorting of, 163
CAT, *see* Chloramphenicol acetyltransferase
Cathepsin
  lysosomal sorting of, 161
  phosphorylation of, 138
  processing of, 161
cel genes and products, cellulose degradation and, 115
Cell fractionation, 15
Cell fusion assays, 37, 144
Cellulosome, 115
Chaotropic agents, peripheral membrane proteins and, 9
Chaperonin (BP) (chloroplast), 87, 194
Chloramphenicol acetyltransferase (CAT)
  chloroplast routing of, 176
  hybrid proteins, 52, 63, 176
  mitochondrial routing of, 176
  as reporter protein, 24
  transport into ER, 63

Chloroplast
  energy for protein import, 182
  evolution of nuclear-encoded proteins and, 170
  organization, 4, 169
  phosphorylation in protein import, 183
  routing signals (transit peptides), 174
Chloroquinone, effect on regulated exocytosis, 157
Cholera toxin
  assembly and secretion, 122
  uptake, 224
cis Golgi
  enzymes in, 141
  transport of secretory proteins into, 133, 136
Clathrin
  assembly proteins, 165, 216
  composition of, 165
  endocytosis and, 216
  gene for (yeast), 165
  lysosomal enzyme sorting and, 161
  mutations affecting, 165
  in trans-Golgi network, 148, 156
  uncoating of, 156, 165
Clostridium thermocellum, protein secretion by, 115
Coat protein precursor (bacteriophage f1/M13)
  insertion into liposomes, 66
  processing site, 57, 62
  signal peptide of, 67
Coated pits, 213
Colchicine, effect on polarized sorting, 155
Colicins
  release of, 124
  uptake of, 226
Collagen, in polarized cells, 151
Colloidal gold
  immunocytochemistry and, 16
  import into nuclei, 203, 208
Competence in protein translocation, 79, 137
Competence factors
  bacterial, 85, 89
  eukaryotic secretory, 82
  in mitochondrial protein import, 183
Condensing granules in regulated exocytosis, 157
Constitutive secretion, 149
Cotranslational translocation
  in bacteria, 85
  in chloroplast protein import, 170, 221
  in eukaryotic secretory pathway, 79
  in mitochondrial protein import, 170
Cross-linking, 18, 28, 42, 76, 203
Crystallography, use of to determine protein structure, 19
Cu^{2+}
  effect on import of metallathionin into mitochondria, 184
  requirement for prepeptidase/transit peptidase activity, 176
Cysteine, fatty acylation of, 104, 109, 166
Cytochemistry, localization of Golgi enzymes and, 142
Cytochrome b_2, sorting in mitochondria, 191
Cytochrome b_5, signal sequences of, 70
Cytochrome c_1, import and sorting of in mitochondria, 191
Cytochrome *c* oxidase subunit II, secondary processing of in mitochondria, 192
Cytochrome *c* oxidase subunit IV (COX4)
  organelle-specific routing of, 176
  prepeptide of, 175
Cytochrome *c* oxidase subunit VIIa, mitochondrial routing of, 179
Cytochrome *c* peroxidase, processing of, 192
Cytochrome *f,* thylakoid routing of, 193
Cytoplasmic membrane (bacterial), 5, 45
  domains of, 15, 96
  separation of, 15
Cytoskeleton
  exocytosis in yeasts and, 155
  Golgi organization and, 147
  polarized secretion and, 154
  regulated exocytosis, 159
Cytosolic factors
  in mitochondrial protein import, 183
  in protein export in bacteria, 85
  in protein–RERM interactions, 82

D

Default sorting
  bacterial secretory proteins, 112
  in eukaryotic secretory pathway, 149
Deoxynojirimycin, 137
Detergent solubilization of proteins, 15

Dextran, uptake into nuclei, 201, 208
Dihydrofolate reductase (DHFR)
hybrid proteins, 25, 52, 173, 197
mitochondrial routing of, 173
peroxisomal routing of, 197
as reporter protein, 24
Diphtheria toxin, uptake of, 224
Disulfide bridges
effect on protein translocation, 84
formation in endoplasmic reticulum, 130
DNA K protein (*E. coli*), 89
Docking protein, 76
Dog pancreas microsomes, 40
Dolichol, 129
Domains, mapping of in proteins, 20

E

E1 protein (coronavirus), retention in Golgi, 147
E1a protein (adenovirus), karyophilic signal of, 204
E19 protein (adenovirus), retention in endoplasmic reticulum, 134
Egasyn, 135
Electron diffraction, use of in protein structure determination, 20
Elastase (*Pseudomonas*), activation and secretion of, 125
Electron microscope, 16
end genes and products (yeast), roles in endocytosis, 34, 226
Endocrine cells, 155
Endocytic receptors, 212
Endocytosis 5, 211
fluid phase, 211, 225
mutations affecting, 34, 226
in vitro, 40, 217
in yeasts, 225
Endoglycosidase resistance, 18, 40, 143
Endoplasmic reticulum
protein modification in, 128
protein retention in, 131, 133
protein routing to, 69
recycling to, 133
ribosomes attached to, 69, 77
Energy
in mitochondrial and chloroplast protein import, 182
in peroxisomal protein import, 196
requirement for, 42
in secretory protein translocation, 97
sources of, 42
5-Enolpyruvylshikimate-3-phosphate synthase, import into chloroplasts, 184
Enzyme complexes, secretion of, 115
Epidermal growth factor endocytosis, 221
Epitope
mapping, 20
surface exposition in *E. coli* (applications of), 234
ER, *see* Endoplasmic reticulum
Erwinia chrysanthemi, extracellular proteases of, 123, 125
Ester-linked fatty acids, 104, 109
Excretion, definition of, 8
Exocrine cells, 155
Exocytic vesicles, 150
Exocytosis, 5
events in, 159
Exotoxin A
secretion of, 122
uptake of, 224
Export, applications in bacteria, 231
Exported proteins, definition of, 8
Expression systems, 28

F

F_0-ATPase subunit 9, import into mitochondria, 183
F_1-ATPase β-subunit, import into mitochondria, 184
f1 (M13) bacteriophage
precoat protein, 57, 62, 66
protein VIII membrane anchor, 100
as secretion vehicle, 234
Farnecyl in lipoproteins, 166
Fatty acids (in lipids), 36
Fatty acylation
in bacterial lipoproteins, 90, 104
detection of, 18
in eukaryotic lipoproteins, 107, 109
polarized VSV G protein and, 154
Fatty acyl transferases, 106, 109, 135
Ferredoxin, sorting of in chloroplasts, 193
Ferritin, endocytosis of, 222
Fimbriae, 6
processing of, 92
secretion of, 124
Flagellae, 6
secretion of, 124
Fluid-phase endocytosis, 222

Fractionation of cells, 15
FTS I protein, lipoprotein nature of, 104
Fucosyltransferase, 139
Fusion pore in regulated exocytosis, 160

G

G protein (GTP binding protein), 167
in exocytosis, 159
in secretory pathway, 83, 168
in transport through Golgi, 145
G protein (trypanosomes), PI lipoprotein, 107
G protein (vesicular stomatitis virus)
glycosylation of, 130, 133
Golgi transport of, 144, 189
human growth hormone hybrids, 156
lipoprotein nature of, 109
membrane anchor of, 101
polarized export of, 153
sorting signal in, 155
trimerization of, 131
Galactokinase, as reporter protein, 24
Galactosyltransferase, 139, 148
Gap junction in polarized cells, 151
Gene fusions, 22
uses of, 22, 30, 33, 50, 52
GERL, *see* Golgi
Globomycin, 91, 104
Globulin P, sorting to plant vacuole, 164
GLS1 gene and product (yeast α-1-2 glucosidase), 130
Glucose trimming, 130
Glyceride
in bacterial lipoproteins, 106
in PI lipoproteins, 108
Glycosylation
of bacterial proteins, 112
detection of, 18
in endoplasmic reticulum, 128
in Golgi, 137
inhibition of, 18
Glyoxisomes, *see* Peroxisomes
Golgi
cisternae, 141
endoplasmic reticular lysosomes (GERL), 147
organization of, 3, 141
protein modification in, 137
retention of proteins in, 147
Gram stain in bacterial classification, 5
GRO EL protein, 87, 89, 194
GTP
in protein–endoplasmic reticulum interactions, 83
in regulated exocytosis, 159
in secretory pathway, 168
in transport through Golgi, 145

H

Hemagglutinin (HA) (influenza virus)
activation of, 148
membrane anchor, 101
polarized sorting of, 152
signal peptide, 65
sorting signal of, 153
trimerization of, 131
HDEL
receptor for, 136
role in ER retention, 136
Heat labile enterotoxin
hybrid proteins, 52
receptor for, 223
secretion of, 122
uptake of, 224
Heat shock proteins, 83, 89, 96, 132, 185, 194
chloroplast transit peptide of, 174
Heat shock response (bacterial), 89
Helical hairpin model, 67
Hemolysin (*Aeromonas*), activation of, 125
Hepatitus B precore protein, signal peptidase action on, 93
Hepatocytes, polarity of, 153
Herbicide resistance, chloroplast genes for, 239
Histoincompatibility antigen class II invariant chain, CAT hybrids and signal sequence, 63
hly genes in α-hemolysin secretion, 123
Horseradish peroxidase, endocytosis of, 148, 222
HSP70 proteins, 83, 96, 185
relation to clathrin dissociation, 165
Human growth hormone, sorting of in regulated exocytosis, 156
Hybrid proteins
biotechnology and, 231
uses of, 22
Hydrophobicity
membrane anchors and, 101

secretory routing signals and, 46, 53, 58, 62, 101
Hydroxylamine, effect on lipoproteins, 18, 109
Hypercholesterolemia, 213

I

IgA and maternal IgG, transcytosis of, 220
IgA protease, secretion of, 121
Immunofluorescence, 16
Immunoglobulin, processing of, 45
Immunogold cytochemistry, 16
Inner membrane (mitochondrial), sorting of proteins to, 190, 192
Insertion loop model, 66, 95
Insulin, exocytosis and processing of, 159
Interblocks in transit peptides, 174
Interleukin 1, secretion of, 166
Intermembrane space (mitochondrial), sorting of proteins to, 190
Invertase
  β-galactosidase hybrids, 51
  export and secretion of, 83
  as reporter protein, 24
  signal peptide, 58
  vacuolar sorting of (in hybrids), 163
In vitro assays, 39

K

Karyophilic signal, 10, 202
  common features of, 205
  cross-linking of, 203
  in hybrid proteins, 203
  multiple copies of, 205
  mutations affecting, 205
  receptor for, 207
Karyoskeleton, 199
KDEL, as endoplasmic reticulum retention signal, 135
3-Ketoacyl CoA thiolase, mitochondrial routing of, 179
Kex peptidases, role in protein processing in Golgi, 40
Killer factor, processing of, 140

L

Lactose permease, membrane insertion and topology of, 64
LAC Z, *see* β-galactosidase
LAM B protein
  outer membrane sorting of, 117
  topology of, 116
Lamellae, in thylakoids, 169
Laminin, polarized secretion of, 152
Laminin (nuclear), 199
laxA gene (*Pseudomonas*), role in elastase secretion and activation, 124
Leader peptidase, *see* Signal peptidase I
Leader peptide, 10
Lectins
  inhibition of nuclear targeting by, 209
  use of to detect sugar residues, 142
lep gene (*E. coli*), 91
Light-harvesting chlorophyll a/b protein (chloroplast), processing and sorting of, 194
Lipids
  nonbilayer structure of, 97
  in nonsecretory organelles, 190, 198
  polarized distribution in plasma membrane, 151
  role in protein translocation, 97
Lipopolysaccharide (LPS), 6
  role in outer membrane protein folding and sorting, 119
Lipoprotein signal peptidase, 91, 104
Lipoproteins
  bacterial, 91, 104
  eukaryotic, 107, 109
Live vaccines, 234, 238
Low-density lipoprotein
  endocytosis of, 221
  receptor, 212, 221
Low temperature, effects on secretory pathway, 35, 157, 161
LPP lipoprotein, 104
  signal peptide, 57
lsp gene (*E. coli*), 104
Luciferase, peroxisomal routing of, 197
Lysis proteins
  fatty acylation of, 104
  role in colicin release, 124
Lysosomal enzyme sorting signal, 138, 160
Lysosome
  endocytosis to, 221
  function of, 218
  secretory enzyme sorting to, 160
  secretory membrane protein sorting to, 162

M

M13 phage, *see* f1 phage
MAL E signal peptide, 52
MAL E-β-galactosidase, aborted export and toxicity of, 52
Maltose binding protein, *see* MAL E signal peptide
Manine Darby canine kidney (MDCK) cells, 39, 152
Mannan proteins (yeast), 139
Mannose
  phosphorylation of, 138
  suicide selection, 33
  trimming, 130, 138
Mannose-6-phosphate (M6P), 138, 160
  lysosomal enzyme sorting and, 160
  receptors, 160, 275
  yeast vacuolar proteins and, 163
Mannosidase II, 138
Mannosylation, 128
MAS1 gene (yeast), 177
MATα2 protein (yeast), nuclear targeting of, 203
Matrix export signal (in mitochondrial proteins), 10, 192
Membrane anchor, 11, 63, 99
Membrane contact sites
  in bacteria, 107, 154
  in chloroplasts and mitochondria, 186
Membrane fluidity, 36
Membrane perturbants, 36
Membrane potential, 45
  colicin uptake and, 227
  diphtheria toxin uptake and, 223
  protein translocation and, 97, 182, 186, 195
Membrane protein topogenesis and topology, 8, 11, 63, 99, 113
  amphiphatic membrane anchor, 103
  determination of, 99
  exceptions, 102
Membrane proteins, types of, 8
Membrane trigger hypothesis, 65
Methotrexate, effect on DHFR import into mitochondria, 183
Microfilaments, 154
Microinjection, 37
Microsomes, 39
Mitochondria, organization of, 4, 169
Mitochondrial proteins
  energy for import, 182, 236
  hybrid proteins, 172
  origin of nuclear-encoded, 169
  outer membrane routing signal, 179
  routing signals, 171
Mitoplast, 181
Monesin, 35
Mutations
  affecting routing signals, 26
  affecting targeting pathway, 26, 30
  suppressors of, 30, *see also sec* and *SEC*
Myristate, in lipoproteins, 110, 116

N

N-acetylgalactosamine transferase, 139
N-acetylglucosamine transferase, 129
Neomycin phosphotransferase, as reporter protein, 24
Neuraminidase (influenza virus), export of, 65
N-glycosylation, 127
Nitrous acid, action on PI lipoproteins, 108
N-linked oligosaccharides, processing of in Golgi, 137
NLP A protein (*E. coli*), 107
Nuclear protein, processing of, 207
Nuclear protein import
  energetics of, 208
  receptor dependence of, 209
  regulated, 206
  role of pore in, 207
Nuclear protein retention signal, 201, 206
Nuclear targeting of nonnuclear proteins, 203, 206
Nucleoplasmin, routing of, 202
Nucleus, 199

O

O-glycosylation, 130, 137
Oligosaccharyltransferase, 128
OMP A protein (*E. coli*)
  export of, 87
  topology of, 116
Opsin, membrane anchor and signal sequences in, 64, 101
Outer membrane (bacterial), 6
  intermediate in protein secretion, 121
  separation of, 15

Outer membrane proteins
  β-galactosidase hybrids and, 117
  sorting of, 117, 155
  topology of, 116
  use in live vaccines, 235
Ovalbumin
  absence of glycosylation, 133
  signal sequence of, 62
3-Oxo-acyl CoA thiolase, peroxisomal routing of, 197

P

PAGE, *see* Polyacrylamide gel electrophoresis
Palmitate
  in chloroplast proteins, 194
  in secretory lipoproteins, 104
Parathyroid hormone, signal peptide of, 60
Passenger proteins, 22
Patch signal, 12, 167
Penicillin G acylase, processing of, 113
Penicillinase, *see* β-lactamase
Peptidase IV, signal peptide peptidase and, 94
Peptidoglycan
  location in bacteria, 6
  restriction of protein diffusion, 117
Percoll gradients, 15
Periplasm, 5
  intermediate in protein secretion, 122
  sorting of proteins to, 112
Permeabilized cells, 38
Peroxisomal proteins
  receptor on peroxisomes, 196
  routing signals in, 197
Peroxisomes
  protein import into, 195
  routing of membrane proteins to, 198
PET genes (yeast), 192
pH
  effect on protein targeting, 35, 157, 159, 161, 218
  measurement in organelles, 35
Pheromones, fatty acylation and secretion of, 166
PHO A, *see* Alkaline phosphatase
Phophatidylinositol (PI) lipoproteins
  identification of, 18
  modification of, 107
Phosphoglycerate kinase, peroxisomal routing of, 197
Phospholipase A, colicin release and, 124
Phospholipase C, action on PI lipoproteins, 18, 107
Phosphorylation
  in chloroplast protein import, 182
  of chloroplast thylakoid proteins, 195
  in nuclear assembly, 201
Photosystem II in thylakoids, 170
Phytohemagglutinin (plant vacuole), sorting of, 163
"Piggyback" import of nuclear proteins, 206
Pili, *see* Fimbriae
Pinocytosis, 212, 222
Planar crystals of membrane proteins, 20
Plasminogen activator inhibitor 2, signal sequence, 62
Plastocyanin
  sorting of in chloroplasts, 192
  transit peptide of, 174
Polarity of protein sorting, 38, 151
  receptors and, 153
Polyacrylamide gel electrophoresis (PAGE), 13
Polysomes, membrane association of, 69, 170
Pores
  in nuclear envelope, 199, 207
  in peroxisome membrane, 196
Porin (bacterial), structure and outer membrane sorting, 115
Porin (mitochondrial), insertion into outer membrane, 179, 187
Posttranslational modification, 17
  in bacteria, 112
  in chloroplasts, 183, 195
  in endoplasmic reticulum, 128
  in Golgi, 137
  in mitochondria, 192
  in nucleus, 207
Posttranslational translocation
  in bacteria, 85
  into chloroplasts and mitochondria, 170
  in eukaryotic secretory pathway, 79
  into peroxisomes, 195
Precursor proteins, 9
Prelysosome, 162, 217
Prepeptidase, 176

Prepeptide (mitochondrial), 10, 171
  artificial, 175
  complex, 178, 190
  interaction with membrane, 172
  mutations affecting, 174
  processing sites in, 176
  receptor, 180, 186
  sequence and structure, 172
  two-step processing of, 178
Prepro-α-factor, signal peptide, 60
Primary endosome, 216
prlA mutations in *secY* gene (*E. coli*), 88
Prohormone processing, 140, 158
Proofreading of secretory proteins, 131
Propeptide, 10
  in bacterial secretory proteins, 133
  in regulated exocytosis, 156
  in vacuolar protein sorting, 163
Proteases
  activation of, 113, 124
  role in vesicle fusion (regulated exocytosis), 159
  secretion by *Bacillus,* 113
  secretion by *E. coli,* 123
  uses of, 20, 39
Protein A
  as export vehicle, 235
  in immunocytochemistry, 16
Protein 34 (phage T4), export of in hybrid proteins, 51
Protein VIII (phage f1), membrane anchor of, 100
Protein coat
  clathrin, 148, 164, 213
  secretory shuttle vesicles and, 144
Protein detection *in situ,* 16
Protein disulfide isomerase, 130
Protein import, 9, 211
Proteinase A (yeast vacuole)
  autoactivation of, 163
  default secretion of, 162
  propeptide of, 163
Proteolytic processing
  in bacteria, 113
  detection of, 17
  in Golgi, 140
  in regulated exocytosis, 158
  sites of in polypeptide, 17
Protein structure
  determination of, 19
  prediction of, 21
Pullulanase
  fatty acylation of, 104
  secretion and micellation of, 107, 121
Puromycin, effects on ribosome–membrane association, 77
Pyruvate kinase, nuclear targeting of in hybrid proteins, 24, 203

R

RAS proteins, myristoylation and export, 166
Regulated exocytosis, 155
Rennin, phosphorylation and lysosomal sorting of, 138
Reporter proteins, 22
RER, *see* Rough endoplasmic reticulum
Respiratory syncytial virus, surface proteins (applications of), 239
Retention signal, 11
  in ER proteins, 133
  in Golgi proteins, 147
  in nuclear proteins, 202, 206
Ribonuclear protein
  import into nuclei, 206
  in secretory protein translocation (SRP), 72
Ribophorins, 77, 96
Ribosome receptor (in ER), 77
Ribosome-coupled translocation (of secretory proteins), 82
Ribosomes associated with membranes, *see* Polysomes
Ricin uptake, 223
RNA, 76
  4S (bacterial), 76
  6S (bacterial), 76
  7S, in SRP, 76
Rough endoplasmic reticulum (RER), 3, 45, 71, 77, 92, 96, 126, 136
Routing signal
  chloroplast, 171
  definition, 9
  mitochondrial, 171
  nuclear, 202
  peroxisomal, 195
  secretory, 46, 62

RUBISCO
binding protein, 87, 194
small subunit
chloroplast routing of, 176
transit peptide of, 173
subunit assembly of, 194

S

S complex (*Bacillus*), 71, 76, 89
Salvage signal
definition, 11
retention of proteins in ER and, 133
SEC4 gene and product
fatty acylation of, 166
role in yeast secretory pathway, 150, 166
SEC11 signal peptidase gene (yeast), 93
SEC18, 137
secA (*E. coli*), 30, 87
secB (*E. coli*), 87
secD (*E. coli*), 30, 87
secE (*E. coli*), 30, 71, 97
secY (*E. coli*), 30, 87, 97
Sec^- mutants
bacteria, 30
yeast, 32
Secondary endosome, *see* Prelysosome
Secondary structure
effect on protein translocation, 79, 183
prediction of, 21
Secretagogues, 155, 159
Secreted proteins
bacterial, 112
definition, 8
Secretion, applications of, 229
Secretory component (in transcytosis), 220
Secretory granules, 3
proteolytic processing in, 148, 158
Secretory pathway
eukaryotes, 1, 45, 166
homologies between eukaryotic and prokaryotic, 98
prokaryotic, 5, 112
Secretory proteins (definition), 4
Shuttle vesicles (secretory pathway), 136, 144
Sialic acid transferase, 135, 148
Signal hypothesis, 66
Signal peptidase I (bacterial), 90
export and topogenesis, 64, 102
Signal peptidase II (bacterial), 91
Signal peptidase III (bacterial), 92
Signal peptidase (eukaryotic), 92
Signal peptidase in fimbrillin/pilin processing, 92
Signal peptide, 10, 46
applications of, 229
artificial, 59
basic changes in, 56, 60, 66, 75
comparisons of, 49
effects on protein conformation, 65
hydrophobic core of, 47, 52
in gene fusions, 49, 231, 235, 238
interaction with membranes, 65
location in secretory protein precursors, 61
mutations affecting, 31, 52, 58
processing of, 46, 90
processing site, 46, 53, 90, 105
receptor, 69
structure, 46, 66
toxicity and degradation of, 93
universality of, 49
Signal peptide peptidase, 93
Signal recognition particle (SRP), 71
in bacteria, 76
independent translocation, 79
receptor, 76
structure, 74
translation arrest and, 72
in yeast, 76
Signal sequence, 10, 46, 62
in gene fusions, 63
mutations affecting, 64
Sorting signal (definition), 11
SRP, *see* Signal recognition particle
SSA genes and proteins (yeast), 83
STE2 gene and product (α-factor receptor), 225
STE13-encoded peptidase, 141
Stop transfer signal, 11, 100, 190
Storage granules (secretory), 159
Subtilisin, autocatalytic processing and secretion of, 114
Sucrose gradient, 15
Sulfation, 140
Superoxide dismutatase, export by bacteria, 51, 232
Surface layers in bacteria, 5, 112
Surface proteins (bacterial), glycosylation of, 112

T

T antigen, nuclear targeting of, 203
tRNA methylase, prepeptide of, 172
Targeting signals, types of, 9
Tetanus toxin, receptor, 223
TGN, *see* Trans-Golgi network
Thioester linkage of fatty acids in lipoproteins, 109
Thylakoid, 170
  peptidase, 192
  protein sorting to, 192
Tight junction, 152
TON B protein, colicin uptake and, 226
Topogenesis (membrane protein), 100
Toxins
  receptors for, 21, 223
  as therapeutic agents, 240
  uptake of, 223
Transcytosis, 4, 153, 220
Trans-Golgi network (TGN), 147, 156
  condensation of proteins in, 157
  protein sorting in, 147
Transferrin
  endocytosis of, 221
  receptor
    endocytosis of, 221
    lipoprotein nature of, 109
    polarity of, 152
    recycling of, 221
Transit peptidase, 178
Transit peptide, 10, 171
  applications of, 239
  structure and sequence of, 174
Transitional element, 131, 134, 137
Translation arrest in secretory pathway, 72
Translocase (secretory), 96
Translocation channel (secretory protein), 68, 96
Translocation competence, 79, 183
Trigger factor, 85
Trigger hypothesis, 65
Triose phosphate isomerase, hybrid protein export in bacteria, 51
Trypsinogen, sorting into regulated pathway, 157
Tunicamycin, 36, 129
  effect on polarized protein sorting, 154
  effect on vacuolar protein sorting, 163
Tyrosine sulfation, 140, 148

U

Uncoatase (coated vesicles), 164
Unfoldase, 79, 96, 183
Urea, precursor denaturation by, 84, 183

V

Vacuolar protein sorting, 35, 163
Vesicles
  in chloroplast protein transport, 192
  fusion of, 147, 159
Viruses
  endocytosis of, 211
  receptors for, 20
VP7 protein (rotavirus), retention in ER, 134
VPL/VPT genes and products in vacuolar protein sorting, 35, 163

W

Wheatgerm agglutinin, effect on nuclear protein import, 209
Wheatgerm lysate, use in *in vitro* assays, 39, 72

X

X-ray diffraction, protein structure determination and, 19

Y

YPTI gene and product (yeast), protein secretion and, 150

Z

Zeeweger syndrome, peroxisomal protein targeting in, 198
Zymogens (proteases), 113, 124